A Report of the Institute for

INSURMOUNTABLE RISKS

BRICE SMITH

The Dangers of Using Nuclear Power to Combat Global Climate Change

published jointly by
IEER Press
Takoma Park, Maryland
and
RDR Books
Muskegon, Michigan / Berkeley, California

Insurmountable Risks

published jointly by

RDR Books	IEER Press
1487 Glen Avenue	6935 Laurel Avenue, Suite 201
Muskegon, MI 49441	Takoma Park, MD 20912
Phone: (510) 595-0595	Phone: (301) 270-5500
Fax: (510) 228-0300	Fax: (301) 270-3029
E-mail: read@rdrbooks.com	E-mail ieer@ieer.org
www.rdrbooks.com	www.ieer.org

Copyright 2006 by Institute for Energy and Environmental Research
All Rights Reserved

ISBN: 1-57143-162-4
After Jan. 1, 2007: 978-1-57143-162-2

Library of Congress Control Number: 2006928939

Cover photo © Corbis Corporation. All Rights Reserved
Cover design and production: Richard Harris

Distributed in the United Kingdom and Europe by
Roundhouse Publishing Ltd., Millstone, Limers Lane,
Northam, North Devon EX39 2RG, United Kingdom

Printed in Canada

*Dedicated to my nephew Grant Turner Koh
and the world he will grow up in.*

Acknowledgements

Work on this report began nearly three years ago when I first became aware of the nuclear power study conducted at the Massachusetts Institute of Technology. Over that time I have benefited from many discussions on the topic of nuclear power and global climate change from formal conferences to casual dinners. First and foremost among those that I would like to thank is Dr. Arjun Makhijani, President of the Institute for Energy and Environmental Research (IEER). His help and guidance in framing this study was invaluable and I greatly benefited from his review of its contents as well. This work would not be what it is today without Dr. Makhijani's significant intellectual contributions. For example, his historical analysis of U.S. nuclear mythology, originally presented in *Nuclear Power Deception* by Dr. Makhijani and Scott Saleska, was one of the central themes around which I wrote the introduction to this work. In addition, Dr. Makhijani was the first to recognize the significance of our economic projections in which each of the alternatives for reducing greenhouse gas emissions tended to have a cost between six and seven cents per kilowatt-hour when projected over the near to medium-term. The recognition that these alternatives would be economically competitive with new nuclear power became one the central conclusion of my current analysis. Dr. Makhijani was also instrumental in framing my discussion of the link between nuclear disarmament and non-proliferation and my discussion of long-term nuclear waste management.

I would also like to thank Melissa Kemp, a policy analyst and organizer for Public Citizen's Energy Program, for her help in framing my discussion of renewable energy resources. In particular, I am grateful for her encouragement in examining the potential for economically competitive solar power to be developed in the coming years.

I am extremely grateful to Lois Chalmers, the IEER librarian, for all of her tireless work in obtaining the information and resources that I required to complete this analysis and for having the patience to so carefully fact check the citations in this report. Her attention to detail has added much to the strength of this work. In addition, I would also like to thank Michael Borucke, now a member of the Science and Technology Team at Northeast States for Coordinated Air Use Management (NESCAUM), for his help in checking many of the calculations in this report while working for IEER. As always, however, I remain solely responsible for the contents of this report, including any omissions or errors.

Outreach work on *Insurmountable Risks* is part of IEER's ongoing program on energy and the environment. In this light, I would like to thank Sara Barczak, Safe Energy Director of the Southern Alliance for Clean Energy, for her encouragement in completing this work, for reviewing its contents, and for helping IEER to conduct outreach on these issues throughout the Southeastern United States.

Finally, I would like to gratefully acknowledge the foundations that have generously supported IEER during the preparation of this report:

> Colombe Foundation, Educational Foundation of America, Ford Foundation, John Merck Fund, Livingry Foundation, New Cycle Foundation, New-Land Foundation, Ploughshares Fund, Public Welfare Foundation, Stewart R. Mott Charitable Trust, and Town Creek Foundation.

I would also like to thank those individuals who have become donors to IEER. Your support is deeply appreciated and helps make our work possible.

Brice Smith
Takoma Park, Maryland
May 11, 2006

Foreword: The return of the nuclear messiahs

Here is a book for the times. Uranium enrichment and reprocessing, once terms reserved for eggheads dealing in nuclear esoterica, are in the headlines everyday. Politicians and diplomats argue about them and the proliferation threats arising from the spread of commercial nuclear power technology.

Yet, strangely, in a parallel universe also being played out on the public stage, is the nuclear industry's claim, amplified by the megaphones of the media, that nuclear power can play a vital role in saving the Earth from another peril – severe climate disruption caused by the anthropogenic emissions of greenhouse gases.

Could it? Could nuclear power really help save the world from what is arguably the worst environmental scourge ever to confront humanity? History would suggest two things: caution about the nuclear industry's messianic proclamations and careful analysis of the problem.

The early promises of the fervent advocates of nuclear energy were of an economic paradise that nuclear energy would usher in for everyone from the needy to the greedy. No whim or need would go unfulfilled. But it was mainly fantasy and propaganda.

Almost two decades ago, browsing through the stacks of a well endowed library, I ran into a 1950 article written by a research engineer by the name of Ward Davidson from Consolidated Edison Company of New York. It was published in the then-nuclear industry journal *Atomics*. Updating an earlier 1947 opinion, he wrote that the technical problems facing nuclear power were even more daunting than he had imagined. For example, the materials requirements would be stringent, given the high temperatures and damage from high neutron fluxes. Testing of the alloys to ensure the quality and uniformity needed would be difficult. All this meant, of course, that nuclear power would be quite expensive.

Reading that prescient 1950 assessment was an eye opener. Like almost everyone else, I believed that the common technical conclusion prevalent in nuclear circles in the 1940s and 1950s was that the nuclear energy would soon be "too cheap to meter." After all, that statement was made by the Chairman of the U.S. Atomic Energy Commission in 1954 and endlessly repeated. I had presumed that it was simply a mistake, but who doesn't make mistakes? This was the first inkling of what further research would decisively show: it was the uniform conclusion of all serious analyses at the time that nuclear electricity would be expensive.[1]

"Too cheap to meter" was part self-delusion, as shown by the florid and fantastic statements made by the most serious people such as Glenn Seaborg, who led the team that first isolated plutonium, and Robert Hutchins, the President of the University of Chicago during the Manhattan Project. And it was part organized propaganda designed to hide the horrors of the hydrogen bomb.

In September of 1953, less than a month after the detonation of the Soviet's first hydrogen bomb, AEC Commissioner Thomas Murray wrote to the commission's chairman that the U.S. could derive "propaganda capital" from a publicity campaign surrounding their recent decision to construct the Shippingport nuclear power plant.[2] Sterling Cole, the chairman of the Joint Committee on Atomic Energy in the U.S. Congress, reached a similar conclusion regarding the importance of demonstrating the "benefits" of nuclear power as a counter balance to the immense destructive force of the hydrogen bomb. This conclusion, in fact, led Cole to worry that the Soviets might beat the U.S. to a functional nuclear power plant, and thus steal the claim to being the true promoters of the "peaceful" atom. In a letter to a fellow Congressman, Sterling Cole wrote

> It is possible that the relations of the United States with every other country in the world could be seriously damaged if Russia were to build an atomic power plant for peacetime use ahead of us. The possibility that Russia might actually demonstrate her "peaceful" intentions in the field of atomic energy while we are still concentrating on atomic weapons could be a major blow to our position in the world.[3]

As early of 1948, the Atomic Energy Commission reported to Congress that "the cost of a nuclear-fuel power plant will be substantially greater than that of a coal-burning plant of similar capacity."[4] In the January 1949 issue of *Science*, Robert Bacher, one of the original members of the AEC and a member of the scientific team at Los Alamos during the war, cautioned that despite the progress that was being made, it was "far too

---

[1] This Foreword is based on Part I of Makhijani and Saleska 1999.
[2] Murray 1953
[3] Cole 1953
[4] AEC 1948

early to make any predictions about the economic feasibility of atomic power."[5]

One of the most direct of the early critiques of the economics of nuclear power came in a December 1950 speech before the American Association for the Advancement of Science by C.G. Suits. At the time, Suits was the Vice-President and Director of Research at General Electric which was then operating the Hanford plutonium production reactors in Washington State and was one of the principal companies developing nuclear reactors for the production of electricity. In his speech, which was reprinted in the industry journal *Nucleonics*, Suits stated bluntly that:

> It is safe to say... that atomic power is *not* the means by which man will for the first time emancipate himself economically, whatever that may mean; or forever throw off his mantle of toil, whatever that may mean. Loud guffaws could be heard from some of the laboratories working on this problem if anyone should in an unfortunate moment refer to the atom as the means for throwing off man's mantle of toil. It is certainly not that!
>
> ...
>
> ... At present, atomic power presents an exceptionally costly and inconvenient means of obtaining energy which can be extracted more economically from conventional fuels... The economics of atomic power are not attractive at present, nor are they likely to be for a long time in the future. This is expensive power, not cheap power as the public has been led to believe.[6]

In 1953, an official AEC study concluded that "no reactor could be constructed in the very near future which would be economic on the basis of power generation alone." Significantly, this language was identical to that in a study published by industrial companies and major utilities including Bechtel, Monsanto, Dow Chemical, Pacific Gas and Electric, Detroit Edison, and Commonwealth Edison.[7]

The dismal assessment of the prospects of nuclear went back to the Manhattan Project. In a star-studded 1948 report, authored by Enrico Fermi, Glenn Seaborg, and J. Robert Oppenheimer, the authors concluded that there was "unwarranted optimism" about the speed with which the technical difficulties facing nuclear power could be overcome. Ironically, the

---

[5] Bacher 1949 p. 6 and LANL Biography
[6] Suits 1951
[7] Makhijani and Saleska 1999 p. 67-68

self-same Glenn Seaborg waxed eloquent about how plutonium fuel could transport everyone into a technical wonderland of "planetary engineering" – which, of course, could only be done if energy were very cheap..

Now, over half a century after the fantasies and the propaganda, and over a quarter of a century since the last reactor was ordered in the United States, the nuclear industry is returning. Then it was the promise of an endless supply of fuel – what Alvin Weinberg, the first Director of Oak Ridge National Laboratory called a "magical" energy source. Uraniun-238, not a reactor fuel, would be turned into fuel in breeder reactors, even as those same reactors consumed plutonium fuel. The net result would be more fuel at the end of the cycle than there was at the beginning. With supplies of uranium-238 being rather vast, the physics of the fantasy was only slightly exaggerated.

But physics is not enough. An energy source must still meet the tests of safety, reliability, and cost. In the case of nuclear energy, there is also the unique problem of nuclear proliferation, in part hidden in the form of the plutonium content of the spent fuel and in part in the form of the spread of know-how. Taken together, these factors made the physics "magic" evaporate the first time around. Breeder reactors and the associated reprocessing have yet to be commercialized after over $100 billion in expenditures worldwide (constant 1996 dollars) and more than fifty years of effort.[8]

It is the same today. As Brice Smith notes, and as several careful analyses have shown, the carbon dioxide emissions from a nuclear electricity system can be kept very small, in fact, an all nuclear-energy-system could theoretically reduce them to zero. But the physics is not the problem now; nor was it then.

The problems are:

1. How much will nuclear energy cost relative to other means of getting rid of carbon dioxide emissions?
2. What kinds of subsides will be required, given that Wall Street is skittish about nuclear power?

---

[8] Makhijani 2001.

3. What will be the risks of catastrophic accidents if we build reactors at the rate of one a week or more, cookie-cutter style around the world?
4. What will happen to the security of power supply in case of terrorist attacks or disastrous accidents on the scale of Chernobyl?
5. What about all the plutonium in the waste?

In *Insurmountable Risks*, Brice Smith carefully analyzes all these questions and more. It is a meticulously researched work that points to the great dangers of attempting to solve the problem of reducing carbon dioxide emissions by resorting to large-scale use of nuclear energy. Were there no alternative, the severity of the threat facing humankind as well as other species from global climate change might well warrant serious consideration of the risks of nuclear energy. But we do have alternatives that will not leave proliferation headaches and risks of radioactive landscapes like the ghostly zone around Chernobyl to future generations.

Before buying into the idea that nuclear energy is going to save us from global climate change because of its theoretical potential for low carbon dioxide emissions, read this book. And then work for the alternatives.

Arjun Makhijani, Ph.D.
President, Institute for Energy and Environmental Research
Takoma Park, Maryland
May 15, 2006

Table of Contents

Foreword: The return of the nuclear messiahs iii
Acronyms and Abbreviations xi
Chapter One: The World of Tomorrow and Yesterday 1
   Section 1.1 - From Peaceful Panacea to Environmental Necessity 2
   Section 1.2 - The Realities of Climate Change 12
   Section 1.3 - Case Study: the MIT Nuclear Power Report 26
Chapter Two: The White Elephant 29
   Section 2.1 – The Projected Cost of Nuclear Power 34
      Section 2.1.1 – Lowering the Capital Cost and Construction Time 37
      Section 2.1.2 – Reducing the Financial Risk Premium 44
      Section 2.1.3 – Impact of Potential Cost Improvements 52
      Section 2.1.4 – Summary of Nuclear Power Economics 53
   Section 2.2 – The Economics of Nuclear Power as a Carbon Mitigation Strategy 54
      Section 2.2.1 – "Carbon-Free" Portfolios 54
      Section 2.2.2 – Direct Taxation of Carbon Emissions 55
   Section 2.3 – Alternatives for the Near-Term (2006 - 2020) 60
      Section 2.3.1 – The Economics of Efficiency 61
      Section 2.3.2 – The Power of Wind 65
      Section 2.3.3 – Summary of Near-Term Options 71
   Section 2.4 – Alternatives for the Medium-Term (2020 - 2050) 72
      Section 2.4.1 – Liquefied Natural Gas and Fuel Switching 73
      Section 2.4.2 – Increased Use of Wind and Other Renewable Energy Resources 79
      Section 2.4.3 – Coal Gasification 86
      Section 2.4.4 – Carbon Capture and Storage 89
   Section 2.5 - Conclusions 96
Chapter Three: Megawatts and Mushroom Clouds 100
   Section 3.1 – Uranium Enrichment 105
   Section 3.2 – Reprocessing and the Plutonium Economy 114
   Section 3.3 – Tritium Production 124
   Section 3.4 – Strengthening Non-Proliferation Efforts 126
      Section 3.4.1 – Enhanced Inspections under the IAEA 130
      Section 3.4.2 – Restricting Access to Fuel Cycle Technologies 138
      Section 3.4.3 – Increased Consequences for Suspected Proliferators 148
      Section 3.4.4 – Disarmament and Nonproliferation 155
   Section 3.5 - Conclusions 160
Chapter Four: A Culture of Safety? 165
   Section 4.1 – The Record of Safety 168
      Section 4.1.1 – The Problems of Youth 171
      Section 4.1.2 – The Problems of Aging 173
      Section 4.1.3 – The Problems of New Reactors 182
   Section 4.2 – The Impacts of A Catastrophic Accident 184
      Section 4.2.1 – Human Consequences of an Accident 188
      Section 4.2.2 – Economic Consequences of an Accident 192

Section 4.2.3 – The Risks from the Nuclear Fuel Cycle 196
Section 4.2.4 – Safety and Public Opinion.. 200
Section 4.3 – Probabilistic Risk Assessments 202
Section 4.3.1 – The Rasmussen Report and the History of the PRA Methodology 205
Section 4.3.2 – Issues of General Completeness................................. 208
Section 4.3.3 – "Human Factors"... 214
Section 4.3.4 – Computers and Digital Control Systems.................... 216
Section 4.3.5 –Expert Judgment and Uncertainties of Methodology. 221
Section 4.4 – Safety of an Expansion of Nuclear Power 224
Section 4.5 - Conclusions... 229
Chapter Five: The Legacy of Nuclear Waste... 233
Section 5.1 –Disposal of "Low-Level" Nuclear Waste.............................. 235
Section 5.2 – Geologic Disposal of Spent Nuclear Fuel and High-Level Waste 237
Section 5.2.1 – General Uncertainties Regarding Geologic Disposal. 241
Section 5.2.2 – The History of Geologic Disposal in the United States 245
Section 5.2.3 – Ready, Fire, Aim… The DOE Strategy at Yucca Mountain.. 250
Section 5.2.4 – Engineered Barriers at Yucca Mountain, the Changing Focus.. 262
Section 5.2.5 – The "Technical" versus "Legal" Limit at Yucca Mountain.. 268
Section 5.2.6 – Additional Concerns Regarding Yucca Mountain 272
Section 5.3 – Transportation of Spent Fuel 274
Section 5.4 – Alternative Waste Management Strategies......................... 280
Section 5.4.1 – Monitored Retrievable Storage (MRS) 282
Section 5.4.2 – Separation, Transmutation, and MOX Fuel............... 284
Section 5.4.3 – Deep Boreholes 287
Section 5.5 – Conclusions 290
Chapter Six: Looking Back, Moving Forward 295
Appendix A: Uranium Supply and Demand... 307
Section A.1 – Estimates of Uranium Resources... 308
Section A.2 – Estimates of Uranium Production Capacity 312
Section A.3 – Stretching Uranium Resources 316
Section A.4 – Estimates for Cumulative Uranium Demand 318
Section A.5 – Impacts of Uranium Supply and Demand on Proliferation 321
References... 325
Endnotes 377
Index.. 417

Acronyms and Abbreviations

AEC	U.S. Atomic Energy Commission
BWR	Boiling Water Reactor
CBO	Congressional Budget Office
CCGT	Combined Cycle Gas Turbine
CO_2	Carbon Dioxide
CRAC-2	Calculation of Reactor Accident Consequences for U.S. Nuclear Power Plants
CTBT	Comprehensive Test Ban Treaty
DOE	U.S. Department of Energy
EAR	Estimated Additional [Uranium] Resources
EIA	U.S. Energy Information Administration
EPA	U.S. Environmental Protection Agency
GAO	U.S. Government Accountability Office (formerly General Accounting Office)
GNEP	Global Nuclear Energy Partnership
GW	gigawatt (one billion watts)
HEU	Highly Enriched Uranium
HOSS	Hardened On-Site Storage
HTGR	High-temperature Gas-cooled Reactor
IAEA	International Atomic Energy Agency
IEA	International Energy Agency
IEER	Institute for Energy and Environmental Research
IGCC	Integrated Gasification Combined Cycle
IPCC	Intergovernmental Panel on Climate Change
LANL	Los Alamos National Laboratory
LEU	Low Enriched Uranium
LLNL	Lawrence Livermore National Laboratory
LLW	Low-Level Waste
LNG	Liquefied Natural Gas
MIT	Massachusetts Institute of Technology

MMBtu	Million British Thermal Units
MOX	Mixed-Oxide Plutonium Fuel
MT	metric tons
MTIHM	Metric Tons Initial Heavy Metal
MW	megawatt (one million watts)
MWD	megawatt-day (86.4 billion joules)
NASA	National Aeronautics an Space Administration
NATO	North Atlantic Treaty Organization
NEA	Nuclear Energy Agency
NPT	Treaty on the Non-Proliferation of Nuclear Weapons
NRC	U.S. Nuclear Regulatory Commission
NREL	National Renewable Energy Laboratory
NWPA	Nuclear Waste Policy Act
NWTRB	Nuclear Waste Technical Review Board
OECD	Organization for Economic Cooperation and Development
PFS	Private Fuel Storage, LLC
PRA	Probabilistic Risk Assessment
PUREX	Plutonium Uranium Refining by Extraction
PWR	Pressurized Water Reactor
QA/QC	Quality Assurance / Quality Control
RAR	Reasonably Assured [Uranium] Resources
SR	Speculative [Uranium] Resources
START	Strategic Arms Reduction Treaties
THC	Thermohaline Circulation
TMI	Three Mile Island
TRU	Transuranic Waste
TSPA	Total System Performance Assessment
TVA	Tennessee Valley Authority
WCS	Waste Control Specialists, LLC
WIPP	Waste Isolation Pilot Plant

Chapter One: The World of Tomorrow and Yesterday

> It is not too much to expect that our children will enjoy in their homes electrical energy too cheap to meter, --will know of great periodic regional famines in the world only as matters of history, -- will travel effortlessly over the seas and under them and through the air with a minimum of danger and at great speeds, - and will experience a lifespan far longer than ours, as disease yields and man comes to understand what causes him to age. This is the forecast for an age of peace.[1]
>
> - Lewis Strauss, Chairman of the U.S. Atomic Energy Commission (1954)

If an enterprise as enormous and costly as nuclear power can have an anniversary, than 2004 would have been its 50th. In the last half a century, the promotion of nuclear power has undergone many changes, but today it remains a technological pariah in many parts of the world despite the potential interest of countries like India and China and the growing interest of the Bush Administration. Following the Second World War, with the rapid expansion in the production of consumer goods in the United States and other countries, nuclear power was held out as the hope for a new and better world run by cheap and plentiful electricity. The widespread failure of nuclear power in the U.S. from an economic perspective, coupled with the unique problems of safety, waste disposal, and nuclear weapons proliferation has forced its proponents to look for new rationales in their attempt to revive the "nuclear option."

Over the last 20 years, the prospect of global climate change driven by the emission of greenhouse gases[2] has provided just such a rationale. Compared to the other major energy sources now utilized around the world to generate base load electricity (coal, oil, and natural gas), nuclear power plants emit far lower levels of greenhouse gases even when mining, enrichment, and fuel fabrication are taken into consideration.[3] Today, the argument that nuclear power is essential to preventing the potentially catastrophic consequences of global warming is the dominant theme uniting arguments to keep existing plants open beyond their intended operational lifespan and for new power plants to be built. These arguments are achieving new prominence, even among some environmentalists, following the entry into force of the Kyoto Protocol on February 16, 2005.[4]

However, just as the claim that nuclear power would one day be "too cheap to meter" was known to be a myth well before ground was broken on the first civilian reactor in the United States, this book will show that a careful examination today reveals that the expense and unique vulnerabilities associated with expanding nuclear power would make it a very risky option for trying to address the problems of climate change. These costs and vulnerabilities are well known and, in fact, are essentially the same set or problems that led to the failure of nuclear power the first time around. As summarized by the Intergovernmental Panel on Climate Change in 2001

> In spite of this advantage [negligible greenhouse gas emissions], nuclear power is not seen as the solution to the global warming problem in many countries. The main issues are (1) the high costs compared to alternative CCGTs [Combined Cycle Gas Technologies], (2) public acceptance involving operating safety and waste, (3) safety of radioactive waste management and recycling of nuclear fuel, (4) the risks of nuclear fuel transportation, and (5) nuclear weapons proliferation.[5]

It has been more than 50 years since ground was broken on the first civilian nuclear power plant and more than 25 years since the last reactor order was placed in the United States. It is time for nuclear power to finally retire so that the global community can move on to focusing its efforts on developing more rapid, effective, and sustainable options for addressing the most pressing environmental concern of our day.

Section 1.1 - From Peaceful Panacea to Environmental Necessity

The year 1954 was indeed a banner year in the birth of the nuclear power industry. On January 21, 1954, the *U.S.S Nautilus* set sail becoming the world's first nuclear powered submarine. In September, ground was broken by President Eisenhower at the site of the Shippingport nuclear power plant in Pennsylvania. The reactor there used a similar design to that which had been developed for the *Nautilus*. Earlier that year Congress, under pressure from the Atomic Energy Commission and its industrial partners, had amended the Atomic Energy Act to encourage greater participation of private companies in the nuclear enterprise. When the 60 megawatt-electric (MW)[6] reactor at Shippingport opened three years later, it would become the first commercial nuclear reactor in the U.S. Outside the United States, things were also moving rapidly in 1954. Three months before ground was broken at Shippingport, the Soviet Union connected a small 5 MW reactor to the electric grid at Obninsk, and by that time work had already been underway for more than a year on the 50 MW Calder Hall reactor in England that would produce both electricity and weapons grade plutonium.

However, 1954 was also an auspicious year for nuclear power in which Lewis Strauss, then Chairman of the Atomic Energy Commission, uttered the single most famous promise for the future of the then infant technology

> It is not too much to expect that our children will enjoy in their homes electrical energy too cheap to meter, --will know of great periodic regional famines in the world only as matters of history, -- will travel effortlessly over the seas and under them and through the air with a minimum of danger and at great speeds, - and will experience a lifespan far longer than ours, as disease yields and man comes to understand what causes him to age. This is the forecast for an age of peace.[7]

Strauss was hardly alone in making such pronouncements of the revolutionary potential of "cheap" and "inexhaustible" nuclear power. For example, shortly after World War II, Robert Hutchins, then Chancellor of the University of Chicago where scientists had conducted the initial research on nuclear reactors during the war, predicted that

> A very few individuals working a few hours a day at very easy tasks in the central atomic power plant will provide all the heat, light, and power required by the community and these utilities will be so cheap that their cost can hardly be reckoned.[8]

David Lilienthal, the head of the Tennessee Valley Authority and the first Chairman of the Atomic Energy Commission, wrote of the "almost limitless beneficial applications of atomic energy."[9] Finally, Glenn Seaborg, the co-discoverer of plutonium, one of the leading chemists in the Manhattan Project, and Chairman of the Atomic Energy Commission from 1961 to 1971, was perhaps one of the most poetic of the early nuclear proponents. He envisioned a world where a scientific elite would "build a new world through nuclear technology." He saw nuclear power as opening the way for "planetary engineering," enabling humans to irrigate the desert, move mountains, and redirect rivers. He saw millions of homes receiving heat and light from a single nuclear plant, and an expanding industrial sector powered by cheap and plentiful electric power. Seaborg even envisioned such novel applications as nuclear powered artificial hearts and SCUBA suits heated by plutonium.[10] Connected to these fantastic prophecies, however, was also a powerful Cold War backdrop that sought to make use of nuclear power as a propaganda tool in the conflict between the United States and the Soviet Union.[11] At stake in the push to develop nuclear power was the ability to lay claim to the mantle of the "peaceful" atom in the context of growing public anxiety over a nuclear arms race that was rapidly escalating.

On July 16, 1945, the U.S. detonated the world's first nuclear device at the Alamogordo test site in New Mexico. The core of this "gadget" was made from plutonium that had been produced in the nuclear reactors at the Hanford Engineer Works in southeastern Washington State. Less than a month later, the world was formally introduced to the power of nuclear energy when the cities of Hiroshima and Nagasaki were destroyed in a pair of blinding flashes that resulted in more than 100,000 immediate deaths as well as untold suffering among many of the survivors. On August 29, 1949, the Soviet Union tested its first atomic bomb at a site in Kazakhstan shattering the U.S. nuclear monopoly, and rendering vulnerable the U.S. population to the same type of fate that had been inflicted on the citizens of Hiroshima and Nagasaki. On October 3, 1952, the British joined the "nuclear club" by testing their first atomic device in Australia. Less than a month later, the U.S. tested the world's first thermonuclear device in the Pacific. Following close behind, the Soviet Union tested its first thermonuclear device on August 12, 1953. These tests opened the way for weapons a thousand times more powerful than those dropped on Japan. As the year 1953 came to an end, the twin images of the mushroom cloud and the devastation of Hiroshima and Nagasaki were the faces of nuclear energy to the people of the world.

In early 1954 the U.S. nuclear weapons complex suffered one of its most serious and public accidents which highlighted a third face of nuclear energy, namely fallout from weapons testing. While it was known within the weapons complex that atmospheric nuclear tests could result in fallout over vast areas, the March 1, 1954 test of the first "weaponized" hydrogen bomb provided dramatic evidence of these dangers to the public. The discussion of this test, carried out openly in the pages of U.S. newspapers, was particularly impactful given that it was coupled to a weapon of almost unimaginable destructive power. The bomb detonated in the Bravo test resulted in a blast of 15 megatons, more than twice the expected yield and nearly 1,000 times more powerful than the bomb that destroyed Hiroshima. The mushroom cloud from the blast rose 40 kilometers into the sky and had a diameter of 120 kilometers after 10 minutes. The explosion was described by then Secretary of Defense Charles Wilson as unbelievable.[12] President Eisenhower told a news conference that he believed "the scientists must have been surprised and astonished at the results," and Representative Chet Holifield, a member of Congress and witness to the Bravo test, described the explosion to the media as "so far beyond what was predicted that you might say it was out of control."[13]

The fallout cloud from the Bravo test heavily contaminated a Japanese fishing vessel called the *Lucky Dragon* sparking an international incident. In addition, the fallout also affected the inhabited islands of Rongelap, Ailinginae, Rongerik, and Utirik in the Marshall Islands. Of the 239 Marshallese who were exposed to the fallout from Bravo, the 86 people on Rongelap received the highest doses. By the end of October 1954, one of the 23 crew members of the *Lucky Dragon*, Aikichi Kuboyama, had died from radiation poisoning and the rest remained in intensive care.[14] Approximately 3,000 people attended Kuboyama's funeral which took place just weeks after ground was broken at Shippingport.[15] Finally, despite assurances by AEC Chairman Lewis Strauss that the "only contaminated fish discovered were those in the open hold of the Japanese trawler [the *Lucky Dragon*]" roughly one out of every eight boats inspected (683 all told) were, in fact, found to have radioactive fish onboard. All together, Japanese monitors found 457 tons of tuna contaminated by the fallout.[16]

Converting the image of nuclear power from a destructive and potentially apocalyptic force into that of a constructive and beneficial force was a major goal of those within the U.S. nuclear establishment right from its very beginning.[17] This effort culminated in President Eisenhower's December 8, 1953 address before the General Assembly of the United Nations in which he announced the launch of an initiative to spread the use of civilian nuclear technology around the world. By the time of Eisenhower's "Atoms for Peace" address, the U.S. had already conducted more than 40 nuclear tests, including the detonation of the world's first thermonuclear device. Eisenhower noted that since WWII, the U.S. nuclear arsenal alone had already grown to the point where it exceeded "by many times the explosive equivalent of the total of all bombs and all shells that came from every plane and every gun in every theatre of war in all of the years of World War II."[18] He painted a grim picture of a nuclear war between the United States and the Soviet Union and concluded that to allow this situation to remain the only reality of nuclear energy was "to confirm the hopeless finality of a belief that two atomic colossi are doomed to malevolently eye each other indefinitely across a trembling world."[19] The alternative he proposed was to convert the face of nuclear energy to one of civilian use. In his speech to the U.N. Eisenhower concluded that

> It is not enough to take this weapon out of the hands of the soldiers. It must be put into the hands of those who will know how to strip its military casing and adapt it to the arts of peace.[20]

Even in the wake of the deadly contamination that followed the Bravo test, the "peaceful" justification of nuclear power was not far removed. In his official statement on the H-Bomb tests in the Pacific, AEC Chairman Lewis Strauss claimed that not only had the tests been "of major importance to our military strength and readiness," but they had also moved the world closer to finally benefiting from the utilization of nuclear power. Strauss concluded his prepared remarks by saying:

> Finally, I would say that one important result of these hydrogen bomb developments has been the enhancement of our military capability to the point where we should soon be more free to increase our emphasis on the peaceful uses of atomic power – at home and abroad. It will be a tremendous satisfaction to those who have participated in this program that it has hastened that day.[21]

While there were intensive programs to develop reactors for generating electricity that began well before December 1953, it was Eisenhower's proposals for the formation of an International Atomic Energy Agency under the United Nations and the launch of an international cooperative effort on the development of nuclear energy that were the true watershed events leading to the rapid expansion of nuclear power around the world. Following Eisenhower's UN address, the first conference at which Soviet and U.S. scientists were able to interact in a meaningful way since the 1930s opened in Geneva for the express purpose of discussing the "peaceful" application of nuclear power.[22]

Understanding these political motivations behind the drive to develop nuclear power helps to explain the significant disconnect between the pronouncements from scientists and public officials and the engineering assessments from those most closely involved with the development of reactors. As originally discovered by Dr. Arjun Makhijani and detailed in *Nuclear Power Deception: U.S. Nuclear Mythology From Electricity "Too Cheap to Meter" to "Inherently Safe" Reactors* by Dr. Makhijani and Scott Saleska, the internal engineering assessments of those within the nuclear complex were in many cases in direct opposition to the public pronouncements of nuclear power's proponents. Specifically, the claim of "cheap" nuclear power was repeatedly and sharply challenged by experts in both the government and industry as discussed in Dr. Makhijani's foreword to the present work.[23] This pessimistic view of nuclear power was not limited to the United States either. In the Soviet Union, the construction of the electricity producing reactor at Obninsk was opposed by some because it was thought to be uneconomical. In fact, the project was seen among some of the leading figures within the Soviet nuclear establishment as little more than a "concession" to Igor Kurchatov, the head of the Soviet nuclear weapons program and strong supporter of the "peaceful utilization of the energy of the atom."[24] Despite the pessimistic outlook for the economic viability of this technology, the nuclear establishment decided to push ahead and aggressively promote the commercial use of nuclear power.

The intense promotion by the government and reactor developers, notwithstanding, most utilities remained hesitant to invest in nuclear power. In 1957, however, two significant events occurred that helped to spark the development of nuclear power. The first was the opening of the Shippingport reactor in Pennsylvania, while the second was the passage in Congress of the Price-Anderson Act limiting the liability of operators of nuclear facilities in the event of a catastrophic accident. Pushed along by large government subsides, nuclear power developed slowly throughout the rest of the 1950s and early 1960s. By 1964, there were only five plants online, with a combined capacity well below that of one large nuclear plant today.[25] All of these facilities had been built with some level of financial support from the government. However, with reactor manufacturers like General Electric and Westinghouse offering to build a number of "first-mover" plants below the actual construction cost, coupled with highly optimistic assessments from the AEC about the economics of future nuclear plants, the 1960s saw what has come to be known as the "Great Bandwagon Market." By the end of 1967, 45 percent of all US nuclear capacity ever brought online had already been ordered.[26]

On the 13 so-called "turnkey" contracts that GE and Westinghouse made in order to stimulate the market, it is estimated that their losses totaled between $875 million and $1 billion.[27] The utilities that did not have fixed cost contracts had to absorb the full extent of cost overruns as the price of nuclear power plants continued to exceed predictions. Over time, the economics of nuclear power grew progressively worse as average capital costs and construction times continued to climb. The average overnight cost of nuclear plants escalated from $817 per kW (in 1988 dollars) for the 13 plants that began operation between 1971 and 1974 to $3,133 per kW (in 1988 dollars) for the 10 plants that started operation in 1987-88. In fact, the least expensive of the 25 plants connected to the grid in 1985-88 was *more expensive* than the most expensive of the 25 plants that entered operation between 1971 and 1976 (see Figure 1.1).[28]

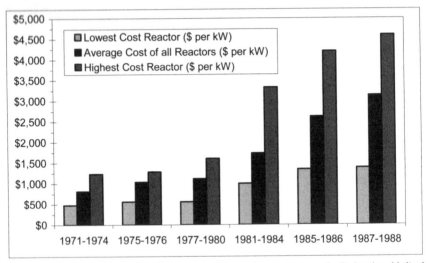

Figure 1.1: Overnight cost of commercial nuclear reactors built in the United States grouped by the date in which they entered operation (all figures in 1988 dollars). The difference between the cost of the most expensive and least expensive reactors completed in the 1980s was more than three and a half times the average cost of the reactors completed in the 1970s.[29]

Coupled to the increase in capital cost, the construction times for nuclear plants also increased significantly. The average lead time for plants that opened prior to 1975 was 5.4 years, while reactors that came online between 1985 and 1989 required an average of approximately 12.2 years to build. The shortest construction time for any of the 47 plants that became operational between 1980 and 1989 was longer than the average time needed to build the 40 plants that came online prior to 1975.[30]

In addition to the rapid increases in both the construction time and construction cost of nuclear power plants, which publicly shattered the myth of "too cheap to meter," events have demonstrated the falsehood of the "peaceful atom" as well. It is significant that, just as the notion that nuclear power would be cheap and plentiful was clearly contradicted well before Strauss's famous speech, the link between nuclear power and the potential for nuclear weapons proliferation was acknowledged well before Eisenhower's "Atoms for Peace" proposal was unveiled. The fact that it is a short step from the development of nuclear power to the development of nuclear weapons was discussed by General Leslie Groves, the head of the Manhattan Project, as early as November 28, 1945, in testimony before the U.S. Congress.[31] In addition, a study commissioned by then Under Secretary of State Dean Acheson in 1946 and chaired by David Lilienthal, concluded that "[t]he development of atomic energy for peaceful purposes and the development of atomic energy for bombs are in much of their course interchangeable and interdependent."[32] The link was again highlighted in the first report of the U.S. Atomic Energy Commission, which was reprinted in the February 21, 1947 issue of *Science*. In that report, the Commissioners acknowledged that, because uranium was the raw material for both nuclear bombs and nuclear reactors, "[t]here is accordingly a very deep and basic relation between weapons and the peacetime uses of atomic energy."[33]

It was these types of concerns over the potential connection between civilian and military uses of nuclear facilities that led to the negotiation of the 1968 Treaty on the Non-Proliferation of Nuclear Weapons.[34] In exchange for aid in developing commercial nuclear power, and a commitment from the nuclear weapons states to pursue "good faith" negotiations on disarmament, the NPT required states that did not already possess nuclear weapons at that time to promise not to manufacture or otherwise acquire them and empowered the International Atomic Energy Agency to be a watchdog on proliferation tasked with conducting inspections of "civilian" nuclear facilities.

Despite the warnings that had been sounded years earlier, the world appeared shocked in May 1974 when India conducted a surprise test of what it called a "peaceful" nuclear device. The plutonium used in this device had been created in a Canadian supplied heavy water research reactor and separated in a facility whose design was based on information received from the U.S. Atomic Energy Commission under the "Atoms for Peace" program. The revelation of the Indian nuclear weapons program caused President Ford and later President Carter to turn the U.S. away from its previous support of using plutonium as a reactor fuel.[35]

The world again appeared startled in June 1981 when Israel chose to invade Iraqi airspace and bomb the French built Osirak reactor, about 15 miles east of Baghdad, due to fears that Saddam Hussein would use the reactor as part of a secret nuclear weapons program.[36] At the time, the bombing was "strongly condemned" by all the members of United Nations Security Council, including the United States. In fact, the Security Council unanimously passed a resolution which stated that the air strike was "in clear violation of the United Nations charter and the norms of international conduct."[37] Five years later the tables were turned somewhat when a former Israeli nuclear technician named Mordechai Vanunu revealed to the world in the October 5, 1986, issue of the London *Sunday Times* that Israel itself had had a secret nuclear program for more than two decades at the Dimona nuclear facility in which its own French built research reactor was being used to create plutonium for atomic bombs.[38]

As the realities of nuclear power displaced the early mythologies of "cheap" and "peaceful" energy, the nuclear industry in the United States and many other countries ground to a halt. A total of 115 reactors were canceled between 1972 and 1984 costing the utilities more than $20 billion.[39] By 1992, a total of 121 reactors had been canceled. A number of other orders were also rescinded before any money changed hands.[40] The last order for a reactor that was actually completed and connected to the grid in the U.S. was placed more than 30 years ago in 1973, and no new reactor orders have even been placed since the two that were signed in 1978 and later canceled.[41] The last nuclear plant to be brought online in the U.S. was the TVA's Watts Bar 1 in Tennessee. This reactor began operation nearly a decade ago in early 1996 after a total of 23 years of construction.[42] In 1985, approximately one year before the Chernobyl disaster in the Soviet Union, a cover story in the business magazine *Forbes* succinctly summed up the economic history of the U.S. nuclear industry as follows:

> The failure of the U.S. nuclear power program ranks as the largest managerial disaster in business history, a disaster on a monumental scale. The utility industry has already invested $125 billion in nuclear power, with an additional $140 billion to come before the decade is out, and only the blind, or the biased, can now think that most of the money was well spent. It is a defeat for the U.S. consumer and for the competitiveness of U.S. industry, for the utilities that undertook the program and for the private enterprise system that made it possible.[43]

Following such a "monumental" failure to live up to the public prophesies of cheap and inexhaustible power, combined with the Indian nuclear test, the Israeli attack on Iraq, as well as the Three Mile Island and Chernobyl accidents, nuclear power in the late-1980s, as in the early 1950s, was in serious need of a face lift.

It just so happened that it was also during the 1980s that it began to be more widely recognized that human activities, particularly the burning of fossil fuels, was adversely impacting the planetary climate. This awareness was heightened by the publicity that followed the 1987 publication by James Hansen, the Director of NASA's Goddard Institute of Space Studies, and Sergej Lebedeff of their groundbreaking analysis showing a sustained anomalous warming of the Earth's surface.[44] Capitalizing on the fact that nuclear power has a low level of carbon emissions compared to fuels like coal, oil, or natural gas, the nuclear power industry turned to climate change in hopes of motivating a nuclear revival. In much the same way that early nuclear proponents sought to highlight a "peaceful" role for nuclear energy in an attempt to overcome concerns about the awesome destructive capacity of nuclear weapons, these later advocates of nuclear power sought to highlight an "environmentally beneficial" role for nuclear energy in their attempt to overcome concerns relating to reactor safety and nuclear waste disposal. As early as 1988, while other major industries such as automobile manufacturers and oil companies were preparing to fund efforts to deny the very existence of global warming, nuclear proponents were writing that "concern about the earth's rising temperature could turn a technological pariah into a savior - if new reactor designs overcome worries about atomic safety."[45] In 1991, the World Bank noted that "growing concerns about the impacts of fossil fuel combustion on the earth's atmosphere have led the nuclear industry to promote nuclear power as a more benign alternative."[46] Today, the rationale of "clean" and "environmentally friendly" nuclear power has come to dominate the discourse over the technology's future.

For example, at the website for the Nuclear Energy Institute, a powerful nuclear industry trade group, one of the most prominent links is entitled "Nuclear the clean air energy."[47] The Department of Energy's *Roadmap to Deploy New Nuclear Power Plants in the United States by 2010* highlights the fact that nuclear power is a "clean" source of electricity and that it "does not emit greenhouse gases."[48] The DOE roadmap goes on to point out that nuclear power's lack of carbon emissions is "an attribute of increasing importance in the U.S. and around the world."[49] In May 2003, New Hampshire became the first state to offer credits for new nuclear capacity as part of its efforts to reduce greenhouse gas emissions.[50] In the ongoing British debate over climate change, Prime Minister Tony

Blair, Former Environment Ministers Michael Meacher and John Gummer, and environmentalist James Lovelock have all declared their belief that nuclear power plants should play a role in reducing carbon dioxide emissions.[51] Even the U.S. *National Energy Policy* crafted by Vice President Dick Cheney and his energy task force linked their promotion of nuclear power with the fact that these plants "discharge no greenhouse gases into the atmosphere."[52]

This last example of promoting nuclear power on environmental grounds is particularly noteworthy because, despite campaign promises to regulate carbon dioxide as a pollutant, the Bush administration has consistently refused to act on the Kyoto Protocol. In fact, the Administration actually pulled out of negotiations in Morocco in the fall of 2001 in which 38 of the 39 countries required by Kyoto to limit emissions agreed to collectively achieve a 5.2 percent reduction over 1990 levels. On November 10, 2001, the agreement reached in Morocco, without U.S. participation, was signed by a total of 178 countries.[53] However, it was not until June of 2002 that the Bush administration even acknowledged publicly the scientific consensus regarding greenhouse gas emissions and global climate change. It is telling that, despite their record on the issue of climate change, the energy policy crafted by the Bush/Cheney White House echoes the modern promotional theme for new reactors by highlighting the low carbon emissions of nuclear power as one of its main selling points.

Over the past 20 years, a great deal has been learned about the functioning of the Earth's climate and about the relationship between local, regional, and global changes. Before examining the potential role envisioned for nuclear power by its proponents in reducing the emissions of greenhouse gases, we will first briefly review the current consensus surrounding the extent of future climate disruption, as well as the potential consequences of those changes in order to better understand the scale and urgency of the reductions in emissions that will be needed.

Section 1.2 - The Realities of Climate Change

The threat of global climate change is perhaps the largest single vulnerability associated with the world's current energy system. Outside of full scale thermonuclear war, it is perhaps the largest single environmental vulnerability of any kind facing the world today. It has been more than 175 years since Fourier first theorized that the Earth's atmosphere works in the same way as glass surrounding a hot-house, and more than 100 years since Svante Arrhenius, the Nobel Prize winning chemist, first put forth the theory that CO_2 in the atmosphere, including that released from the burning of fossil fuels, could trap enough heat to cause a widespread warming of the Earth's surface.[54] Today, after more than a century of rapid industrialization based on ever expanding fossil fuel usage,[55] there is an overwhelming consensus among the world's scientific and political communities that human impacts on the Earth's climate are both real and a cause for serious concern.

Worries over the potential consequences of climate change and the ineffective actions taken by the federal government[56] has prompted a number of U.S. cities and states to take matters into their own hands. For example, in June 2003, three states (Connecticut, Maine, and Massachusetts) sued the Environmental Protection Agency for failing to regulate CO_2 emissions under the Clean Air Act.[57] In July 2004, eight states (California, Connecticut, Iowa, New Jersey, New York, Rhode Island, Vermont, and Wisconsin) as well as the city of New York filed a lawsuit against five of the largest power providers in the U.S. in a bid to force them to reduce their carbon emissions by approximately 3 percent per year over the next 10 years.[58] As of December 8, 2005, 195 mayors from cities in 38 states and the District of Columbia representing an estimated 40 million people had joined the U.S. Mayors' Climate Protection Agreement which commits their cities to meeting or exceeding the limits set by the Kyoto Protocol.[59] Finally, in June 2005, Governor Schwarzenegger signed Executive Order S-3-05 setting a goal of reducing the greenhouse gas emissions from the State of California to their year 2000 levels by 2010, to their 1990 levels by 2020, and to 80 percent below their 1990 levels by the year 2050.[60]

The Intergovernmental Panel on Climate Change (IPCC), an organization of the world's leading climate scientists chartered through the World Meteorological Organization and the United Nations Environment Program in 1988, is the world's leading authority on the scientific consensus surrounding the influence of human activities on the planetary climate. In their 2001 assessment report, the panel concluded that "most of the observed warming over the last 50 years is likely to have been due to the increases in greenhouse gas concentrations."[61] The current level of carbon dioxide in the atmosphere is higher than it has been in at least the last 420,000 years, and is likely the highest it has been in the past 20 million years. By 2100, the IPCC predicts that the concentration of atmospheric CO_2 will increase by 55 to 300 percent over the current level, and that by the end of this century the Earth's average temperature will have increased by 1.4 to 5.8 °C.[62] The panel concluded that "[t]he projected rate of warming is much larger than the observed changes during the 20th century and is very likely to be without precedent during at least the last 10,000 years."[63] Accompanying this increase in average temperature, the IPCC predicted that the world's oceans will rise between 9 and 880 cm by 2100. Their median estimate of nearly a half meter rise is two to four times larger than the increase in ocean levels recorded during the 20th Century.[64] These estimates, however, did not include any significant melting from the Greenland ice sheets. Unfortunately, these glaciers are now thought to be at potential risk if carbon emissions continue to grow unchecked.

The magnitude of potential temperature increases presented by the IPCC in their last assessment is not an absolute range. Work conducted since 2001 has found that even larger climate changes than those reported by the IPCC are possible. For example, the *climateprediction.net* experiment, which is being run by a collaboration of British scientific institutions, used distributed computing techniques and the combined power of a large number of individual personal computers to carry out more than two thousand unique global warming simulations. From this ensemble, the researchers identified realistic models with climate sensitivities much higher than those identified by the IPCC in their 2001 assessment. The models used in the ensemble had a response to a doubling of CO_2 over pre-industrial levels which ranged from an increase in average temperature of less than 2 °C to an increase of more than 11 °C.[65] While these results highlight the possibility that larger climate sensitivities may need to be considered, particularly in light of the potential consequences from such large global temperature changes, it is still generally believed that the most likely value for the actual climate sensitivity lies within the range cited by the IPCC in 2001.[66]

When considering the relative costs, benefits, and risks of strategies proposed to help mitigate the damage caused by greenhouse gas emissions, it is important to have a clear idea of how significant the impacts of climate change might be and what the costs of doing nothing could amount to. With global warming comes such possibilities as an increased intensity and frequency of storms and natural disasters including floods and droughts, additional pressures on ecosystems and agricultural capacity, and changing patterns of disease. While a full review of all the potential impacts from climate change is far beyond the scope of this report, a brief review will suffice to illustrate the seriousness of the problem we now face and highlight the immediacy of the need for action.

In the past, for example, when the global average temperature was up to 2 °C higher than today due to natural causes, there were catastrophic floods that are estimated to have been four to ten times more severe than typical floods seen today. To put this into perspective, the flooding in 2004 caused by the monsoon season in India, Bangladesh, Nepal, and Pakistan destroyed crops over a vast area and left more than 2,000 people dead and millions homeless.[67] According to the IPCC

> The frequency and magnitude of many extreme climate events increase even with a small temperature increase and will become greater at higher temperatures (high confidence). Extreme events include, for example, floods, soil moisture deficits, tropical and other storms, anomalous temperatures, and fires. The impacts of extreme events often are large locally and could strongly affect specific sectors and regions. Increases in extreme events can cause critical design or natural thresholds to be exceeded, beyond which the magnitude of impacts increases rapidly (high confidence).[68]

In particular, one consequence of global warming is likely to be a gradual increase in the intensity of hurricanes and typhoons as measured both by the amount of precipitation and by the maximum wind-speed. If such increases in average intensity were to occur, it would also necessarily lead to an increase in the likelihood of highly destructive category-5 storms.[69] The 2004 and 2005 hurricane seasons provided a stark illustration of what unusually severe storms can do. Between August 13 and September 25, 2004, Florida was struck by four separate hurricanes (one category-2, two category-3, and one category-4) causing a combined total of $26.8 to $30.8 billion in insured damage in that state alone. Just one of these storms, Hurricane Jeanne, killed more than 1,500 people in Haiti and left many more homeless.[70] In late August 2005, Hurricane Katrina devastated the U.S. Gulf Cost flooding large parts of the city of New Orleans and other areas. This storm drove more than 270,000 people into temporary shelters and as of October 14, 2005 more than 160,000 evacuees were still living in hotels at the government's expense. By early October an estimated 1,242 people had died in the states of Louisiana (1,003 people), Mississippi (221 people), Florida (14 people), Georgia (2 people), and Alabama (2 people) as a result of Katrina. Overall, the estimated costs from this hurricane are expected to total more than $200 billion.[71]

Although it is often called simply "global warming", the changes to the Earth's climate will be far from uniform warming. The local and regional variations in the climate's response to increased atmospheric CO_2 will result in significant impacts on ecosystems, and will likely add to the already rapid loss of biodiversity. The United Nations 2005 Millennium Ecosystem Assessment summarized the current situation as follows[72]

> Over the past 50 years, humans have changed these ecosystems more rapidly and extensively than in any comparable period of time in human history, largely to meet rapidly growing demands for food, fresh water, timber, fiber and fuel. This has resulted in a substantial and largely irreversible loss in the diversity of life on Earth.[73]

The Millennium Assessment went on to conclude that

> By the end of the century, climate change and its impacts may be the dominant direct driver of biodiversity loss and changes in ecosystem services globally... The balance of scientific evidence suggests that there will be a significant net harmful impact on ecosystem services worldwide if global mean surface temperature increases more than 2° Celsius above preindustrial levels or at rates greater than 0.2° Celsius per decade (*medium certainty*).[74]

Similar conclusions were reached by the IPCC in their review prepared for the United Nations Convention on Biological Diversity. In this assessment, the IPCC concluded that due to the impacts of climate change

> The risk of extinction will increase for many species, especially those that are already at risk due to factors such as low population numbers, restricted or patchy habitats, limited climatic ranges, or occurrence on low-lying islands or near the top of mountains.... Without appropriate management, rapid climate change, in conjunction with other pressures, will cause many species that currently are classified as critically endangered to become extinct, and several of those that are labeled endangered or vulnerable to become much rarer, and thereby closer to extinction, in the 21st century.[75]

Unique or threatened ecosystems such as coral reefs and mangrove forests are particularly vulnerable to the detrimental impacts of climate change. In addition, one of the areas most likely to feel the earliest and largest overall impacts from global warming will be the Arctic. Already the impacts from the recent warming trends are beginning to be felt with the reduction in floating ice and earlier thaws throughout the region. Large expanses of the Alaskan and Siberian permafrost have begun to melt, leading to significant changes in the landscape and ecosystem. The melting of the permafrost will also likely result in further warming as large amounts of previously trapped methane and carbon dioxide are released from the ground into the atmosphere and the loss of snow and ice cover leads to greater amounts of heat being absorbed from the sun.[76]

The likely impacts from continued warming in the Arctic have been extensively investigated by the Arctic Council, an intergovernmental organization composed of representatives from the arctic states, Indigenous Peoples organizations, as well as other interested governments, scientific bodies, and non-government organizations.[77] The Arctic Assessment concluded that ice-dependent seals such as the ringed, ribbon, and bearded seals are "particularly vulnerable to the observed and projected reductions in arctic sea ice because they give birth to and nurse their pups on the ice and use it as a resting platform." Additional negative impacts are likely for other animals in the marine ecosystem as well. For example, the population of ivory gulls in Canada has decreased by 90 percent over the past twenty years as sea ice has retreated from costal nesting grounds. The largest impacts, however, are expected to be on the polar bear populations since they are at the top of the Arctic food chain and are also dependent on the sea-ice. The impacts of climate change will be exacerbated by the fact that polar bears are already being adversely affected by the bioaccumulation of heavy metals and other chemical pollutants present in their environment.[78] In light of these facts, the Arctic Assessment concluded that

> Polar bears are unlikely to survive as a species if there is an almost complete loss of summer sea-ice cover, which is projected to occur before the end of this century by some climate models. The only foreseeable option that polar bears would have is to adapt to a land-based summer lifestyle, but competition, risk of hybridization with brown and grizzly bears, and increased human interactions would then present additional threats to their survival as a species. The loss of polar bears is likely to have significant and rapid consequences for the ecosystems that they currently occupy.[79]

It will not be only sea-ice dependent species that will likely be affected. The increase in freezing rain compared to snow, the earlier spring thaws, the more frequent freeze-thaw cycles, and the northward shift of vegetation under sustained warming will all act to displace tundra vegetation and impair foraging condition necessary for reindeer and caribou populations. These impacts could result in a decline in the animals' populations with dramatic impacts on the way of life for many arctic residents, primarily Indigenous Peoples.[80]

Already an estimated 25 percent of the entire world's mammals and 12 percent of the world's birds are believed to be facing "a significant risk of global extinctions" making them more vulnerable to the impacts of climate change and other stressors.[81] A recent analysis of more than 1,100 plant and animal species found that an average of 18 to 35 percent of these species could be "committed to extinction" over the next century in the event of temperature increases ranging from just 0.8 °C to more than 2.0 °C. For temperature changes in the range of 1.8 to 2.0 °C, the researchers found that an average of nearly one-fourth of all species they examined could be committed to extinction by the century's end.[82] While there are many unknowns with such assessments,[83] the authors concluded that

> Despite these uncertainties, we believe that the consistent overall conclusions across analyses establish that anthropogenic climate warming at least ranks alongside other recognized threats to global biodiversity. Contrary to previous projections, it is likely to be the greatest threat in many if not most regions. Furthermore, many of the most severe impacts of climate-change are likely to stem from interactions between threats, factors not taken into account in our calculations, rather than from climate acting in isolation. The ability of species to reach new climatically suitable areas will be hampered by habitat loss and fragmentation, and their ability to persist in appropriate climates is likely to be affected by new invasive species.[84]

Beyond the increases in global temperature, the increased amounts of CO_2 that are being dissolved in the oceans create an additional threat to marine ecosystems. Between 1800 and 1994, it is estimated that the oceans absorbed approximately 48 percent of all anthropogenic CO_2 emissions amounting to approximately 118 billion additional tons of carbon being dissolved in the oceans. This is estimated to be approximately 30 percent of the total capacity of the sea to absorb CO_2.[85] By the mid-90s, this additional carbon, which is dissolved in part as carbonic acid, had already led to an decrease in the ocean's pH by 0.1 units.[86] The CO_2 currently dissolved in the ocean is concentrated near the surface, and has a relatively slow mixing time with deeper water. Nearly half of all dissolved carbon dioxide is found at depths shallower than 400 meters. The vertical mixing of the thermohaline ocean circulation, however, allows the North Atlantic to store a larger fraction of the CO_2 than its surface area would imply.[87] The vertical transport of CO_2 in the ocean is dominated by biologically mediated processes. These processes include passive transport mechanisms such as the sinking of shells, bones, fecal matter, and dead organisms as well as active transport mechanisms such as the migration of zooplankton to the deep ocean. Significantly, these processes themselves may be disrupted by the impacts of climate change on marine ecosystems.[88] Under one of the future emission scenarios considered by the IPCC, the ocean pH would be predicted to decrease by a further 0.3 units by 2100, leading to a net increase in hydrogen ion concentration of two and half times over the pre-industrial level. Even larger increases in the ocean acidity are possible if CO_2 emissions were to continue on their current business-as-usual trend.[89]

Experiments have shown that reduced pH and increased levels of dissolved carbon dioxide in water can lead to, among other effect,

- reduced growth rates in prawns and juvenile white sturgeon
- decreased egg size and delayed hatching in perch
- reduced sperm motility in white sturgeon
- decreased calcification rate for organisms with shells or other exoskeletal structures (such as coral) and a dissolution of calcified structures that have already formed
- reduced growth and reproduction in mollusks
- impaired cardiac function in fish
- reduced egg production and hatching rate in marine copepods
- reduced body length and fertilization rates in sea urchins
- morphology changes and malformed skeletogenesis in sea urchins.[90]

Additional concerns are raised by (1) the fact that experiments have shown that, when tested at the same pH levels, seawater acidified by elevated dissolved CO_2 was significantly more toxic than water acidified by the addition of hydrochloric acid[91]; (2) the fact that the tolerable pH range for marine plankton has been shown to decrease with increasing exposure time[92]; (3) the fact that the susceptibility of fish species to acute lethality from dissolved CO_2 was found to vary significantly with developmental stage being largest in the early egg and juvenile stages and least in the early larval stage[93]; and (4) that there are likely to be synergisms between the deleterious impacts of increased water temperature and increased CO_2 on many marine species.[94]

While marine populations have traditionally been viewed as less prone to extinction from human impacts than terrestrial populations, recent work has shown this belief to likely be "too general and insufficiently conservative."[95] A number of fisheries in both the Atlantic and Pacific are threatened and are currently the source of international efforts to try and restore their health. For example, after a century of exploitation, the Atlantic cod fisheries collapsed in 1992 resulting in a suspension of fishing.[96] In fact, a number of marine species have already been pushed to almost certain local or regional extinction as a result of human influences.[97] With respect to the impacts of global warming and increased levels of dissolved CO_2, the species likely to be most heavily impacted are those that rely on calcified exoskeltons such as coral.[98] Warm water coral reefs around the world are already suffering from bleaching due to increased temperatures and damage from other causes. In fact, since 1980, nearly 30 percent of the warm-water coral reefs have already disappeared. The impact of rising ocean temperatures, increasing acidification, and alterations to ocean cycles such as the El Nino Southern Oscillation will add further to this stress, increasing concerns over local or regional extinctions.[99] As summarized by the U.K. Royal Society

> Organisms will continue to live in the oceans wherever nutrients and light are available, even under conditions arising from ocean acidification. However, from the data available, it is not known if organisms at the various levels in the food web will be able to adapt or if one species will replace another. It is also not possible to predict what impacts this will have on the community structure and ultimately if it will affect the services that the ecosystems provide. Without significant action to reduce CO_2 emissions into the atmosphere, this may mean that there will be no place in the future oceans for many of the species and ecosystems that we know today. This is especially likely for some calcifying organisms [such as coral].[100]

As noted, the changing climate will have many direct and indirect effects on human health. These impacts will go far beyond the likely increase in heat related deaths and the decrease in cold related fatalities that would accompany an overall warming of the planet. In its 2001 assessment, the IPCC concluded that

> Global climate change will have a wide range of health impacts. Overall, negative health impacts are anticipated to outweigh positive health impacts. Some health impacts would result from changes in the frequencies and intensities of extremes of heat and cold and of floods and droughts. Other health impacts would result from the impacts of climate

> change on ecological and social systems and would include changes in infectious disease occurrence, local food production and nutritional adequacy, and concentrations of local air pollutants and aeroallergens, as well as various health consequences of population displacement and economic disruption.[101]

Similar conclusions were reached in a 2003 analysis by the World Health Organization.[102] In fact, the existing models suggest that climate change may already be detrimentally affecting human health and increasing the global burden of disease. While this effect is currently small in comparison to other major risk factors such as pollution, the risks posed by climate change are believed to be "increasing rather than decreasing."[103] Like many of the impacts of climate change, the health impacts will hit hardest in areas of the Global South where the heightened vulnerability of these populations will hinder their efforts at adaptation.[104]

On top of the impacts addressed above which relate to relatively gradual climate changes that will allow humans and ecosystems some time to adapt, there are additional concerns raised by the possibility that gradual warming could spark rapid and dramatic changes in the Earth's climate. Such rapid shifts from one climate state to another have been observed in the past and are still only partially understood. These rapid changes in climate conditions are believed to have played an important role in past changes to human lifestyles, including the complete collapse of some ancient societies.[105] In 2002 the National Research Council of the U.S. National Academy of Sciences concluded that

> Abrupt climate changes were especially common when the climate system was being forced to change most rapidly. Thus, greenhouse warming and other human alterations of the earth system may increase the possibility of large, abrupt, and unwelcome regional or global climactic events. The abrupt changes of the past are not fully explained yet, and climate models typically underestimate the size, speed, and extent of those changes. Hence, future abrupt changes cannot be predicted with confidence, and climate surprises are to be expected.[106]

In a 2003 White Paper on the future of the British energy system, the Department of Trade and Industry painted the following picture of the potential impact of rapid climate change

> Worldwide, the consequences [of global warming] could be devastating, especially in the developing world where many

millions more people are likely to be exposed to the risk of disease, hunger and flooding. In addition, there is a risk of large scale changes such as the shut-down of the Gulf Stream or melting of the West Antarctic ice sheet, which although they may have a very low probability of occurring, would have dramatic consequences.[107]

The models examining the possibility that increased global warming may lead to a long-term weakening or even total shutdown of the thermohaline circulation (Gulf Stream) have shown that the likelihood of these outcomes are strongly affected by many assumptions, including the choice of climate sensitivity, the rate of average global warming, and the rate at which Greenland's ice sheets melt. These models have consistently shown that such a shutdown is possible (particularly when the feedback from potential melting of the Greenland ice-sheets is included), and that such a shutdown could lead to sustained, long-term cooling in the North Atlantic and over Europe as a result.[108]

As noted, the models considered by the IPCC in 2001 did not include any significant contribution from the melting of the Greenland or Antarctic ice-sheets.[109] Recent modeling results and experimental measurements have begun to raise questions about the long-term stability of these ice-sheets, and thus about the IPCC's conclusion that a shutdown of the thermohaline circulation is not likely to occur within the next century. For example, in February 2002, the 3,250 km^2 Larsen B ice shelf broke up over the period of just a few weeks releasing approximately 500 billion tons of ice as icebergs. Following this collapse, the movement of several of the West Antarctic glaciers behind the former ice shelf has significantly accelerated and these glaciers have begun to thin more rapidly. This loss of ice is already contributing to annual sea level rise and raises serious concerns regarding the long-term stability of the Antarctic ice sheets.[110] Professor Chris Rapley, Director of the British Antarctic Survey and former Executive Director of the International Geosphere-Biosphere Programme at the Royal Swedish Academy of Sciences, summarized the importance of these recent finds as follows

> Satellite measurements tell us that a significant part of the West Antarctic ice sheet in this area is thinning fast enough to make a significant contribution to sea level rise, but for the present, our understanding of the reason for this change is little better than hypothesis. The last IPCC report characterised Antarctica as a slumbering giant in terms of climate change. I would say that this is now an awakened giant. This is a real cause for concern.[111]

The Greenland ice sheets are showing similar phenomena, although for different reasons. Recent measurements have shown that the loss of ice from Greenland has accelerated recently compared to the 1990s. These additional losses are due to both higher summer melting and accelerated glacier movements. Earlier models have also shown that significant long-term melting is possible for scenarios with large, sustained increases in atmospheric CO_2 concentrations. If a major portion of the ice sheets on Greenland were to melt, it would alone be sufficient to raise sea levels by several meters.[112]

A weakening or total shutdown of the thermohaline circulation would have wide reaching impacts on weather patterns throughout the world. The shutdown would also impact marine ecosystems and coastal fisheries and reduce the ocean's ability to absorb atmospheric carbon dioxide. Examples of the possible impacts on humans from a sustained cooling in Europe can be found in the history of the Middle Ages. In 1250 the climate in northern Europe rapidly cooled over the period of a decade. A second cooling trend after 1600 led to what is known as the Little Ice Age.[113] Edward Bryant, a professor of Geosciences and advisor to the Australian government on climate change, describes the impact of these periods of cooling as follows:

> Society in the thirteenth century had progressed to such an extent in Europe that the Industrial Revolution almost took place. However this breakthrough was interrupted by the climate deterioration that began about 1250. Not only did climate changes of only 1°C require a major agricultural adaptation, but they also had a major impact upon social fabric, human health and survival.[114]

Heavy rains between 1315 to 1317 in Europe destroyed crops resulting in a widespread famine that killed approximately 10 percent of the population.[115] The impacts of such rapid climate change on human health could also be far more dramatic than those considered for gradual change. For example, changes in climate are believed to have helped facilitate the spread of the Black Plague which ultimately killed roughly 50 percent of the Chinese population between 1200 and 1400, and led to the deaths of 15 to 60 percent of the population in various regions of Europe between 1350 and 1550.[116]

Considering the potential consequences arising from rapid climate changes, Andrew Marshall, the head of an influential think-tank within the Department of Defense since 1973, commissioned an official Pentagon study to examine the national security implications of global warm-

23

ing. The two authors Marshall tapped for this work were Peter Schwartz, the former head of strategic planning at Royal Dutch/Shell, and Doug Randall, from the Monitor Group's Global Business Network.[117] The Pentagon study did not attempt to conduct a complete risk analysis, nor did it seek to make a prediction about what is most likely to happen, but it did provide important insight into the potential impacts from a very rapid shift in the global climate.

The scenario considered by the author's of the Pentagon study was a continued warming trend through 2010 followed by a decade long cooling similar to the one that occurred 8,200 years ago. The period of cooling would be triggered by the shutdown of the thermohaline circulation described above. Relying on the historical analog, they considered a drop in temperatures throughout Asia and North America of 2.8°C, while those in Northern Europe were estimated to drop by 3.3 °C. These changes were assumed to spark decade long droughts in a number of agricultural regions throughout the world as well as in major population centers in Europe and the eastern United States. Winter storms and winds increased in magnitude adding to the impacts of the overall climate change. Globally, crop yields were assumed to be reduced by 10 to 25 percent, and that they would become less predictable from year to year due to climate fluctuations.[118]

The authors concluded that the societal pressures caused by these ecological changes might lead more resource poor countries to "initiate in struggles for access to food, clean water, or energy."[119] Refugees displaced by changing land use patterns, fishing rights along coastal and inland waters, and access to fresh water are all possible sources of conflict between countries identified in this case study. Given the already limited supplies of fresh water available today, the authors point out the potential for conflict surrounding some of the 200 river basins that touch multiple countries. These rivers include some of the world's most important water ways such as the Danube (touching ten countries in Europe), the Nile (touching nine countries in Africa), and the Amazon (touching six countries in South America).[120]

While the Pentagon analysis concluded that the likely impacts within the U.S. and western Europe would be manageable given their large economic and resource bases, it is important to recall that the Great Depression of the 1930s was made far more severe by the drought that struck the U.S. west. The mass internal migration caused by people leaving the "Dust Bowl" to look for work in places like California is well documented. Even relatively minor losses in economic growth triggered by increases in food and energy costs resulting from climate impacts could have significant impacts on millions of people in the U.S. where the gap between rich and poor continues to grow ever more pronounced and many families already live below the poverty line. In 2005, Hurricane Katrina showed the potential for a natural disaster to lead to the displacement of several hundred thousand people even within the United States. Finally, there are already internal conflicts brewing within the United States over access to fresh water supplies. These conflicts are not limited to the Southwest where much of the attention on water issues is typically focused. Inter and intra-state disputes over access to fresh water resources have begun to arise in many areas of the U.S. including in the South, along the East coast, and in the Pacific Northwest. These conflicts would likely increase in at least some areas as climate change further alters the local and regional hydraulic cycle.

Finally, perhaps one of the most important conclusions presented in the Pentagon study with respect to the present discussion on the future of nuclear power, is the authors belief that the heightened potential for resource wars could make further nuclear proliferation "inevitable."[121] In addition to the nine current states that are known or believed to possess nuclear weapons (the United States, Russia, China, France, Britain, India, Pakistan, Israel, and North Korea) the Pentagon case study included nations such as Japan, South Korea, Germany, Iran, and Egypt among those that may potentially acquire nuclear weapons in response to the national security pressures brought about by the changing climate. Importantly, the authors note that the spread of nuclear weapons would be facilitated by the expansion of nuclear power "as countries develop enrichment and reprocessing capabilities to ensure their national security."[122]

While the uncertainties surrounding the potential impacts of global warming are often large and numerous, the possible outcomes are so varied and so severe in their ecological and human impacts that no time should be lost in trying to mitigate the damage that humanity has done and continues to do to the Earth's climate. Definitive proof will only come following a catastrophe, and by then it will be too late to effectively take action. The potential impacts highlighted above combined with our rapidly evolving understanding of the climate system provides a strong motivation to prioritize mitigation strategies that are economically reasonable and that will have the largest likelihood of making significant contributions in the near to medium term (the next one to four decades) while not jeopardizing the future implementation of sustainable long-term strategies. It is in this light that we examine the question of what role, if any, nuclear power might play in strategies to combat global warming.

Section 1.3 - Case Study: the MIT Nuclear Power Report

There are many ways in which an expansion of nuclear power might be envisioned as part of efforts to reduce greenhouse gas emissions. In order to make our current investigation tractable, therefore, we chose to consider in detail two representative examples of nuclear growth scenarios. Arguably one of the most important analyses supporting the use of nuclear power in the future is a 2003 study from the Massachusetts Institute of Technology entitled *The Future of Nuclear Power*.[123] Given the importance of MIT in the scientific and technical communities, and the broad interdisciplinary nature of the Institute's investigation, we chose to make this report our primary focus.

The MIT report begins with the important acknowledgement that

> The generation of electricity from fossil fuels, notably natural gas and coal, is a major and growing contributor to the emission of carbon dioxide - a greenhouse gas that contributes significantly to global warming. We share the scientific consensus that these emissions must be reduced and believe that the U.S. will eventually join with other nations in the effort to do so.[124]

In order to contribute to a reduction in future greenhouse gas emissions, the MIT report envisions a "global growth scenario" with a base case of 1,000 GW of nuclear capacity installed around the world by 2050. It is interesting to note that 1,000 gigawatts of capacity is what the nuclear industry expected, in 1970, to be online in the U.S. alone by the year 2000. All of the reactors in operation today would be shutdown by mid-century, and therefore the net increase represented by the global growth scenario over today's effective capacity would be roughly 230 percent.[125] To give a sense of the scale of this proposal, the MIT scenario would require a new reactor to come online somewhere in the world every 15 days on average between 2010 and 2050. In the US alone, where 300 reactors are envisioned under the MIT growth scenario, there would have to be one new power station completed every 50 days on average. For comparison, the actual rate of plant construction around the world over the last 25 years has averaged roughly one plant beginning operation every 23 days.[126] The authors of the MIT report point out the difficulty this level of growth would pose, and note that "[t]he implied construction rate near the mid-century endpoint of the global growth scenario would be challenging and exceed any rate previously achieved."[127]

One might imagine that a tripling of installed nuclear capacity would have a significant impact on reducing carbon emissions, however, this turns out not to be the case for the MIT projections. This is due to the fact that the overall electricity demand in their scenario would increase nearly as rapidly as the installed nuclear capacity. In the MIT base case, nuclear power would account for a total of 19.2 percent of the world's electricity in 2050 compared to 16.3 percent in the year 2000.[128] Therefore, even with all the time and money that would have to be spent on the revival of the nuclear power industry, the carbon emissions from the electricity sector would be predicted to continue increasing over the next four decades. In the United States, where the largest share of nuclear construction is assumed to occur, the carbon emissions from electricity production in 2050 would be predicted to *increase* by approximately 13 to 62 percent over their year 2000 levels under the global growth scenario.[129]

In order to consider the implications of a more aggressive scenario in which a more serious effort is made to limit carbon emissions through the expanded use of nuclear power, we have developed our own proposal which we call the steady-state growth scenario. Using the same level of projected growth in global electricity consumption as assumed by the authors of the MIT analysis (i.e. 2.1 percent per year), we calculated the number of nuclear reactors that would be required in 2050 in order to simply *maintain* the carbon emissions from the electricity sector at their year 2000 levels.[130] Much steeper reductions in overall emissions will likely be required in order to avoid the more serious consequences of climate change, but we have used this set of assumptions as an illustrative example. Making a range of assumptions about the contribution of renewables (20 to 40 percent of total generation in 2050) and about the contribution of natural gas fired plants (1.5 to 2.5 times the current level of generation), we find that between 1,900 and 3,300 gigawatts of nuclear capacity would be required world wide. For simplicity we will use a value of 2,500 gigawatts of nuclear power online in 2050 as our alternative case study for a revival of nuclear power.[131]

The following chapters will examine the implications of pursuing the global or steady-state growth scenarios. We will begin by examining the economics of nuclear power and compare the cost of these carbon mitigation strategies to those of alternatives including increased energy efficiency and the use of renewables, liquefied natural gas, and coal gasification with carbon sequestration. Each of these options is capable of being commercialized in the near to medium term and they are all capable of making significant contributions to a reduction in greenhouse gas emissions. The last three chapters will address the unique vulnerabilities of nuclear power in some detail. First, we will consider the potential impact on nuclear weapons proliferation that would accompany the spread of fuel cycle facilities around the world. Second, we will examine the issue of reactor safety and the risks posed by the potential for a core meltdown or cooling pool fire to release large amounts of long-lived radionuclides into the environment causing significant loss of life and long-term contamination of vast areas. Third, we will conclude by discussing the issue of nuclear waste and the difficulties that have and will continue to be experienced in efforts to safely manage this very dangerous material over unprecedented timescales. Throughout this discussion we will find that nuclear power does not make sense from a security or a human and environmental health perspective. Trading the vulnerabilities of climate change for the vulnerabilities of nuclear power is not a sound energy policy technologically and economically viable alternatives are available.

Chapter Two: The White Elephant

> The bottom line is that with current expectations about nuclear power plant construction costs, operating cost and regulatory uncertainties, it is extremely unlikely that nuclear power will be the technology of choice for merchant plant investors in regions where suppliers have access to natural gas or coal resources. It is just too expensive.[132]
> - *The Future of Nuclear Power* (2003)

> The failure of the U.S. nuclear power program [through the mid-1980s] ranks as the largest managerial disaster in business history, a disaster on a monumental scale. The utility industry has already invested $125 billion in nuclear power, with an additional $140 billion to come before the decade is out, and only the blind, or the biased, can now think that most of the money was well spent. It is a defeat for the U.S. consumer and for the competitiveness of U.S. industry, for the utilities that undertook the program and for the private enterprise system that made it possible.[133]
> - James Cook in *Forbes Magazine* (1985)

Concerns over reactor safety, nuclear weapons proliferation, and radioactive waste management have had a powerful impact on the development of nuclear power and are significant vulnerabilities associated with any future revival of this technology. However, the failure of nuclear power to date has been due in large part to its inability to compete economically with other sources of base load generation such as coal and natural gas. The near-term future of nuclear power, therefore, rests heavily on the predictions for the cost of building and operating the next generation of reactors compared to the cost of competing technologies. When considering what those costs might be, a distinction has to be made between building reactors with designs similar design to those already in existence and building new types of reactors that have not yet been widely commercialized. The first category includes the evolutionary and advanced light-water reactors, while the second category includes the Generation-IV designs such as the pebble bed reactor being developed by an international consortium in South Africa.

Given the need for significant reductions in carbon emissions to be made over the next few decades in order to avert the more serious potential consequences of global warming (see Section 1.2), the reactor designs most likely to play a role would be those in an advanced state of development, namely the light-water reactor designs. Examples of these designs include:

> Evolutionary Light-Water Reactors
> - Advanced Boiling Water Reactor (ABWR) developed by General Electric
> - System 80+ PWR developed by Combustion Engineering
> - SP-90 advanced PWR developed by Westinghouse
>
> Advanced Light-Water Reactors
> - The Simplified Boiling Water Reactor (SBWR) developed by General Electric
> - The Safe Integral Reactor (PWR) developed by a joint effort led by Combustion Engineering
> - Advanced Passive PWR (AP-600 and AP-1000) developed by Westinghouse

In early 1996 the first ABWR from General Electric to be completed began generating electricity in Japan. In the United States, both the General Electric and Combustion Engineering designs noted above received certification from the Nuclear Regulatory Commission in 1997, while the Westinghouse AP-600 received design certification from the NRC in 1999, and the AP-1000 received design certification in January 2006. The evolutionary designs typically have a large capacity (on the order of 1,100 MW) and do not have safety features fundamentally different than those fielded today. The advanced designs, on the other hand, are generally smaller (between 320 and 750 MW) with a lower power density in the reactor cores, a simplified design, and passive safety features such as gravity fed emergency cooling water supplies that do not require as many pumps or as much electric power to operate in the event of an accident. Some advanced designs, however, like the AP-1000, have larger capacities.[134]

In addition to new light-water reactors designs, one of the fundamentally different designs that is believed to offer some potential for making a significant contributing to new nuclear construction before 2050 is the high temperature gas cooled reactor (HTGR). Assuming "success at every turn" the authors of the MIT study believe that HTGRs could "make up as much as one-third" of the capacity envisioned in their global growth scenario. However, due to the smaller size compared to light-water reactors (the HTGR designs are typically proposed in the 125 to 350 MW range), there would have to be between 1.4 to 4 times as many gas cooled reactors built as light-water reactors in order to achieve a total combined capacity of 300 GW in the U.S.[135]

With respect to the cost of these new reactor designs, the authors of the MIT study noted that "past operating experience with HTGR plants, at Peachbottom, at Fort St. Vrain, and in Germany is mixed and there is no reliable basis on which to estimate the economics of HTGR plants relative to LWR plants."[136] Even this, however, is a somewhat optimistic reading of history. In 1959, the U.S. decided to build a 30 MW experimental gas-cooled reactor near Oak Ridge National Laboratory that would produce electricity for the government owned Tennessee Valley Authority. With its costs escalating rapidly, the project was eventually canceled in 1965, leaving the reactor only half finished. As remembered by Alvin Weinberg in 1994, the director of Oak Ridge during the time of this reactor's development, "the whole abandoned project stands to remind me, as I drive along Bethel Valley Road, that in those earliest days, we made expensive mistakes."[137]

The first HTGR to be successfully operated in the U.S. was Peach Bottom 1, a small 40 MW reactor in Pennsylvania. However, this reactor was shut down in November 1974 after less than seven and a half years in service. The only other HTGR reactor operated in the U.S. was the 330 MW Fort St. Vrain reactor in Colorado. This plant was permanently closed in August 1989 after 10 years in commercial operation. During this time, the Fort St. Vrain reactor achieved a lifetime capacity factor of just 14.5 percent and had a forced outage rate of nearly 61 percent making it the worst performing reactor in United States history after Three Mile Island Unit 2 which suffered a partial core meltdown just one year and one day after first achieving criticality.[138] Given uncertainties surrounding the possibility of the HTGRs being successfully developed and fielded commercially in the next few decades, the economics of new nuclear construction will be dominated by the costs of light-water reactors. The poor economics and operating history of gas-cooled reactors is also relevant to the possible use of nuclear power as a source of hydrogen, since a variety of the HTGR is one of the four options currently being given serious consideration for high-temperature hydrogen production.[139]

Finally, despite the recent increased interest in reprocessing among some elements of the U.S. government, we will only consider in detail the economics of the once-through fuel cycle in which the spent nuclear fuel is disposed of directly in a deep geologic repository. This choice is driven by the serious proliferation risks that accompany the use of reprocessing technologies (see Section 3.2), as well as by the economic failure of the plutonium economy to date. In addition, spent mixed-oxide plutonium fuel (MOX) fuel from light-water reactors is not typically reprocessed due to its isotopic makeup. Given that it is both thermally hotter and more radiotoxic than spent uranium fuel, it is more difficult to dispose of in a repository.[140] As with our decision to focus on light-water reactor designs, the focus on the once-through fuel cycle was also the position taken by the authors of the MIT study. Based upon their analysis, the authors concluded that

> *The once through cycle has advantages in cost, proliferation, and fuel cycle safety*, and is disadvantageous only in respect to long-term waste disposal; two closed cycles have clear advantages only in long-term aspects of waste disposal, and disadvantages in cost, short-term waste issues, proliferation risk, and fuel cycle safety.[141]

Specifically, the authors of the MIT study estimated that reprocessing the spent fuel and using the separated plutonium in MOX would be more than four times as expensive as making fresh fuel from uranium and disposing of it directly in a geologic repository. A previous estimate from the U.S. National Academy of Sciences found that, even if the plutonium was assumed to be provided to the utilities free of charge (such as would be the case for plutonium turned over from the surplus military stockpile), then MOX fuel would still be more than one and half times as expensive as fuel from fresh uranium.[142] Other than the high cost of reprocessing itself, the excess cost of plutonium fuel is due to the fact that fuel fabrication costs are much higher for MOX than low-enriched uranium (LEU) fuel. This is due to the stricter health and safety requirements as well as the cost of adding stricter materials accounting systems to safeguard the weapons usable plutonium. Similar estimates for the cost of MOX fuel have been made internationally as well. For example, a study in 1997 estimated that the cost of reprocessing spent fuel and manufacturing MOX fuel in Japan would be 5.6 to 7.3 times as expensive as making fuel from fresh uranium. The study also found that, even if the plutonium was free, MOX fuel would remain 1.7 to 2.5 times as expensive as LEU fuel.[143] The Japanese estimates are somewhat higher that those from the U.S. due to the higher costs of reprocessing and MOX fuel fabrication in Japan.[144]

Due to the greater difficulties encountered with the use plutonium fuel, MOX is typically limited to making up less than one-third of the fuel in the core of a light-water reactor. Table 2.1 summarizes the cost estimates for the MOX fuel cycle from two U.S. studies and two studies from countries that are actively pursuing the use of plutonium fuel (France and Japan). In each case the use of MOX fuel was projected to lead to an increase in the total cost of generating electricity from nuclear power compared to the once-through fuel cycle. Significantly, the French estimates were presented as part of an official government study prepared for the Prime Minster by members of that country's nuclear establishment.

Table 2.1: Estimated increase in the cost of electricity from the use of mixed-oxide plutonium fuel (MOX) in the United States, France, and Japan compared to the use of low-enriched uranium fuel.[145]

	Estimated Increase in Cost of Nuclear Electricity (cents per kWh)	Percentage of Reactor Fuel Supplied by MOX	Percentage Increase in Cost of Electricity
United States (MIT)	0.276	16.0	4.1
United States (University of Chicago)	0.165	not specified	2.3 to 3.1
France	0.131	10.3	5.3
Japan[(a)]	0.988 to 1.31	33.3	10.3 to 13.7

(a) The higher costs in the Japanese study are due in large part due to the assumption of a larger percentage of MOX usage as well as the higher costs associated with the shipment of spent-fuel to France for reprocessing or for the use of the domestic Rokkasho Reprocessing Plant which has suffered extensive cost overruns and delays.[146]

In light of these added expenses, we will only examine the cost of electricity from new light-water reactors operating on the once-through fuel cycle. If the current proposals for reprocessing spent commercial fuel in the United States (see Section 3.2) were to come to fruition, the economics of nuclear power would only grow worse as can be seen in Table 2.1. In the subsequent sections of this chapter, we will turn to a consideration of whether the projected cost for electricity from new nuclear power plants is likely to decrease significantly in the near to medium-term through improvements in the construction or financing of the plants. While it is traditional to then compare these costs for new nuclear development with those from competing baseload generation such as coal or natural gas fired plants, this does not, in itself, adequately address the question of nuclear power's potential role in strategies to address the threat of climate change. In the present context it is more relevant to compare the costs of nuclear power to available alternatives with similar or greater potential to reduce greenhouse gas emissions from the electricity sector. In this light, we will consider proposals to directly subsidize nuclear power due to its low emissions as well as proposals to tax carbon emissions from fossil fuel plants.

In the final sections of this chapter we will compare the economics of nuclear power to that of alternative emission reduction strategies such as increasing energy efficiency, expanding the use of wind power and other renewable technologies like advanced hydropower and thin-film solar cells, stabilizing long-term natural gas prices through an increased reliance on liquefied natural gas, the development of combined cycle power plants using coal gasification technology, and the direct capture of CO_2 from fossil fuel plants for sequestration in geologic repositories. Some of these alternatives carry their own potentially serious environmental and security impacts, but we have found that each are likely to be economically competitive or advantageous compared to the expanded use of nuclear power while simultaneously allowing us to avoid the unique vulnerabilities associated with nuclear weapons proliferation (Chapter Three), the risks of catastrophic reactor accidents or large scale terrorist attacks (Chapter Four), and the long-term difficulties encountered with managing long-lived and highly radioactive nuclear waste (Chapter Five).

Section 2.1 – The Projected Cost of Nuclear Power

Despite the improvements that have been made to the economics of operating reactors in the U.S. over the past 15 years, nuclear power remains an expensive option for the future. In fact, the authors of the MIT study themselves acknowledge that, for new construction, "nuclear power is *much more costly* than the coal and gas alternatives even in the high gas price cases."[147] In the MIT reference case, the assumptions most favorable to nuclear power are those corresponding to a 40 year operational lifetime for the power plants and an effective capacity factor of 85 percent. Under these assumptions electricity from nuclear power was found to be nearly 60 percent more expensive than coal and between 20 and 75 percent more expensive than natural gas depending on the assumptions made for the future fuel price. Table 2.2 summarizes the main results from the MIT economic analysis.

Table 2.2: Levelized cost of electricity from new nuclear power, pulverized coal, and combined cycle natural gas fired power plants as estimated by the 2003 interdisciplinary study entitled "The Future of Nuclear Power" conducted at the Massachusetts Institute of Technology.[148]

Generation Type	40 Year Levelized Cost (¢/kWh) 85% Capacity Factor	40 Year Levelized Cost (¢/kWh) 75% Capacity Factor
Pulverized Coal (fuel cost: $1.30 per MMBtu)[a]	4.2	4.6
Natural Gas (CCGT) (fuel cost: $3.77 to $6.72 per MMBtu)[b]	3.8 to 5.6	3.9 to 5.7
Nuclear (overnight capital cost: $2,000 per kW)[c]	6.7	7.5

(a) Levelized cost of coal over 40 years with a 0.5% rate of cost escalation.
(b) Levelized cost of natural gas over 40 years with a 0.5% to 2.5% rate of cost escalation. The price for natural gas has fluctuated a great deal in recent years and has been both above and below the range considered here at times. For comparison, the average price of natural gas sold to the electric power sector was $6.11 per thousand cubic feet in 2004 and $8.45 per thousand cubic feet in 2005.[149] However, long-term gas prices can be expected to remain within the range considered by the authors of the MIT study if policies on efficiency, conservation, and an increased reliance on liquefied natural gas are pursued (see Sections 2.3.1 and 2.4.1).
(c) The overnight capital cost is the amount of money that it would cost to build the plant if it could be completed instantly. This value takes into account the cost of labor and materials, but does not include such things as interest payments on money borrowed by the utility to build the plant.

Compared to coal or natural gas, the capital costs of nuclear power are a much more significant portion of the total cost of generating electricity, and therefore the comparison grows less favorable to nuclear power if the assumptions regarding the plant life or capacity factor are reduced. Given these estimates for the cost of generating electricity from new power plants, the authors of the MIT study summarized their conclusions as follows

> The bottom line is that with current expectations about nuclear power plant construction costs, operating costs and regulatory uncertainties, it is extremely unlikely that nuclear power will be the technology of choice for merchant plant investors in regions where suppliers have access to natural gas or coal resources. ***It is just too expensive.***[150]

The principal conclusions from the MIT economic analysis are consistent with those from a major study on the economics of nuclear power conducted at the University of Chicago in 2004. The U Chicago study focused on the cost of power plants that could be put into service by 2015 and, like the MIT study, calculated a levelized cost of electricity from nuclear, pulverized coal, and natural gas fired plants assuming an annual capacity factor of 85 percent.[151] Table 2.3 summarizes the estimated costs from this study.

Table 2.3: Levelized cost of electricity from new nuclear power, pulverized coal, and combined cycle natural gas fired power plants as estimated by the 2004 study entitled "The Economic Future of Nuclear Power" conducted at the University of Chicago.[152]

Generation Type	Levelized Cost of Electricity (¢/kWh) 85% Capacity Factor
Pulverized Coal (fuel cost: $1.02 to $1.23 per MMBtu)[a]	3.3 to 4.1
Natural Gas (CCGT) (fuel cost: $3.39 to $4.46 per MMBtu)[a]	3.5 to 4.5
Nuclear (overnight capital cost: $1,200 to $1,800 per kW)	5.3 to 7.1

(a) Average price of coal or natural gas over the lifetime of the plant calculated by IEER from the information presented in the University of Chicago study. The estimated price of natural gas in the University of Chicago study was derived mainly from projections by the DOE's Energy Information Administration.

Despite the differences in their models, it is significant that the conclusions of the U Chicago study regarding the cost of nuclear compared to coal or natural gas are generally consistent with those of the MIT report; namely that electricity from new nuclear plants will be more expensive than electricity from coal (29 to 115 percent more expensive compared to 60 percent in the MIT study) and more expensive than electricity from natural gas (18 to 103 percent more expensive compared to 20 to 75 percent in the MIT report). The authors of the University of Chicago study noted that such results should not come as any surprise because

> No observers have expected the first new nuclear plants to be competitive with mature fossil power generation without some sort of temporary assistance during the new technology's shake-down period of the first several plants.[153]

This conclusion, of course, fails to recognize that nuclear power is also a "mature" technology that has been commercialized for more than 50 years with nearly 440 plants online around the world, and that it has already been the recipient of far more extensive government subsidies than any other source of electricity in U.S. history.[154]

Finally, it is important to note that the MIT and U Chicago estimates (which are focused primarily on light-water reactors that could be built in the United States) are broadly consistent with the findings of studies in other countries and for other types of reactors. For example, it is estimated that, at the typical discount rate used by the Central Electric Authority, electricity from new heavy water reactors in India would be 37 percent more expensive than electricity from coal fired plants.[155] Similarly, estimates from Canada are that electricity from either the heavy water CANDU-6 or the light-water ACR-700 would be 24 to 49 percent more expensive than electricity from coal.[156]

In each of these analyses, the primary focus was on comparing the cost of new nuclear plants to those of competing base load fossil fuel plants. However, the more important comparison in the present context is between the costs of nuclear and those of other strategies for reducing greenhouse gas emissions. Before making such a comparison, however, it is necessary to examine the possible improvements to the generating cost of nuclear power that have been proposed. Given the importance of construction costs to the overall cost of electricity from nuclear power plants, the main improvements considered relate to lowering the capital cost of the reactors, shortening the plant's construction time, and improving the conditions under which the plant's construction is financed.

Section 2.1.1 – Lowering the Capital Cost and Construction Time

The base case in the MIT study estimates that new nuclear power plants will be four times as expensive to build as natural gas fired plants and 54 percent more expensive than coal plants. In addition to the larger capital costs, the MIT reference case also assumes that nuclear plants will take two and a half times longer to build than natural gas fired plants and 25 percent longer than coal plants. On the other hand, the base case of the University of Chicago study assumes a lower range of capital costs for new nuclear plants but a longer lead-time. In fact, the U Chicago study surprisingly assumes that the capital cost of new nuclear plants could fall as much as 16 percent below the cost of new coal fired plants (see Table 2.4).[157]

Table 2.4: Comparison of the assumptions for overnight capital cost and lead-time for construction used in the MIT and University of Chicago studies.[158]

Generation Type	MIT Study		University of Chicago Study	
	Overnight Capital Cost ($ per kW)	Lead-Time (years)	Overnight Capital Cost ($ per kW)	Lead-Time (years)
Natural Gas	500	2	500 to 700	3
Coal	1,300	4	1,182 to 1,430	4
Nuclear	2,000	5	1,200 to 1,800	7

The middle of the U Chicago range for capital costs is 20 percent higher than the MIT base case for natural gas, roughly equal to the MIT value for coal, and 25 percent lower for nuclear. Despite these differences, which all favor the economics of nuclear power relative to fossil fuels, Tables 2.2 and 2.3 show that the levelized cost of electricity from new nuclear plants estimated by the MIT and University of Chicago models are reasonably consistent (6.7 cents per kWh in the MIT study versus 5.3 to 7.1 cents per kWh in the U Chicago study). This consistency between the two estimates is due, in large part, to the increased financial penalty incurred by nuclear plants in the U Chicago analysis due to its assumption of a longer construction time and higher financial risk premium (see Section 2.1.2).

The potential improvements over the base case considered by the authors of the MIT study were a 25 percent reduction in the capital cost to $1,500 per kW and a 20 percent reduction in lead-time to 4 years. The U Chicago study did not consider any further improvements in the capital cost of new reactors, but they did consider a reduction in lead-time from seven to five years. Interestingly, the improved capital cost estimate considered in the MIT study is equal to the middle of the base case range used in the U Chicago study, while the improved lead-time considered in the U Chicago study is equal to the base case of the MIT study. While we agree that the MIT base case estimates for the capital cost and lead-time are reasonable assumptions for future nuclear plants, they should be considered already optimistic and not subject to further significant improvement.

One reason to regard the MIT base case as a reasonable, but optimistic, starting point is that the values for the capital cost and lead-time are significantly lower than what would be expected from historical experience in the United States. While it is difficult to compare experiences between different periods in time, just as it is difficult to compare experience between different countries, historical experience remains relevant as a guide to the types of uncertainties that may arise in long-term projections of future costs and as a caution against overly optimistic assumptions about future cost reductions. As summarized by the Energy Information Administration

> Thus, it is important to study the historical information on completed plants, not only to understand what has occurred but also to improve the ability to evaluate the economics of future plants. This requires an examination of the factors that have affected both the realized costs and lead-times and the expectations about these factors that have been formed during the construction process.[159]

In its analysis, the EIA found that the real overnight costs of nuclear plants went up by more than 340 percent between those plants which started construction in 1966-67 (with an average cost of $700 per kW in 1982 dollars) and those that entered construction in 1974-75 (with an average cost of $3,100 per kW in 1982 dollars). In its analysis, the EIA attributed three-fourths of the escalation in costs to "increases in the quantities of land, labor, material, and equipment used to build a nuclear power plant" and the remaining one-fourth to "increases in the real financing charges, escalation in the rate of increase in the prices of land, labor, material, and equipment during the construction period, and increases in construction lead-times."[160]

The National Research Council of the U.S. National Academy of Sciences, examining the same historical data in a somewhat different way, reached similar conclusions to those of the EIA. The National Research Council Committee grouped reactors by their completion date instead of by the date their construction began and found that from 1971 through 1988 there was a sustained increase in average plant costs as well as average lead-times. The NRC found that the average cost of nuclear plants that entered operation between 1971 an 1974 was $817 per kW (in 1988 dollars) while those that entered operation between 1987 and 1988 averaged $3,133 per kW (in 1988 dollars), an increase of 280 percent. Perhaps even more telling is the fact that the least expensive plant that examined by the NRC from the period 1987 to 1988 was more expensive than the most expensive plant that entered operation between 1971 and 1974.[161] The authors of the MIT study themselves recognize this history of cost escalations, and acknowledge that, if they had used the cost data for plants completed in the late 1980's and early 90's, that their resulting overnight cost estimates "would have been much higher."[162]

The National Research Council found a similar trend in relation to nuclear plant construction times as well. Overtime, the NRC found a constant increase in average lead-times growing from 5.4 years for reactors that came online prior to 1975 to 12.2 years for those coming online from 1985 through 1989. In this case, it found that the minimum construction time achieved for any of the 47 plants that became operational between 1980 and 1989 was longer than the average time needed to build the 40 plants that came online before 1975. Taking the 110 reactors under consideration into account, the NRC Committee found that the average lead-time required to complete construction was 8.4 years, or about two-thirds longer than the MIT base case assumption.[163]

The National Research Council also performed an analysis of global nuclear construction, and found that the average lead-time achieved worldwide between 1978 and 1989 was approximately 7.7 years. The United Kingdom was found to have had the longest construction times listed with an average for this period of 12.8 years compared to 11.1 for reactors built in the United States. The shortest construction times listed over this time period were Japan, with an average of 4.7 years, followed by France, with an average of 5.9 years.[164] Thus, even the shortest average construction times achieved anywhere between 1978 and 1989 were comparable to the base case assumption in the MIT report. Turning to more recent experience, in 2002 the International Atomic Energy Agency reported that, for plants that began construction around the world after 1993, there was an average of 5.3 years between the start of construction and the commencement of commercial power generation.[165] In light of this history, the discussion of estimated construction times for future nuclear plants from the University of Chicago study is worth quoting at length:

> The stated DOE position of a 5-year construction schedule is based on the new streamlined regulatory policy. The base case in the present study is 7 years for anticipated construction time. This is the time period of major financial outlays prior to revenue generation from power sales. The business significance of this period is that it is a time of negative cash flow, during which interest costs accrue on expenditures. This duration is based on the assumption that the business community will form expectations taking account not only of the newer announced regulatory procedures but also of earlier experiences with construction times. The Scully interviews with fi-

40

nancial and utility executives, as well as anecdotal reports, reinforce the importance to the business community of expectations regarding construction time. Deutsche Bank's LCOE [levelized cost of electricity] calculations for new nuclear power in the United States rely on a 7-year construction period.[166]

When making use of estimates for future construction costs or lead-times from the nuclear industry or promotional agencies like the Department of Energy that depart significantly from the historical experience reviewed above, it is important to consider their track records at making such estimates. In its review, the DOE's Energy Information Administration concluded that

> ... although the utilities did increase their lead-time and cost estimates as work on the plants proceeded, they still tended to underestimate real overnight costs (i.e., quantities of land, labor, material, and equipment) and lead-times even when the plants were 90 percent complete.[167]

Table 2.5 shows how little improvement there was at estimating lead-times times between those plants that began construction between 1966 and 1969 and those that began construction between 1974 and 1977, and that the nuclear industry actually grew slightly worse at estimating the final plant cost despite its increase in experience. Specifically, even when past plants were three-fourths complete, the industry was still underestimating the final construction cost by roughly 23 percent and the final lead-time by an average of 12 to 21 percent.

Table 2.5: Comparison of utility estimates made before and during construction to the final cost and total lead-time of the plants. Despite the increase in experience between the late 1960s and the mid-1970s, the utilities continued to underestimate both the cost and amount of time required to complete nuclear plants even when the facilities were three-fourths complete.[168]

Year in Which Construction Began	Percentage of Plant Completed When Estimate Was Made	Percent by Which the Utility Underestimated the Final Construction Cost	Percent by Which the Utility Underestimated the Total Lead-Time
1966-1969	0%	63%	47%
	75%	22%	21%
1974-1977	0%	72%	45%
	75%	24%	12%

As a further check on the reasonableness of the base case assumptions used in the MIT study, we can compare them to a 2003 analysis conducted by the nonpartisan Congressional Budget Office and a 2005 analysis from the International Energy Agency. Based on information provided by the Nuclear Regulatory Commission, the Department of Energy, and "industry sources," the CBO assumed that a new nuclear plant with a capacity of 1,100 MW could be built starting in 2011 at a cost of approximately $1,900 to $2,700 per kW, with a best estimate of $2,300 per kW.[169] The estimate cited by the CBO for the expected construction cost of a competing natural gas fired plant was $536 per kW, while their estimate for a coal plant was $1,367 per kW. The estimates used by the International Energy Agency were generally consistent with those from the CBO and with the base case of the MIT report. Specifically, the IEA estimated that a new nuclear plant could be built in the U.S. for $1,894 per kW, while a new pulverized coal fired plant would cost $1,160 per kW and a new combined cycle natural gas plant could be built for $609 per kW.[170]

Instead of looking just at U.S. history, experience in other countries is often cited to support claims for lower capital costs for new nuclear plants. However, the construction costs for three recent Japanese nuclear plants averaged more than $2,500 per kW, while the construction costs for Japan's two new ABWR units totaled between $1,800 and $2,000 per kW.[171] In addition, the only reactor currently under construction anywhere in western Europe is a Finnish light-water reactor. Converting the Finnish estimate for the cost of this plant from euros to dollars, we find that the overnight cost of this plant would be projected to be more than $1,900 per kW.[172] Thus, both the recent experience in Japan as well as the projections for future costs in Europe are generally consistent with the MIT base case assumption of $2,000 per kW.

One important caution that must be kept in mind, however, is that it is generally quite difficult to compare raw construction estimates between different countries given the differences in labor costs, the costs and availability of materials and transportation, the relationship between the government and the utilities, currency exchange rates, the ability for citizens to participate in the licensing process, and many other factors. An alternative way to compare international experience in light of these difficulties is to consider the relative capital cost of nuclear plants compared to coal fired plants. In 1986, the Nuclear Energy Agency of the Organization for Economic Cooperation and Development (OECD/NEA) analyzed the economics of coal and nuclear plants that could be brought online by 1995. The average cost for nuclear construction in the 12 countries considered was 77 percent more expensive than coal ($1,442 per kW for nuclear versus $812 per kW for coal in 1984 dollars).[173] At the low end of the range were Japan with a nuclear construction costs 32 percent higher than coal and France with a ratio of 34 percent. At the high end were the United States where the average cost of a nuclear plant was 120 percent above that of coal, and Spain with a ratio of 143 percent.[174]

In 1998, the OECD updated this study to consider the economics of plants that could be commissioned in the years 2005 to 2010 at the latest. This update to the 1986 study considered six of the same countries as the older report in addition to six new countries.[175] In the updated study, Japan and France continued to have the lowest ratios of nuclear to coal capital costs, while this time India and Finland were found to have the highest relative costs. For the twelve countries considered in 1998, a nuclear power plant was still more than 50 percent more expensive to build on average than a coal fired plant ($2,210 per kW for nuclear versus $1,452 per kW for coal in 1996 dollars). In no country examined was a nuclear plant estimated to be cheaper to build than a coal plant.[176] The findings of these studies raise very serious doubts about the low end of the capital cost range used in the U Chicago study where nuclear plants were projected to be as much as 16 percent less expensive to build than new coal fired plants.[177]

The MIT base case assumes a capital cost of nuclear that is 54 percent above that of coal 1.54 ($2,000 per kW for nuclear versus $1,300 per kW for coal). This ratio is generally consistent with the international average from the 1986 and 1998 OECD/NEA studies. However, the reduction in nuclear plant costs considered by the authors of the MIT study would result in a cost ratio of just 15 percent ($1,500 per kW for nuclear versus $1,300 per kW for coal). This lower value is highly suspect in that it would surpass the expectations for power plant construction in France, and would be nearly one-fourth less than the international average. Therefore, this analysis of international construction costs further argues against the likelihood of any significant improvement to the cost of new nuclear plants over those considered in the MIT base case.

Even if a small number of initial plants could be built with lower capital costs and/or lead-times, this experience would likely be viewed with a high level of conservatism by utilities and financial institutions considering the past escalation in construction cost and lead-times. This is particularly true in light of the fact that any improvements would have to be maintained under a very demanding timetable set by the global growth scenario in which more than one reactor would have to come online somewhere in the world every 15 days for four decades. Meeting the more aggressive steady-state growth scenario would put an even greater strain on the nuclear industry with more than one reactor having to come online every six days over 40 years.

Section 2.1.2 – Reducing the Financial Risk Premium

The high capital cost and long lead-time of nuclear plants are particularly detrimental to the economic viability of nuclear power due to the fact that the interest rate charged to utilities on loans during construction is higher for nuclear plants than it is for fossil fuel plants. This higher rate is a result of the fact that nuclear plants are considered to be riskier investments that carry a higher degree of uncertainty and a greater risk of failure than investments in conventional generation. In part, this excess risk is due to the fact that nuclear plants have large generating capacities and must be planned further in the future than fossil fuel plants and thus, if demand grows less than expected the nuclear plant may end up being canceled during construction. In addition, there is the potential for public opposition to the construction of a new reactor to lead to delays that could quickly increase the plant's cost. As noted by the authors of the MIT study:

> ... even if investment in nuclear power looked attractive on a spreadsheet, investors must confront the regulatory and political challenges associated with obtaining a license to build and operate a plant on a specific site.... Many planned plants, some of which had incurred considerable development costs, were canceled. Delays and "dry-hole" costs are especially burdensome for investors in a competitive electricity market.[178]

In fact, between 1972 and 1984 (the period immediately following the first energy crisis in which energy demand growth in the U.S. slowed markedly), more than $20 billion was spent on 115 nuclear plants that were later canceled.[179] By 1992, a total of 121 reactors had been canceled, not counting those that had been ordered but were canceled before much money had been spent.[180] By the early 1980s, the choice to continue pursuing nuclear construction was recognized as an important factor in the downgrading of utility credit ratings by Standard & Poor's.[181] No new nuclear plants have been ordered in the U.S. since 1978 (more than a quarter of a century ago), and the two orders that were placed in 1978 were subsequently canceled. In fact, it has been nearly ten years since the last new reactor was brought online in the United States.[182]

Since the late 1980s, there have been a number of efforts undertaken to try to overcome this excess financial risk for nuclear power. The most important of these efforts was the passage of the Energy Policy Act of 1992 which changed the licensing rules to allow utilities to receive combined construction and operating licenses (COL). This new licensing process will allow a reactor to be operated without further regulatory review or possibility for public participation and comment if the Nuclear Regulatory Commission is satisfied that the plant has been built according to the agreed upon specifications. Despite this change in the regulations, the U Chicago study summarized the current financial position of nuclear power as follows:

> The recent combining of construction and operating licenses into a single step gives hope that construction delays and uncertainties encountered in the last generation of nuclear plants can be avoided in new construction, but in the absence of actual experience, there is a perception that nuclear plants are riskier than others, as discussed by Scully Capital [an investment banking and financial advisory service].[183]

Thomas E. Capps, chairman and chief executive officer of Dominion Energy, was somewhat more direct when he told the New York Times in May 2005 that "Standard & Poor's and Moody's would have a heart attack" if Dominion chose to begin construction of a new reactor, and that "my chief financial officer would, too."[184] This is a particularly significant assessment given that Dominion operates seven reactors in three states, and that it is currently seeking an early site permit to allow the possible placement of a new reactor at their North Anna site in Virginia.

The question, therefore, is not whether new nuclear plants will have to be financed with an allowance for the excess risk, but is instead how big the penalty is likely to be. In this context, the authors of the U Chicago study note that

> Financial terms in recent nuclear construction overseas are not a satisfactory guide to a risk premium in the United States because of differences among countries in business practices, differences in business climate, varying degrees of involvement of governments in nuclear projects, and differences in regulatory regimes.[185]

For example, the new nuclear plant that is to be built in Finland assumes that the construction costs will be financed at a real interest rate of just 5 percent, and that the plant would not pay income tax. This plant will also be operated by a non-profit company that sells electricity directly to industrial users under long-term power purchase agreements which reduces the uncertainty of demand growth. Similar cautions are sometimes needed within the U.S. as well. For instance, the government owned Tennessee Valley Authority assumes that the refurbishment and restart of the mothballed Browns Ferry 1 reactor, which has been shutdown since the mid-1980's, will be financed at an effective interest rate just 0.8 percent above that of 10-year treasury notes, and that it would also pay no taxes.[186]

Given that no new plants have been ordered in the U.S. in more than a quarter century, and the fact that experience from other countries is not a reasonable basis for comparison, the MIT and U Chicago studies arrive at somewhat different estimates for the level of the risk premium that would be imposed on the next round of nuclear construction (see Table 2.6).

Table 2.6: Financing rates for debt and equity assumed by the MIT and University of Chicago analyses. The effective interest rate is the weighted average for the assumed mix of debt and equity. The primary difference between these assumptions is the lower debt rate for nuclear used by the authors of the MIT study.[187]

Type of Generation	MIT (2003)			University of Chicago		
	Equity	Debt	Effective Rate	Equity	Debt	Effective Rate
Fossil Fuel Plants	12%	8%	9.6%	12%	7%	9.5%
Nuclear Power Plants	15%	8%	11.5%	15%	10%	12.5%
Financial Risk Premium			1.9%			3.0%

From Table 2.6, we see that, while their assumptions about the financing rate for new fossil fuel plants are similar, the risk premium for nuclear power in the MIT study is more than a third smaller than the risk premium assumed in the U Chicago study. Considering this difference, it is important to note that the authors of the U Chicago study concluded that

Informal conversations with a number of Wall Street analysts corroborated the reasonable magnitude of the 3 percent premium as a **lower bound estimate**.[188]

Thus, even with the modified regulatory process in place in the U.S. and the heightened interest of the Bush Administration, it is unlikely that new reactors could achieve much better financing than that assumed by the MIT base case analysis without a significant and sustained intervention by the federal government.

The influence of the higher interest rate assumed in the U Chicago study on the overall generation cost of electricity from new nuclear plants can be seen in Table 2.7. Despite the fact that the overnight capital cost for the middle of the U Chicago range of base-case costs is 25 percent less than the value in the MIT study ($1,500 per kW versus $2,000 per kW), due to the longer lead-time and higher risk premium, the interest payments in the U Chicago analysis actually contribute about 80 percent more to the final cost of electricity than the interest payments do under the MIT assumptions.

Table 2.7: Comparison of the relative contribution to the total capital cost of new nuclear power plants in the MIT and University of Chicago analyses from interest payments on money borrowed by the utilities. Due to the long lead-time and higher interest rate, these financing costs are a more significant contributor to the total cost of electricity from nuclear power in the U Chicago analysis.

	Overnight Capital Cost	Lead-Time	Effective Interest Rate	Percent of Total Capital Cost Due to Interest[a]	Contribution of Interest to Total Generation Cost
MIT	$2,000 per kW	5 years	11.5 percent	21.5 percent	1.0 cents per kWh
University of Chicago	$1,500 per kW[b]	7 years	12.5 percent	39.3 percent	1.8 cents per kWh

(a) The total capital cost includes all elements covered by the overnight cost as well as interest payments made on money that had to be borrowed during construction.
(b) The mid-point of the estimated range in the U Chicago study.

Given the importance of interest payments to the overall cost of nuclear power, there have been a number of proposals put forth to try to reduce the apparent risks associated with the choice to build a nuclear plant. The most likely government intervention in this respect will be to offer loan guarantees to at least the first several new nuclear plants to be built. These guarantees would allow the utilities to borrow money at the risk free rate for the portion of the loan guaranteed by the government, and thus significantly lower the overall effective interest rate.[189] In 2003, the Congressional Budget Office analyzed the impact of a provision in the Senate version of the Energy Policy Act of 2003 (S.14) which would have allowed the DOE to grant 50 percent loan guarantees to the next seven nuclear plants to be built. In addition, the law would have empowered the Department of Energy to enter into long-term power purchase contracts with these "first-mover" utilities to lessen the risk that demand would not grow fast enough to economically support the added capacity.[190] As a result of its analysis, the CBO concluded that

> [b]ecause the cost of power from the first of the next generation of new nuclear power plants would likely be significantly above prevailing market rates, we would expect that the plant operators would default on the borrowing that financed its capital costs.[191]

It estimated that the net subsidy from the government required to keep the plant running following the utility defaulting on its financing would amount to 15 percent of the plant's total construction cost, or roughly $375 million per plant. This proposal could therefore have resulted in a total subsidy of more than $2.62 billion overall to the seven new nuclear plants.[192] The CBO updated this analysis in 2005 to consider the revised language in the Energy Policy Act of 2005 which allows loan guarantees of up to 80 percent to be granted to a variety of energy efforts including new nuclear power plants. The CBO concluded that such guarantees to new nuclear plants "could be for more than $2 billion [per plant] and carry a significant subsidy cost (perhaps as much as 30 percent)."[193]

A second type of government intervention that has gained significant prominence is to provide direct subsidies to a limited number of "first-movers" in the form of production tax credits. These proposal are similar in nature to the tactic that was used by the nuclear plant manufacturers in the late 1950's and early 1960's in which a limited number of early plants were built on a fixed cost basis to encourage utilities to pursue nuclear power. The loss to these reactor manufacturers like Westinghouse and General Electric on these initial "first-mover" promotions has been estimated at between $875 million and $1 billion on 13 plants.[194]

In the MIT report, the authors recommended that each of the next 10 nuclear plants to be built be given a subsidy worth 10 percent of the total construction cost ($200 million per plant), for a total subsidy of $2 billion. In order to ensure that this money would only go to plants that were actually completed, and not to plants that might be started and then abandoned during construction, the authors recommended that the money be paid out as a production tax credit over the first year and half of plant operation. Assuming a capacity factor of 75 to 85 percent, this would work out to a subsidy of 1.8 to 2.0 cents per kWh over the first 18 months the plant is in operation.[195]

A similar, but larger "first-mover" subsidy was proposed in early 2005 by the House Committee on Energy and Commerce of the U.S. Congress. In its discussion draft of the Energy Policy Act of 2005, the Committee included a 1.8 cents per kWh production tax credit for up to 6,000 MW of new nuclear capacity that entered service before January 1, 2021. This subsidy would be paid for up to eight years for each individual plant, and would be limited to $125 million per plant per year. Overall, therefore, this proposal could result in a total subsidy of $6 billion to the first six reactors.[196] The maximum production credit would be received by any plant achieving a capacity factor equal to or better than 80 percent. Assuming a five year construction time and discounting the payments at 3 percent per year, the present value for this subsidy would be equal to $750 million per plant. In other words, this production tax credit would result in the tax-payers absorbing nearly 40 percent of the total overnight construction cost for each of the first six new nuclear plants to be built.[197] An analysis conducted by the Energy Information Administration using the National Energy Modeling System, a computer program designed to predict the likely impact of energy policies, estimated that the total cost of this proposed production tax credit just through the year 2022 would equal $3.85 billion.[198] Despite the fact that the Bush Administration announced on May 17, 2005, that it would not support any major financial incentives for nuclear power beyond the issuance of regulatory insurance as discussed below, the 1.8 cents per kWh production tax credit for the utilities was included in the final version of the energy bill signed into law in August 2005.[199]

Despite the magnitude of these proposed subsidies, they would still not be large enough to fully overcome the higher costs of nuclear power compared with fossil fuels. Under the MIT proposal, the levelized cost of electricity from the ten "first-movers" would be approximately 6.2 cents per kWh which is still well above its estimate for the price of electricity from coal or natural gas. In addition, this estimate assumed that these "first-mover" plants could be built without longer lead-times or higher capital costs than those assumed in the base case. If technical problems, regulatory uncertainties, or public opposition led to an increase in construction costs on the first few plants, the value of this subsidy would be proportionally diminished. As noted by analysts at Standard & Poor's in their 2006 assessment of nuclear power, "given that construction would entail using new designs and technology, cost overruns are highly probable."[200]

With respect to the larger first-mover subsidy that was eventually included in the 2005 Energy Policy Act, the authors of the U Chicago study found that this tax credit would "achieve competitiveness only for the most optimistic cost outcome," and that the middle of their range for the cost of new nuclear power would remain above even their highest estimate for the price of electricity from coal or natural gas.[201] In addition, the authors of the U Chicago study noted that "the production tax credit helps cash flow only after the plant has been built and does not reduce the heavy drain on near-term dollar requirements during the construction period."[202] This would further limit the impact of such strategies on making nuclear power look financially attractive to utilities and investors.

The third way in which the government is attempting to intervene on behalf of the nuclear industry to lower the financial risk was put forth by President Bush in April 2005. Under this proposal, the government would reimburse the utilities for the cost of any delays in nuclear plant construction that might be caused by the regulatory process. Specifically, the President directed the Department of Energy

> ... to work on changes to existing law that will reduce uncertainty in the nuclear plant licensing process, and also provide federal risk insurance that will protect those building the first four new nuclear plants against delays that are beyond their control.[203]

The Energy Policy Act of 2005 enacted the Bush Administration's proposal, expanding the "insurance" to the first six new reactors in line with the potential number of "first-mover" recipients of the 1.8 cents per kWh production tax credit. The law allows for the Department of Energy to reimburse utilities for costs they incurred due to either

> (A) the failure of the Commission to comply with schedules for review and approval of inspections, tests, analyses, and acceptance criteria established under the combined license or the conduct of preoperational hearings by the Commission for the advanced nuclear facility; or
> (B) litigation that delays the commencement of full-power operations of the advanced nuclear facility.[204]

The first two reactors to be built would have 100 percent of these costs covered up to $500 million per plant. The next four reactors to be built would be eligible for 50 percent of their costs to be covered up to $250 million if the delays extend beyond 180 days.[205] While this type of "insurance" program would help to lessen some of the risks associated with building the first six nuclear plants (specifically the risk of legal challenges to the combined construction and operating license process which has yet to be tested by any utility), it would still not eliminate the risk from such uncertainties as slower than expected demand growth and the effects of potential public opposition outside the regulatory process. Moreover, while the $500 million maximum would represent a significant subsidy to the nuclear utilities, a number of past reactors suffered from even larger cost escalations (see Section 2.1.1). Finally, and most importantly, this "insurance" program effectively punishes the NRC for doing its job well and thus will likely have a negative impact on the safety of the next generation of reactors. Unlike the loan guarantees or production tax credits which would amount to a transfer of public money to private hands, this subsidy will likely have a chilling effect on the regulators and further weaken the oversight provided by the NRC (see Section 4.1.3) and the ability of the public to intervene legally.

The very large federal subsidies and market interventions discussed above could make nuclear power appear more attractive to investors already considering building new reactors. However, as noted by analysts at Standard & Poor's in early 2006

> Although these events create some sort of supportive platform for a nuclear renaissance in the U.S., it may not provide sufficient incentive to pursue new construction. From a credit perspective, these legislative measures may not be substantial enough to sustain credit quality and make this a practical strategy.[206]

Thus, even in the face of billions of dollars in tax breaks, loan guarantees, and "risk insurance," Wall Street financial institutions may remain hesitant in financing new nuclear plant construction on favorable terms. However, even if the financing of the first few plants could be improved through heavy subsidization, the many uncertainties that surround the choice to build a nuclear power plant make it very likely that a risk premium of at least two to three percent will remain the norm for financing future nuclear plants.

Section 2.1.3 – Impact of Potential Cost Improvements

As discussed above, the "first-mover" subsidies would not be sufficient by themselves to make nuclear power economically advantageous compared to coal or natural gas. Before moving on to a consideration of how the costs of nuclear power compare to other climate change mitigation strategies, it is important to consider the other potential cost savings that have been put forth and how the authors of the MIT and U Chicago studies viewed the reasonableness of their own proposals. Specifically, the authors of the MIT study concluded that

> *The cost improvements we project are plausible but unproven.* It should be emphasized, that the cost improvements required to make nuclear power competitive with coal are significant: 25% reduction in construction costs; greater than a 25% reduction in non-fuel O&M costs compared to recent historical experience (reflected in the base case), reducing the construction time from 5 years (already optimistic) to 4 years, and achieving an investment environment in which nuclear power plants can be financed under the same terms and conditions as can coal plants. Moreover, under what we consider to be optimistic, but plausible assumptions, nuclear is never less costly than coal.[207]

In fact, even if all of their proposed cost improvements were somehow achieved (which, as discussed in Sections 2.1.1 and 2.1.2, is highly unlikely), not only would coal remain cheaper, but electricity from natural gas would only exceed the price of nuclear power in the highest fuel price scenario considered. In addition, the MIT analysis assumes that the major reductions in construction time and capital cost envisioned for nuclear power plants could be achieved without similar improvements being made for coal or natural gas fired plants. If equally optimistic improvements were considered for the fossil fuel plants, the cost comparison would grow worse for nuclear.

These conclusions from the MIT study are in line with those from the University of Chicago study as well. In addition to starting with lower overnight capital costs, the U Chicago study considered the following cost reduction strategies for new nuclear plants: a 25 to 50 percent federal loan guarantee, an accelerated depreciation schedule, an investment tax credit of 10 to 20 percent, and a reduced construction time from seven to five years. Achieving any one of these cost reductions alone could still leave the price of nuclear more expensive than either coal or natural gas.[208] Based on their result, the authors of the U Chicago study concluded that

> In summary, with the expectation of a 7-year construction period, no individual financial policy can be counted on unambiguously to bring the LCOE [levelized cost of electricity] of first new nuclear plants within the range of LCOE competitive with fossil generation.[209]

Finally, the U Chicago study also considered possible cost reductions to future plants as a result of the experience gained by the industry during the construction of the first few plants. Under what it considered to be "optimistic" assumptions about the rate of this learning, however, they found that, even for the eighth plant (when learning would be virtually complete), the cost of electricity from nuclear would still remain higher than the range of estimated costs for coal or natural gas. It was only when the "optimistic" assumptions about learning were coupled to further highly optimistic assumptions, including a reduction of the construction time to five years on the third plant and the elimination of excess financial risk on the fourth plant, that the cost of electricity from nuclear fell to within the middle of the competitive range for coal and natural gas.[210] These large cost reductions on future plants are based on assumptions that are not in line with the historic experience in the United States, and are very unlikely to be realized.

Section 2.1.4 – Summary of Nuclear Power Economics

Given the optimistic nature of the assumptions built into the base case of both the MIT and U Chicago analyses, it is unlikely that any significant improvements to the economics of nuclear power could be sustained under the demanding construction schedule of the global growth or steady-state growth scenarios. We will therefore consider 6.0 to 7.0 cents per kWh to be a reasonable range for the future costs of nuclear power, with the MIT value of 6.7 cents per kWh as the best single estimate (see Table 2.8). This range is consistent with the reference case projections from the DOE's Energy Information Administration for the cost of new nuclear power reported in their 2005 *Annual Energy Outlook*.[211] Finally, the fact that the base case estimate for the lead-time for construction in the MIT study is equal to the improved lead-time in the U Chicago study, while the base case capital cost in the U Chicago study is equal to the improved capital cost in the MIT analysis further reinforces the reasonableness of the resulting range for the cost of electricity from new nuclear plants.

Table 2.8: Summary of base case assumptions made in the MIT and University of Chicago studies and the resulting levelized cost of electricity from new nuclear power plants.

	Overnight Capital Cost	Lead-Time	Financial Risk Premium	Capacity Factor	Levelized Cost of Electricity
MIT	$2,000 per kW	5 years	1.9 percent	85 percent	6.7 cents per kWh
University of Chicago	$1,500 per kW[a]	7 years	3.0 percent	85 percent	6.2 cents per kWh

(a) The mid-point of the estimated range of capital costs in the U Chicago study.

Section 2.2 – The Economics of Nuclear Power as a Carbon Mitigation Strategy

In the previous section we showed that electricity from new nuclear power plants is likely to be substantially more expensive than electricity from either coal or natural gas. This comparison, however, is not necessarily the one most relevant to the current discussion of what role nuclear power might play in mitigating the impacts of climate change. The more important question in the present context than how nuclear compares to the business as usual usage of fossil fuels, is how the economics of nuclear power compare to those of alternative strategies for reducing carbon emissions. In the current section we will examine two proposals aimed at directly improving the economics of nuclear compared to the continued usage of conventional fossil fuel plants. The first proposal is to directly subsidize all nuclear plants in recognition of their low carbon emissions, while the second proposal, involves the direct taxation of carbon emissions in order to raise the price of electricity from coal and natural gas.

Section 2.2.1 – "Carbon-Free" Portfolios

One of the more aggressive proposals for incorporating climate change into the economics of nuclear power is to include it in future federal or state "'carbon-free' portfolio" standards.[212] These tax credits, usually set up as "renewable energy portfolios," have traditionally been used to promote the development of energy sources like wind and solar. The federal production credit given to renewable technologies in the U.S. (including new or incremental hydroelectric power added after 2005) amounts to approximately 1.8 cents per kWh for the first 10 years of the plant's operation.[213] If such a credit had been adopted for all electricity from nuclear power (i.e. from both new and existing plants), the cost to the government would have been roughly $13.75 billion in 2003.[214] Under the MIT global growth scenario, the cumulative cost of this subsidy between 2005 and 2050 would amount to roughly $358 billion, with an average annual payment to the utilities of nearly $7.63 billion per year.

The payment of this "carbon-free" credit to the nuclear utilities over the next 45 years would amount to nearly two and a half times the estimated spending by the government in both direct and indirect subsidies to promote nuclear power's development over the first 50 years of its development.[215] Assuming a five year lead-time, an 85 percent capacity factor, and discounting the "carbon-free" subsidy at 3 percent per year results in this subsidy having an overnight value equal to approximately half of the total estimated cost of building the plants in the first place. A further way to appreciate the scale of this subsidy, is to note that the average annual payout by the government between 2005 and 2050 would be nearly as large as the <u>total amount of all taxes</u> (including income tax) paid by all large, share-holder owned utilities combined in 2002.[216]

It clearly does not make economic sense to discuss a tax break for nuclear utilities that would be comparable to the entire tax burden of all large, share-holder owned utilities combined. However, even if this kind of subsidy were instituted, the electricity produced by these new nuclear plants would still be more expensive than electricity from coal under the base case assumptions of the MIT model and roughly equal to the cost from natural gas even in the highest gas price scenario if a modest improvement in the heat rate of combined cycle plants is achieved in the coming years.

Section 2.2.2 – Direct Taxation of Carbon Emissions

Up to this point we have primarily considered proposals for money flowing from the taxpayers to the nuclear industry in the form of tax breaks or other direct subsidies. An alternative proposal is to tax greenhouse gas emissions from fossil fuel burning power plants. The idea of a so-called "carbon tax" has been widely considered, and is viewed by many as likely to be a part of any market based strategies for combating carbon emissions. Imposing such a tax is, in effect, an effort to internalize a portion of the potential economic costs of climate change within the cost of generating electricity from fossil fuels.

While the concept of taxing emissions is straightforward, the setting of a "price" for carbon is considerably more difficult. Putting a dollar value on the potential impacts of climate change is highly uncertain given the complexity of both the ecological changes that are possible as well as the complexity of how those changes could impact human health and society. In addition, such schemes for setting the price of carbon are not well suited to address losses that are less quantifiable in economic terms such as the general loss of regional and global biodiversity or the impacts of ecosystem changes on the culture of Indigenous Peoples. An alternative approach is to set the price of carbon based on how significant a reduction in emissions is desired. This approach includes an implicit assumption about what level of climate change would be acceptable through its choice of a stabilization target. The greatest difficulty with this second approach is that it is very sensitive to the assumptions used in modeling the impact of the carbon tax on the choices that would be made regarding energy supply and demand as well as on the assumptions used for the sensitivity of the climate to increased concentrations of greenhouse gases.

Instead of approaching the issue of carbon taxes solely from these more traditional angles, the authors of the MIT study chose to instead focus on the level of taxes that would be necessary to raise the cost of electricity from coal and natural gas to equal that from new nuclear power plants. We will retain this approach in the present analysis since it is the most strait forward way to examine the cost of nuclear power as a CO_2 mitigation strategy. Since it is unlikely that any corporation would undertake a program with as many long-term commitments and as many financial and environmental risks as the building of a new reactor on the basis of a single economic projection, Table 2.9 summarizes the range of carbon taxes that would be required to raise the cost of electricity from fossil fuels to within the estimated range for the cost of nuclear power from new plants.

Table 2.9: Values for the tax per ton of carbon emitted that would be required in the MIT analysis to raise the price of electricity from coal or natural gas fired plants to the level of electricity from new nuclear power.[a]

Generation Cost of Electricity	Natural Gas[b]			Coal
	High Gas Price	High Gas Price (improved heat rate)	Moderate Gas Price	
6.0 cents per kWh	$40	$95	$175	$75
7.0 cents per kWh	$130	$190	$270	$115

(a) In this table, we have retained the assumptions used in the MIT study for the amount of carbon emitted per kWh of generation from coal and natural gas fired plants, and have rounded the values for the resulting carbon taxes to the nearest five dollars. To convert these figures into a tax per ton of carbon dioxide (CO_2) divide by 3.7.
(b) The three natural gas scenarios shown correspond to the high and moderate gas price cases from the MIT study (i.e. a levelized cost over 40 years of $6.72 per MMbtu and $4.42 per MMbtu respectively). Under the high gas price scenario, the authors of the MIT study considered two cases; namely the costs with today's technology and a case in which natural gas plants achieved a 10 percent improvement in their heat rate over the coming years.[217]

Table 2.9 shows that the likely range of carbon taxes required to equalize the price of electricity from coal or natural gas plants with that from new nuclear plants in the MIT analysis falls within the range of about $40 to $270 per ton of carbon. This range compares well to other recent estimates. For example, the study conducted at the University of Chicago found that a carbon tax of $65 to $180 per ton of carbon would be required to raise the cost of electricity from coal or natural gas to within the six to seven cents per kWh range of electricity from new nuclear plants (see Table 2.10).

Table 2.10: Values for the tax per ton of carbon emitted that would be required in the U Chicago analysis to raise the price of electricity from coal or natural gas fired plants to the level of electricity from new nuclear power.

Generation Cost of Electricity	Pulverized Coal	Natural Gas
6.0 cents per kWh	$65 to $90	$80 to $130
7.0 cents per kWh	$100 to $125	$130 to $180

(a) In this table, we have retained the assumptions used in the U Chicago study for the amount of carbon emitted per kWh of generation from coal and natural gas fired plants, and have rounded the values for the resulting carbon taxes to the nearest five dollars. To convert these figures into a tax per ton of carbon dioxide (CO_2) divide by 3.7.

Other estimates have found similar ranges for the required carbon tax. For example, the Intergovernmental Panel on Climate Change (IPCC) concluded in 2001 that carbon allowances of approximately $100 to $250 per ton would be required for nuclear power to break even with new natural gas fired plants. In arriving at this estimates, the IPCC assumed that the reactors were of a state-of-the-art design and that the technology lived up to the cost and lead-time improvements that the nuclear industry envisions.[218] In addition, in 2002, an analysis by the British Performance and Innovation Unit of the Cabinet Office estimated that building nuclear power plants in place of natural gas plants would displace carbon at a cost of £70 to £200 per ton ($100 to $300 per ton).[219] While natural gas prices have increased since these estimates were made, their results are generally consistent with the projected ranges from the MIT and U Chicago analyses.

The imposition of carbon taxes in these ranges would have dramatic economic consequences. For example, in the year 2000, the total carbon emissions from the U.S. electricity sector were estimated to have been nearly 619 million metric tons of carbon.[220] If a carbon tax of $50 to $200 per ton had been imposed in that year, it would have resulted in a total tax of $30.9 to $123.8 billion on the utility industry as a whole. For comparison, the total gross income of all major investor-owned utilities in that same year was just over $233 billion. After expenses, the net income of these utilities was estimated at $13.3 billion.[221]

Given that the total carbon emissions from the electricity sector are projected to rise under the MIT global growth scenario, the annual cost of this tax would also rise despite the construction of 300 hundred nuclear plants throughout the United States. Table 2.11 shows our predictions for the cost to the electricity sector as a whole from carbon taxes in the range of $50 to $200 per ton. These estimates include both a lower carbon growth scenario (i.e. a scenario in which the 300 gigawatts of new nuclear capacity replaces coal) and a higher carbon growth scenario (i.e. a scenario in which the new nuclear plants replace natural gas). It should be recalled that a tax of approximately $100 per ton would equalize the cost of coal, natural gas, and nuclear power under the base case economic assumptions of the MIT model for the high gas price scenario.

Table 2.11: Summary of the estimated costs of imposing a tax of $50 to $200 per ton of carbon under the MIT global growth scenario. The lower carbon growth corresponds to nuclear power displacing only coal fired plants while the higher carbon growth corresponds to nuclear power displacing natural gas. If some of the nuclear plants displaced renewable resources, the resulting carbon emission would be higher than indicated in this table.

	Year	Carbon Emissions[a] (million metric tons)	Total Annual Carbon Tax (billions)	Tax per kWh of Total Generation (¢ per kWh)	Cumulative Carbon Tax (2005 through 2050)
Lower Carbon Growth	2005	663	$33.2 - $133	0.84 - 3.4	$1.62 - $6.47 trillion
	2050	700	$35.0 - $140	0.42 - 1.7	
Higher Carbon Growth	2005	663	$33.2 - $133	0.84 - 3.4	$1.91 - $7.63 trillion
	2050	1,000	$50.0 - $200	0.60 - 2.4	

(a) Estimated by IEER from the assumptions presented in the MIT study and starting from the fact that carbon emissions from the electricity sector were approximately 619 million metric tons in the year 2000 (see Section 1.3).[222]

One important point to keep in mind when considering the results presented in Table 2.11, however, is that we are reporting the value of the tax to the entire electricity sector and the average increase in cost per kWh of total generation regardless of how that electricity was generated. While it is true that many utilities are diversified in the types of generating capacity they operate, the variation among different companies and among different regions in the U.S. as to how much of this tax burden they would share would be significant. For example, in 2003 the states of Illinois, Indiana, Iowa, Kansas, Michigan, Minnesota, Missouri, Ohio, Nebraska, North Dakota, South Dakota, and Wisconsin received 73.6 percent of their total electricity from coal, 2.7 percent from natural gas, and 1.6 percent from hydroelectricity while the states of California, Oregon, and Washington received just 4.7 percent of their electricity from coal, 26 percent from natural gas, and 48.6 percent from hydroelectricity.[223] A further concern is that energy taxes are almost always highly regressive in nature, with the highest proportional impacts falling on the lowest income individuals. This raises serious economic justice concerns with any proposal for large carbon taxes imposed in isolation.

Finally, there is a problem with the implicit assumption in the MIT analysis that such large market interventions on behalf of nuclear power could occur without slowing the rate of demand growth or affecting the choices for generating technology made by utilities outside of the pulverized coal, natural gas, and nuclear paradigm. Already wind power at very favorable sites is economical compared to natural gas fired plants, and carbon taxes in the range of $100 to $200 per ton would be likely to make technologies like coal gasification and carbon sequestration economically attractive (see Sections 2.3 and 2.4).[224] The increase in electricity costs from such a tax would also likely lead to an increased focus on efficiency and a resulting reduction in the rate of demand growth which favors more flexible types of generation than large nuclear plants. In fact, it is this effect on demand that is one of the main reasons that carbon taxes are viewed by many as such an effective market based option for reducing greenhouse gas emissions. Surprisingly, the authors of the MIT provide no discussion of whether their assumption is reasonable that the <u>per capita</u> electricity consumption in the U.S. would increase by more than three-fifths between 2000 and 2050 no matter what the cost or environmental impact of that electricity was.[225]

Section 2.3 – Alternatives for the Near-Term (2006 - 2020)

In order to achieve large reductions in greenhouse gas emissions currently under consideration such as the 40 percent reduction by 2020 proposed by Germany, the 60 percent reduction by 2050 adopted by Britain, or the 80 percent reduction by 2050 committed to by the State of California, significant actions will be required both in the near-term (the next 15 years) as well as the longer term (the next 15 to 45 years).[226] While deep reductions in emissions from other sectors like transportation will be vital to meeting these goals, the emissions from electricity generation will also have to be cutback dramatically.[227] The primary selection criteria for near-term options in the electricity sector are that:

1. they must be capable of making a significant contribution to the reduction in greenhouse gas emissions;
2. they must be either already commercialized or far enough along that they could be ready for the commercial market within a short time;
3. they should be capable of competing economically with current sources of generation to allow their rapid entry into the market;
4. they should, to the extent consistent with the goals of reducing the threat from climate change, minimize other environmental and security impacts; and

5. they should be compatible with the medium and longer term options for the electricity sector.

The two available options that best meet these five selection criteria are efforts to increase efficiency in both the generation and use of energy and the large scale expansion of wind power at favorable sites. We will discuss both of these options in the following sections.

Section 2.3.1 – The Economics of Efficiency

Improvements to the efficiency of energy use as well as a reduction in demand through conservation are widely recognized as robust options that can have significant benefits throughout the industrialized world. Such programs are not only effective at curbing carbon dioxide and other dangerous emissions, but they are also typically more economical and more sustainable than other types of efforts. Unlike programs focused on increasing supply, these demand side options can result in low or negative cost reductions in greenhouse gas emissions by reducing consumption while simultaneously providing new jobs and opening up new avenues of economic growth. Many possibilities for improved energy efficiency are well known and are either ready or nearly ready for full scale commercialization. These programs could be brought online very rapidly in many sectors and could help to reduce the rate of demand growth significantly in the near-term.[228]

In the wake of the Arab oil embargo, the Energy Policy Project of the Ford Foundation produced an analysis in 1975 that discussed how economic expansion could be de-coupled from the growth in primary energy usage.[229] A study in that same year published by the Conference Board, a nonprofit organization which provides information and analysis to the business community, summarized the situation as follows

> Energy use and economic growth are certainly not independent of one another, but the link between them is more elastic than is commonly assumed.[230]

In releasing the company's estimates for future energy consumption and economic growth in 1978, Sheldon Lambert, the Manager for Energy Economics at Shell USA, stated more bluntly that "[w]e have found that we could decouple the two."[231]

In fact, the ability of energy efficiency programs to provide economic, environmental, and energy security benefits has been widely recognized for decades. For example, a 1979 study by Energy Project at the Harvard Business School concluded that

> Conservation may well be the cheapest, safest, most productive energy alternative readily available in large amounts. By comparison, conservation is a quality energy source. It does not threaten to undermine the international monetary system, nor does it emit carbon dioxide into the atmosphere, nor does it generate problems comparable to nuclear waste. And contrary to conventional wisdom, conservation can stimulate innovation, employment, and economic growth.[232]

In the early 1990s, the Congressional Office of Technology Assessment found that

> There is general consensus among energy analysts that we can cut electricity demand growth further and maybe even produce a net reduction in electricity demand over the next several decades. Doing so clearly offers substantial benefits. We believe with wise implementation of cost-effective measures, they likely will outweigh the costs and risks inherent in this strategy.[233]

The OTA went on state that

> Improvements in energy efficiency through the electric utility sector offer the promise of savings for ratepayers and electric utilities, profits for shareholders, and societal benefits to energy security, international competitiveness, and environmental quality.[234]

In 1997 the President's Committee of Advisors on Science and Technology (PCAST), a panel of academics and industry executives set up under the Clinton Administration, concluded that

> R&D investments in energy efficiency are the most cost-effective way to simultaneously reduce the risks of climate change, oil import interruption, and local air pollution, and to improve the productivity of the economy.[235]

The President's Committee recommended doubling the DOE budget for energy efficiency programs between FY1998 and FY2003. It estimated that this increase would yield a forty to one return on the investment by helping to achieve reductions in energy expenditures amounting to $15 to $30 billion by the year 2005 and as much as $30 to $45 billion by the year 2010.[236]

In, a broad review of DOE research and development spending between 1978 and 2000, the National Research Council of the U.S. National Academy of Sciences found that the

> DOE made significant contributions over the last 22 years to the well-being of the United States through its energy efficiency programs. These programs led to important realized economic benefits, options for the future, and a bank of scientific knowledge. The benefits substantially exceeded their costs and led to improvements to the economy, the environment, and the energy security of the nation...[237]

In this study, the National Research Council Committee reviewed a representative sample of 17 programs amounting to about 20 percent of the DOE R&D funding for energy efficiency programs. The estimated net economic benefits of these programs outweighed the costs by nearly nineteen to one (approximately $30 billion in savings versus $1.6 billion in expenditures). In fact, the savings on just these 17 programs outweighed then entire cost of all DOE R&D spending on all energy efficiency programs by more than four to one ($30 billion in savings versus $7.3 billion in expenditures). These numbers do not include the significant environmental and energy security benefits that also accompanied the DOE research efforts. The NRC Committee estimated the value of these additional benefits to be between $3.2 and $21 billion. For comparison, the R&D spending on fossil fuel programs analyzed by the committee had direct economic benefits that were less than 3 percent higher than their costs over the same 22 year time-span ($10.8 billion in savings versus $10.5 billion in costs). Even narrowing the time frame considered in order to eliminate the rush spending on risky programs that occurred in the wake of the first Arab oil embargo, the economic benefits from fossil fuel spending between 1986 and 2000 still only outweighed costs by less than 65 percent as compared to the more than 1,775 percent return on investment achieved by energy efficiency programs.[238]

Finally, the DOE's Energy Information Administration has also been tracking utility spending on Demand Side Management programs since 1991. These utility sponsored programs include both efforts to reduce end use electricity consumption as well as efforts to establish more efficient load management strategies. For the entire period of 1991 through 2002, the utility Demand Side Management programs saved a total of 596.1 billion kWh at an average annual cost of just 3.9 cents per kWh.[239] For comparison, the average amount of electricity saved each year by these programs was equivalent to a reduction in generating capacity of nearly 6,700 MW, assuming a capacity factor of 85 percent.[240] Another way to appreciate the magnitude of these savings is to note that the average annual reduction in the U.S. was greater than the combined electricity consumption for a total of 15 of the 17 "least developed" countries identified in the MIT study in the year 2000.[241]

Implicit in this discussion of improved energy efficiency is the idea that it is energy services such as heating, cooling, and lighting that are of interest and not the raw amount of electricity consumed. In total, it has been estimated that, without the improvements in energy efficiency that have occurred since 1973, U.S. customers would have spent an additional $430 billion in the year 2000 alone on energy services, a 72 percent increase.[242] Technologies that are capable of providing such services directly therefore offer an opportunity to further reduce demand for electricity. One such technology that is mature and already in widespread commercial use is combined heat and power (CHP) for large industrial users. With combined heat and power systems, a portion of the waste heat created as part of generating electricity is used to directly heat co-located facilities. If properly optimized, the effective efficiency of such systems can be as high as 80 percent.[243] While the long-term potential for large CHP systems is limited by the number of industries that are large enough to efficiently operate their own generators, the near-term expansion of such systems would offer a further economic means to reduce the rate of demand growth and lower greenhouse gas emissions.[244]

Another energy service technology that offers the potential for wider commercial deployment in the near to medium-term is direct solar heating. There are currently 1.2 million buildings using solar water heaters in the U.S. and an additional 250,000 solar heaters for swimming pools. It is estimated that up to 29 million existing single-family homes in the U.S. have suitable orientation and sunlight to make use of solar water-heating systems while as many as 70 percent of new homes could be oriented to enable the use of such system. Additional growth potential exists for swimming pool heating systems. Current solar water heating systems can supply 40 to 70 percent of the residential needs, while solar pool heaters can supply between 50 and 100 percent of the required energy services. The cost of these systems have declined by 50 percent over the past decades, and further reductions are projected in the near-term that would likely make them competitive for water heating with the projected range of costs for running an electric water heater using electricity from new nuclear power plants.[245]

In addition to improvements in the efficiency of energy use, there are a number of ways that demand can be reduced with little to no added spending through conservation. Simple efforts, such as ensuring that building are not over lit, making wider use of motion sensitive light switches, buying locally grown foods, air drying clothes when and where possible, and making sure that thermostats are not set so low that sweaters are worn indoors during the summer and then so high that sweaters are not worn in the winter, are all important parts of any overall approach to achieving a more sustainable energy system. Areas for such zero or low cost savings abound in our society. So called "phantom loads", such as the energy drain of televisions, VCRs, and computers when they are turned off, could easily be eliminated by using power-strips that can be switched off or by unplugging the devices when they are not being used. These "phantom loads" can be quite significant, with nearly a quarter of the total energy usage for televisions in a typical household and more than half of the electricity usage for VCRs coming when they are not in use.[246] Coupling improvements in energy efficiency to decreases in wasted and wasteful consumption enhances their impact and adds to their already strong economic viability.

Such benefits from conservation and increased energy efficiency are, of course, not limited to the United States. For example, a 2002 study by the British government's Cabinet Office Performance and Innovation Unit (PIU) found that the cost through 2020 of energy efficiency efforts in the domestic, service, and industrial sectors ranged from a <u>savings</u> of £640 ($960) per ton of carbon to a maximum <u>cost</u> of £130 ($195) per ton of carbon. Pursuing these programs would result in estimated reductions of approximately 28 million metric tons of carbon through 2020 with the savings increasing to 65 million metric tons by 2050. For comparison, the PIU concluded that the use of nuclear power would <u>cost</u> between £70 and £200 ($105 to $300) per ton of carbon and result in savings of just 7 million metric tons of carbon by 2020 with potential reductions increasing to more than 20 million metric tons by 2050.[247] Following the publication of this report, the British government's Department of Trade and Industry concluded that "the cheapest, cleanest and safest way of addressing all our goals is to use less energy."[248]

Section 2.3.2 – The Power of Wind

Combined with energy efficiency efforts to reduce demand, the use of renewable energy, particularly wind power, offers the most economically competitive way to provide near-term incremental growth in supply without increasing carbon emissions. By any measure, the available wind resources are enormous. The Intergovernmental Panel on Climate Change estimated potential global wind resources to be between 5.3 to 14 times the world's total primary energy usage in the year 2000. Even the lower estimate of developable wind energy potential cited by the IPCC was between two and four times its estimate for global primary energy use in 2050.[249] In the United States, it was estimated by the Pacific Northwest Laboratory that the top 12 states in terms of wind potential have a combined annual capacity more than two and half times the total U.S. electricity consumption in 2000.[250] The state with the lowest potential from this group of twelve (New Mexico) has nearly five and a half times more potential wind capacity by itself than the entire amount of generation by all non-hydro renewables from all states in 2000.[251] Using a different estimate for wind potential, researchers at Stanford University, found that more than one-fifth of the monitoring stations in the U.S. had winds averaging Class 3 or higher at a height of 80 meters, and that one in seven had wind speeds averaging Class 4 or higher.[252] The two geographic regions with the greatest wind potential estimated by this method included the states of Arkansas, Iowa, Kansas, Louisiana, Minnesota, Missouri, Nebraska, North Dakota, Oklahoma, South Dakota, and Texas. In these eleven states more than two-fifths of the existing monitoring stations had greater than Class 3 winds, while one in four had Class 4 winds or higher.[253]

Despite this great potential, just 0.3 percent of U.S. electricity production came from wind power in 2003.[254] In addition, while the U.S. led the world in the early exploitation of wind power, the United States currently lags far behind Europe in developing this resource (see Table 2.12). In 1985, the U.S. accounted for nearly 95 percent of the world's installed wind capacity with nearly 18 times more wind than in all of Europe. The U.S. share fell to 76 percent in 1990 and by 2003, the U.S. accounted for just 16 percent of installed wind capacity. Europe, on the other hand, rose from just over 5 percent of global wind capacity in 1985 to 22.5 percent in 1990 and 73 percent in 2003.[255] To put this comparison another way, in 2003, Germany, a country with less than 5 percent of the United States' land area and less than 15 percent of its electricity production, has more than two and a quarter times as much wind capacity installed as the entire U.S. In addition, Spain, a country with just over 6 percent of the U.S. land area and 6 percent of its electricity production, had nearly the same amount of installed wind capacity as in all of the continental United States.[256] In 2003 alone, the European Union added

15 percent more wind capacity to its grid than the cumulative amount installed in the entire United States through the end of 2002.[257]

Table 2.12: Total amount of grid-connected wind capacity (in megawatts) in the Unites States, Europe, and the world between 1980 and 2003. A rapid and sustained rise in installed capacity has been occurring in Europe since 1995.[258]

Region	1980	1985	1990	1995	2000	2003
United States	10	1,039	1,525	1,770	2,554	6,374
Europe	5	58	450	2,494	12,961	28,706
Other Countries	0	0	27	623	2,138	4,214
World	15	1,097	2,002	4,887	17,653	39,294

The impressive growth in Europe not withstanding, the amount of wind power installed continues to lag far behind its economic potential. For example, including offshore installations, the wind energy industry estimated in 1998 that it was feasible for installed wind capacity to have risen to 844 GW by 2010 (nearly 21.5 times the actual level installed through 2003). As a further example, a joint study published by Greenpeace and the European Wind Energy Association in 1999 estimated that a total of 1,200 GW could be installed around the world by 2020 providing 10 percent of global needs.[259]

At a level of wind penetration of 15 to 20 percent, the issue of land usage would be unlikely to pose any significant obstacle to development. Many of the areas with high wind potential in the U.S. are located in rural areas and farmers are able to utilize the land right up to the base of wind mills for crops or grazing with the added benefit of receiving an additional steady source of income. The Nuclear Regulatory Commission itself estimates that just three acres of land would be occupied by wind turbines for every megawatt of effective generating capacity. The remaining land area required for the wind farms would remain available for agricultural uses.[260] Improvements to the capacity factor of wind turbines since the estimate cited by the NRC was made could reduce this land usage by as much as 40 percent.[261] The equivalent land use that the NRC estimated for new nuclear power plants was between two and three acres per megawatt when the full fuel cycle was included. For natural gas fired plants, the NRC estimated that a total of 3.7 acres per megawatt of effective generating capacity were required.[262] Beyond the fact that these levels of land usage are comparable to begin with, the land occupied by wind turbines would require very little remediation to be restored to a green-field condition, but some of the land used for generating nuclear electricity would remain dangerously contaminated with radioactive waste for hundreds to hundreds of thousands of years. Given the large geographic distribution of favorable sites in the U.S., the siting of wind farms in the near-term could be done so as to avoid major bird migration corridors, sensitive ecosystems, and important scenic areas without imposing any meaningful limitations on the increase in wind power.

Estimates for a renewable contribution well beyond 15 to 20 percent have also been seriously considered. For example, in Shell International's 1995 *Long-Term Energy Scenarios*, the world's largest oil company estimated that, achieving the maximum growth rate for renewable resources, could allow up to half of the world's total primary energy demand to be met by renewables in 2050. In its 2001 *Dynamics as Usual* scenario, Shell predicted that, even under a less aggressive development path, as much as one-third of the world's primary energy demand in 2050 could be met by renewable resources.[263] These predictions of what is possible are significantly higher than even the current goals that have been set in Europe where, for example, Britain is planning to generate 10 percent of its electricity from renewables by 2010 and 20 percent by 2020, while the European Union has a announced a goal of generating more than 22 percent of its electricity from renewable energy sources by 2010.[264]

Aside from the question of land use, the main argument made against the expanded use of renewable energy technologies (other than hydroelectricity and certain types of biomass) is that their intermittency makes them unsuitable as a replacement for base load resources like coal or nuclear power, and that the costs of grid integration, as well as transmission losses, make them too expense to develop on a large scale. While these are very complex questions that will require an extensive and sustained effort to address (including developing more robust regional transmissions grids and improving the rules governing grid access and integration), a careful consideration of the current situation reveals that none of these concerns is likely to pose an insurmountable limitation to the economic expansion of wind power in the United States.

The U.S. already has a complex and interconnected grid system that allows electricity to be transmitted over long distances.[265] Many favorable wind sites are located near existing transmission infrastructure allowing for their rapid development, but a significant effort will be required to develop additional transmission lines and strengthened grids in order to reach the level of wind penetration we are considering for the near-term (i.e., 15 to 20 percent of total generation). Significantly, while the existing grid is already under strain as illustrated by the massive Northeast blackout that occurred in the summer of 2003, the DOE's National Renewable Energy Laboratory concluded that

> Initial lower levels of wind deployment (up to 15–20% of the total U.S. electric system capacity) are not expected to introduce significant grid reliability issues.[266]

Similar conclusions regarding grid stability at wind penetrations up to 20 percent have been drawn by a number of utility studies as well.[267] However, the connection of wind to the grid at these levels would impose two related financial penalties.[268] The first penalty relates to the additional efforts required to maintain the stability and reliability of the transmission grid given that there is uncertainty in predictions for the amount of wind power that will be generated on an hour ahead or day ahead basis. The cost of this penalty generally increases with the fraction of electricity supplied by wind since the fluctuations in its generation have a proportionally greater impact on the stability of the overall system. Estimates in the U.S. for these grid integration costs at wind penetrations up to 29 percent of the total supply have been estimated to vary from about 0.1 to 0.55 cents per kWh. A range of costs from 0.2 to 0.5 cents per kWh was used by IEER in its assessment of wind power in New Mexico as a reasonable estimate for near-term penalty for wind development.[269] This range is consistent with estimates from Britain as well. In 2002, the Performance and Innovation Unit of the Cabinet Office concluded that the intermittency cost of renewables like wind and solar is insignificant below penetrations of 5 percent, rises to 0.1 pence per kWh (0.16 cents per kWh) for penetrations between 5 and 10 percent, and reaches 0.2 pence per kWh (0.31 cents per kWh) at penetrations up to 20 percent.[270]

The other financial penalty faced by wind is that the transmission infrastructure connecting the wind farms to the grid must be sized to match the peak generating capacity of the facility, despite the fact that the typical level of electricity generated by the wind farm will only be able to utilize a fraction of this transmission capacity. The size of this financial penalty generally decreases as the capacity factor of the wind farm increases, since less of the transmission infrastructure is left unused. Estimates for the current cost of this penalty range between 0.4 and 0.9 cents per kWh.[271] Therefore, for the near-term expansion of wind power to levels of approximately 15 to 20 percent of total generation, it is appropriate to add between 0.6 and 1.4 cents per kWh to the estimated generation cost of the wind turbines in order to account for the costs of transmission and grid integration.

In light of its potential contribution to the energy system, the economics of wind have been reviewed by a wide range of individuals and organizations (see Table 2.13). The consistent finding among all of these studies is that, at favorable sites, wind is already likely to be far more economic to develop than new nuclear power. The average cost from all these estimates is just 4.8 cents per kWh. Even the very highest estimated costs for wind falls within the range of 6.0 to 7.0 cents per kWh we have been considering for new nuclear power. Significantly, Table 2.13 also shows that, at very favorable sites, the cost of wind power is already comparable to the cost of electricity from new natural gas fired plants as well. The wind farms would have the additional economic benefits of no fuel price uncertainty, no carbon emissions subject to future taxation, and a very high availability.

Table 2.13: Summary of estimates cited for the current cost of electricity from onshore wind power at favorable sites.[272]

Source	Current Cost of Wind (cents per kWh)
Energy Information Administration (2005)	4.5 to 6.0
National Renewable Energy Laboratory (2005)	4.2 to 6.0
Institute for Energy and Environmental Research (2004)	4.1 to 5.6
International Energy Agency (2003)	3.0 to 7.0
Intergovernmental Panel on Climate Change (2001)	3.3 to 5.4
Jacobson and Master (2001)	3.5 to 5.3[a]

(a) Estimated grid connection costs of 0.6 to 1.4 cents per kWh added to the generation cost estimate of Jacobson and Master by IEER to allow comparison to other figures.

Improvements to the cost of wind that are expected over the next 5 to 15 years would add further to the economic advantage of this technology. The capacity factor of wind farms have been rising consistently since 1985, and are expected to continue increasing in the future.[273] These increases, along with other improvements to windmill technology, are expected to continue to lower the generation cost of wind, as well as the penalty it pays for under utilizing its transmission infrastructure. The International Energy Agency estimates that the generation costs of wind power at very favorable sites could drop to just 2.0 to 3.0 cents per kWh by 2010.[274] Similar expectations have been put forth by the National Renewable Energy Laboratory, which sees generating costs at very favorable sites dropping to 2.4 to 3.0 cents per kWh by 2010 and to just 2.2 to 2.7 cents per kWh by 2020.[275]

Section 2.3.3 – Summary of Near-Term Options

While it will require significant public and private effort and investment to implement new energy efficiency programs and to develop the necessary infrastructure to support a large near-term increase in wind power, it is important to keep in mind the costs inherent in simply maintaining the current energy system as well as the difficulties that would be encountered in restarting a nuclear power industry that last hasn't had a new order placed in the U.S. in more than 25 years and hasn't opened a new plant in nearly ten years.

For example, the International Energy Agency estimates that the amount of investment in oil and gas between 2001 and 2030 will total nearly $6.1 trillion with 72 percent of that investment going towards new exploration and development efforts. The amount of investment in the U.S. and Canada over that time is expected to account for about one-fourth of all global investment.[276] With respect to nuclear power, the construction cost of each new nuclear plant, including interest payments, would total nearly $2.6 billion under the MIT base case assumptions, and dozens of such plants would have to be started in the ten to fifteen years in order to remain on track to meet the global or steady-state growth scenarios.[277] The difficulties of restarting the nuclear industry on this scale would be severely aggravated by concerns over nuclear weapons proliferation, the risks of a major reactor accident, and the difficulties in handling the large volumes of radioactive waste that would be generated as the new plants came online (see Chapters Three through Five).

Finally, when considering efforts to expand energy efficiency programs and the use of wind power it is important to note that, unlike the decision to pursue new nuclear power, there is already strong and sustained public support for these programs. In a review of 700 polls conducted between 1973 and 1996, Dr. Barbara Farhar of the DOE's National Renewable Energy Laboratory concluded that

> In summary, the pattern of preferences for using energy efficiency to decrease demand and renewables to supply energy has been consistent in the poll data for at least eighteen years.

> This is one of the strongest patterns identified in the entire data set on energy and the environment.[278]

A 1998 survey conducted by International Communications Research found similar results with three out of every five respondents placing the development of renewable sources of power or improvements in energy efficiency as their highest priority for energy research while only three out of every fifty placed nuclear power first.[279] The public survey conducted as part of the MIT report found that less than 30 percent of people supported any expansion of nuclear power while 77 percent favored an expansion of solar and wind, with more than 50 percent of the respondents favoring a large increase in the use of these renewable resources.[280] Finally, in October 2005, a report from the International Atomic Energy Agency, a body explicitly charged with promoting the spread of civilian nuclear technologies, found that nearly three out of every five people interviewed opposed the construction of any new nuclear plants.[281]

With the proper priorities on investment in transmission and institutional infrastructure, wind power could make a significant contribution to reductions in greenhouse gas emissions while economically displacing natural gas under the high fuel price scenarios.[282] When coupled with aggressive energy efficiency and conservation efforts, the near-term expansion of wind could supply much of the required incremental growth in demand while reducing emissions and providing time for new low-carbon technologies that are nearing commercialization to be brought onto the market. As summarized by the British Department of Trade and Industry

> Energy efficiency is likely to be the cheapest and safest way of addressing all four objectives [i.e. a reduction of greenhouse gas emissions, the maintenance of a reliable energy supply, promotion of competitive markets, and an assurance of adequate and affordable heat to every home]. Renewable energy will also play an important part in reducing carbon emissions, while also strengthening energy security and improving our industrial competitiveness as we develop cleaner technologies, products, and processes.[283]

Section 2.4 – Alternatives for the Medium-Term (2020 - 2050)

The near-term exploitation of wind power and a strong focus on reducing demand through heightened efficiency and conservation are essential elements of sustainable efforts to reduce greenhouse gas emissions. This is not only because of their direct contribution to reducing greenhouse gas emissions, but also because they can provide a buffer time for new technologies and infrastructure to be developed for the transition period in the later part of this half-century. These available transition technologies will often involve a level of compromise in which their environmental and security impacts must be weighed against their ability to reduce greenhouse gas emissions and avoid the unique vulnerabilities associated with nuclear power. Like the near-term options discussed in the previous section, the main criteria upon which to judge the transition technologies are that:

1. they must be capable of making a significant contribution to the reduction in greenhouse gas emissions through 2050;
2. they must be likely to be ready for full scale commercialization by 2020 at the latest;
3. they should be capable of competing economically in a market setting with alternative options such as new nuclear power;
4. they should, to the extent consistent with the goals of reducing the threat from climate change, minimize other environmental and security impacts; and
5. they should be compatible with the ultimate long-term goal of developing an equitable and sustainable global energy system.

The most likely options that can meet these criteria are the continued expansion of energy efficiency programs and the development of wind power and other renewable resources, the continued use of natural gas in combined cycle generating plant with a heightened reliance on LNG imports to help stabilize fuel prices, the use of coal gasification combined cycle power plants, and the integration of carbon capture and storage technologies with fossil fuel plants. Each of these options will be discussed in the following sections.

Section 2.4.1 – Liquefied Natural Gas and Fuel Switching

Per unit of generation, a pulverized coal plant has significantly higher greenhouse gas emissions than a plant using natural gas combined cycle technology. For example, in 2002, coal generated 50 percent of the electricity in the U.S., but was responsible for 83 percent of the total carbon dioxide emissions from the electric sector, while natural gas accounted for 18 percent of total electricity generation and just 13.3 percent of emissions.[284] Although not included by the authors of the MIT study in their list of four "realistic options" for reducing carbon emissions from the electricity sector, fuel-switching has been widely considered as part of the response to global warming during the transition from fossil fuels to a more sustainable, long-term energy system. For example, this option has been considered by the DOE's Inter-laboratory Working Group on Energy-Efficient and Clean Energy Technologies as well as the Intergovernmental Panel on Climate Change. In fact, the IPCC concluded that replacing coal fired plants with more efficient natural gas plants would be likely to make a "relatively large" contribution to carbon reduction if undertaken as part of a larger overall strategy.[285] Switching between fossil fuels would also help to reduce the emissions of other pollutants such as sulfur dioxide and mercury in addition to lowering the emission of particulates.

The main limitation to such strategies, however, is the fact that natural gas supplies have already been stretched thin in the U.S. by the strong growth in combined cycle gas plants throughout the last two decades when natural gas prices remained generally at or below $3.00 per MMBtu. For example, between 1986 and 1999, the amount of electricity generated from natural gas fired plants more than doubled from 248.5 billion kWh to 556.4 billion kWh. Between 1999 and 2003, however, fuel prices rose sharply while the amount of generation from natural gas increased at a lower rate.[286] The 2003 price for natural gas paid by utilities was near the all time high reached in 1982 following the Iranian revolution.[287] As a result of the increased demand for natural gas, the unusually intense Atlantic hurricane seasons in recent years, and heightened political tension in the Middle East following the U.S. led invasion of Afghanistan in 2001 and Iraq in 2003, the average price and price volatility of natural gas has increased sharply in the last few years (see Figure 2.1).

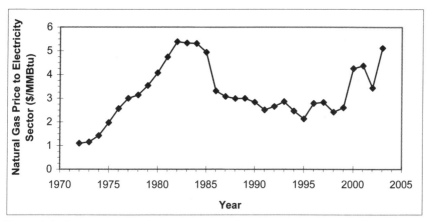

Figure 2.1: Average price of natural gas sold to customers in the U.S. electricity sector between 1972 and 2003 (all figures in constant 2000 dollars). The peak price over this time occurred in 1982 following a sustained increase over the preceding decade. Since the mid-1990s the average price of natural gas has generally been increasing, but the price has shown a greater degree of volatility than at any other time in the past 30 years.[288]

Like petroleum, domestic natural gas production in the United States peaked more than 30 years ago. However, unlike oil, where imports accounted for more than 56 percent of U.S. consumption in 2003, imports of natural gas accounted for less than 18 percent of U.S. consumption in that year. The most significant change to U.S. natural gas imports over the past several years has been the increasing importance of liquefied natural gas (LNG). In 1998, nearly 97 percent of all imported natural gas was brought in via pipelines from Canada while just 2.6 percent was LNG (all of which was imported from Algeria and Australia). In 1999, a new liquidification terminal began operation in Trinidad and Tobago, and, by 2003, LNG was supplying 12.8 percent of the natural gas imports to the U.S. While pipelines from Canada still accounted for 87 percent of imported supply in that year, the contributions of LNG from Trinidad alone had risen to 9.6 percent of total imports.[289] Despite the significant increase in LNG usage, however, it remains a small fraction of total U.S. consumption (just 2.3 percent in 2003), and therefore remains primarily a modest supplement to a much larger pipeline supply.

In order to bring gas prices under better control and allow natural gas to be retained as an economically viable transition fuel for the medium-term it will likely be necessary to combine efficiency and conservation efforts with an increased supply of liquefied natural gas. There are already a few industrialized countries that rely on LNG for the vast majority of their natural gas needs which can serve as models for how a higher penetration of liquefied natural gas might function in the United States. For example, in 2001, South Korea relied on LNG for nearly 100 percent of its consumption while Japan relied on LNG for 97 percent of its natural gas needs, and Taiwan imported 86 percent of its natural gas as LNG.[290]

Between 1990 and 2000, the global gas liquidification capacity grew by 86 percent. It is believed that, assuming a favorable outcome to the current restructuring of the global natural gas market and that demand remains high, it is "easily possible" that the liquidification capacity could increase by another 144 percent by the end of 2012. In fact, increases of more than 200 percent are deemed to be within reason if the conditions for growth were extremely favorable.[291] Liquidification projects, however, are very capital intensive with an expected price of between $1.5 and $2.0 billion per plant, roughly equal to the projected overnight cost of building a new 1000 MW nuclear power plant in the U.S. Achieving the level of growth envisioned in these projections would therefore require a significant near-term investment.[292]

Once the natural gas could be delivered to the wholesale market, the U.S. already has a well developed hub system that would allow the gas to be distributed and traded via pipelines. This infrastructure would help to stabilize supply with the introduction of larger imports of LNG in the future. While not as robust as the U.S. system, the Europeans are also beginning to develop their own system of natural gas hubs, with the United Kingdom further along than other European countries. Improvements to this system will be necessary, but it is not likely to pose a significant limitation to the expanded use of LNG in the medium-term.[293] The primary bottleneck in getting LNG to end users in the U.S., therefore, is likely to be in regasification capacity.

Currently the U.S. has only four regasification terminals, all of which are more than 20 years old and all of which serve the East Coast (Everett, Massachusetts; Cove Point, Maryland; Elba Island, Georgia; and Lake Charles, Louisiana). The terminals in Maryland, Georgia, and Louisiana are expected to complete an expansion of their capacity by 2007. When complete, these expansions will have increased the total baseload capacity of the four existing terminals by nearly 82 percent over their 2003 capacity. In addition, more than two dozen proposals have been put forward for new regasification terminals to be built in North America. Most of these proposals were made following the recent spike in U.S. gas prices (see Figure 2.1). While there has been strong local opposition in some areas to the siting of new regasification facilities due to safety and environmental concerns, and the financial collapse of Enron has changed the nature of companies that are likely to actually carry out plans for building new LNG infrastructure, the Energy Information Administration continues to expect that new regasification facilities will come online in the near to medium-term. Specifically, along with new terminals in the continental United States., the EIA expects that a new regasification terminal will be built in the Bahamas to provide natural gas to Florida and another terminal will be built in Baja California, Mexico to provide gas to California.[294] The EIA assumptions are similar to those from other forecasts in that they all project strong growth in LNG. The LNG imports in 2025 are projected to range from 11.5 to 20.75 times their level in 2003, with the EIA reference case assuming a 16 fold increase in imports over 2003. For comparison, under the EIA reference case, imported liquefied natural gas would amount to more than 20 percent of total projected end-use consumption in 2025, and would be 28 percent greater than the total amount of natural gas consumed by the entire electric power sector in 2003.[295]

A more reasonable growth in LNG imports over the coming decades is possible than that projected by the EIA if the efforts were coupled to a greater emphasis on energy efficiency and conservation, including improvements to natural gas fired power plants and the widespread use of high efficiency heating systems, as well as to an expansion of wind and other renewable sources of power. While there will still likely need to be an increase in LNG imports over the coming decades to help stabilize prices, the pursuit of efficiency and alternative forms of generation would allow time for regasification terminals to be sited with greater deliberation, allowing an opportunity for full and informed participation by affected communities, and allowing the best possible sites to be found (i.e. those that are away from major population centers, particularly fragile or endangered ecosystems, and major shipping routes among other considerations). The siting of terminals in other countries to primarily service the U.S. economy should be carefully scrutinized on both environmental and environmental justice grounds.

As noted above, LNG in the U.S. is currently a small percentage of overall consumption, and therefore its price is strongly coupled to the wellhead and pipeline import price. For example, between 2000 and 2004 the import price of LNG averaged $4.37 per thousand cubic feet compared to $4.19 at the wellhead and $4.52 for pipeline imports.[296] The EIA forecasts assume that demand will grow rapidly enough that, despite a significant increase in LNG imports and domestic production from non-conventional sources, supplies will remain tight and that prices will continue to follow the cost of gas at the wellhead. Between 2003 and 2025, the EIA projects a variable gas price with an average price of approximately $4.26 per thousand cubic feet.[297]

The ongoing liberalization of the international natural gas markets offers a number of opportunities for continued reductions in the cost of LNG. For example, tanker construction costs have decreased by 45 percent since their peak in the mid-1980s and, while the majority of tankers are still tied to specific LNG projects under long-term contracts, the emergence of independent transport companies offers the possibility for further cost savings through greater flexibility and the limiting of unnecessary transport.[298] The biggest influence on recent reductions in cost, however, has been the increase in the train size for liquidification plants. In the 1980s, it was common for trains to have a capacity of 2 to 2.5 million tons while 4 million ton trains are available today. Continued improvements are likely with 5.1 million ton trains under construction and planning currently under way by Exxon Mobil for the construction of twin 7.5 million ton units. Increasing the train size to 4 million tons reduced the liquidification costs by nearly 30 percent relative to the cost of LNG from 2 to 2.5 million ton units. Going to a train size of 7.5 million tons could reduce costs by another 20 percent.[299] Table 2.14 shows the current expectations for the costs of liquidification, transportation, and regasification and their relative contribution to the typical cost of LNG imports.

Table 2.14: Estimated added costs of liquefying natural gas, transporting it via tanker to the eastern United States, and then regasifying it. The largest cost element remains liquefying the natural gas, despite the recent improvements in the size of the trains used.[300]

	Liquidification	Transport	Regasification
Cost Estimates for LNG delivered to the U.S.	$0.97 to $1.09 per MMBtu	$0.34 to $1.46 per MMBtu[a]	$0.30 per MMBtu
Typical Fraction of the Total LNG Import Price	30% to 45%	10% to 30%	15% to 25%

(a) The low end of the transportation costs are those for LNG shipped from Venezuela or Trinidad and Tobago while the upper end of the transportation costs is for shipments from Qatar in the Middle East. Costs for shipments of LNG from Nigeria would fall in the middle of this range at roughly $0.85 per MMBtu.

The moderate gas price scenario considered in the MIT report starts at $3.50 per MMBtu while the high gas price scenario starts at a cost of $4.50 per MMBtu. The prices escalate at different rates under the two scenarios (1.5 percent for the moderate gas price and 2.5 percent for the high gas price) yielding a levelized 40 year cost of $4.42 and $6.72 per million BTU respectively.[301] If the long-term price of natural gas is restrained by the cost of importing LNG, then the moderate gas price is likely to be more correct than the higher price scenario. The levelized gas price in the moderate scenario would be consistent with the recent average price of LNG in the United States noted above ($4.37 per MMBtu between 2000 and 2004), the average price for LNG imports to Japan and South Korea over the past decade (~$4 per MMBtu), and the expected price for future LNG imports to India ($4.10 per MMBtu).[302]

As noted, the long-term cost of natural gas has a strong impact on the economics of combined cycle gas generation. In the MIT and U Chicago models, an increase of $1 per MMBtu in the levelized fuel price leads to approximately a 0.6 cent per kWh increase in the price of electricity.[303] Based on similar considerations regarding the long-term impact of an increased reliance on LNG, the U Chicago study concluded

> Assuming the gas supply infrastructure is stabilized in the near term and new supply options such as LNG are realized, the current cost advantage of natural gas generation [compared to nuclear power] should continue.[304]

Additional development efforts carried out over the coming decade to improve the heat rate and efficiency of natural gas fired plants would lessen the economic impact of higher fuel prices as well.[305] Thus, in light of these considerations, it is reasonable to continue to look on natural gas as an economically viable transition fuel for the medium-term, particularly in its capacity to replace pulverized coal plants and its suitability for being more tightly integrated with high penetrations of wind power as discussed in the following section.

Section 2.4.2 – Increased Use of Wind and Other Renewable Energy Resources

In Section 2.3.2, we discussed the economics of an expansion of wind power in the near-term with penetrations up to 15 to 20 percent of total generation. Further increases in the contribution of wind can be achieved through a reduction in the penalty associated with its intermittency.

There are three main ways that this can be achieved in the medium-term: (1) increased geographic distribution of wind farms connected through strengthened regional transmission grids; (2) integration of wind power with pumped hydro power stations that will make use of more ecologically sound turbine technologies; and (3) closer integration of natural gas fired plants (both single-stage peaking capacity and baseload combined cycle plants) with wind generation.

As noted, sites with favorable wind resources in the U.S. are widely available, and as the development of wind power is expanded, its geographic distribution will naturally increase. This distribution will improve the stability of the wind farm's overall output because while the wind is not blowing in one area, it may be blowing elsewhere.[306] Even over the relatively small distances of a single wind farm, it has been found that increasing the number of turbines will increase the stability of the power output from the site.[307] Over longer distances, a European study found that the correlation between wind speeds at different sites decreased at roughly an exponential rate, and that for separations of approximately 2,500 kilometers (about 1,550 miles) there was no longer any discernable correlation between the wind blowing at one site and that at another.[308] Using wind data from monitoring stations in the United States, researchers at Stanford University found that

> The standard deviation of the wind speed averaged over multiple locations is less than that at any individual location. As such, intermittency of wind energy from multiple wind farms may be less than that from a single farm, and contingency reserve requirements may decrease with increasing spatial distribution of wind farms.[309]

In addition to the distribution of wind speeds becoming more concentrated around the average, the researchers found that the amount of time when the average wind speed at all sites fell below the lower cutoff threshold of the turbines decreased as the number of sites increased.[310] These improvement in the stability of the wind power system viewed as a whole will begin to occur in the near-term, but will become more pronounced as the regional transmission infrastructure and regulatory framework is improved in the longer-term and progressively higher levels of wind penetration are achieved in more varied geographic locations.

The second way in which the intermittency of electricity from wind can be reduced is to make use of pumped hydro power to store the electricity generated by wind farms at periods of low demand and regenerate the electricity during subsequent periods of higher demand.[311] Pumped hydro power is the most effective and widely used form of energy storage available today. The U.S. has an estimated 19.2 gigawatts of pumped hydro power capacity out of a total hydro electric capacity of 99.7 gigawatts. An additional 64 gigawatts of pumped hydro capacity is installed in other OECD countries.[312] The current efficiencies of these systems range between 75 and 80 percent depending on the capacity of the pumps used.[313] While this energy storage capacity is currently available and could be integrated with the near-term expansion of wind power, improvements in the regional transmission grids will be needed and there are additional environmental considerations with hydro power that can be more adequately addressed in the medium-term.

The installed hydro capacity in the U.S. continued to grow through 1995, but it has remained virtually unchanged since then due to concerns over the social and environmental impacts of dams in general, and the impact of the hydro plants on fish and water quality in particular. Advanced hydropower technologies and new operational procedures designed to improve water quality and reduce the detrimental impacts on fish are estimated to be suitable for use at more than 80 percent of the existing hydropower plants in the U.S., as well as suitable for use at existing dams without hydropower facilities, amounting to 15 to 20 GW of new capacity. Many of these technologies, however, still require some level of development before they will be ready for commercialization. For example, the DOE's Advanced Hydropower Turbine System was successfully tested at a pilot scale facility in 2003 and full scale prototype testing of this system, including its effects on the aquatic ecosystem, is expected to be completed by 2010. Field testing of other advanced turbine designs have shown that fish survival up to 98 percent is possible, and the TVA's Lake Improvement Plan has already been able to increase energy production while simultaneously improving downstream fish resources.[314] Adding to the benefits of integrating wind with pumped hydro systems is the fact that the release of water from the upper reservoir could be better timed to ecological needs and then re-pumped by excess wind power at a low marginal cost further reducing the impacts of hydro plants.

Even if as much as one-quarter of the total installed wind capacity was eventually backed up by pumped hydro power, this would increase the effective generation cost of the overall wind-hydro system by just 0.4 to 0.7 cents per kWh which would still leave wind very competitive with the expected range of costs for electricity from new nuclear power.[315] More importantly, this cost increase may actually be reduced or even eliminated if the hydro power backup allows low cost electricity from wind generated at off-peak hours to be resold at times of higher demand. For example, a recent pair of studies considered how a wind farm and pumped hydro system could be optimally integrated into the Portuguese grid.[316] In these studies, the authors examined a 12 MW wind farm coupled to a 3 MW hydro facility, and found that optimizing the system for profit <u>increased</u> the net value of the electricity sold over that of the wind farm alone by 12 to 22.2 percent. In addition, the combination of the two technologies increased the number of hours that the system was able to operate at its nominal power by an average of 5.6 to 19.6 percent over the wind farm alone, demonstrating the benefits of reduced intermittency even in a system optimized for economic gain rather than stability of the electrical output.[317]

A further advantage to the development of cost effective wind-hydro systems in the U.S. or Europe would be their potential for use in smaller electricity systems in the Global South. For example, a pair of recent studies examined the economics of building a pumped hydro backed wind farm on the island of El Hierro in the Canarian archipelago.[318] The authors of these studies found that the wind-hydro system would be economically advantageous over the island's current diesel generators at fuel prices above 0.283 euros per liter ($1.33 per gallon) while allowing renewables to make up 68 percent of the total electricity supply and saving an estimated 21,000 metric tons of carbon emission per year.[319]

A third way in which higher wind penetrations could be achieved would be to improve the integration of wind farms with natural gas fired capacity. As discussed in the previous section, the increased use of LNG in the U.S. could stabilize fuel prices in a range where combined cycle gas plants will continue to be economically competitive compared to other types of base load generation. The capital cost of natural gas plants is a smaller part of the final cost of electricity than for either coal or nuclear plants, and thus a portion of the natural gas plants in operation could intentionally be run below their full capacity without incurring a significant economic penalty.[320] The reserve capacity provided by these operating combined cycle plants could be rapidly brought into use when the output of the wind farms was lower than expected. For larger shortfalls in peak demand, less efficient single cycle gas turbines that can be started rapidly could be used to further stabilize the grid. Combined with a reduction in the intermittency of supply through strengthened regional transmission infrastructure, greater geographic dispersion, and the use of pumped hydro storage, the closer integration of wind with natural gas generation could allow it to reach penetrations well beyond 20 percent in some regions of the United States.

In addition to the expansion of wind power, other renewable resources will likely play a role in the medium-term in diversifying the supply of electricity. As noted by the International Energy Agency,

> ... some renewable technologies actually complement one another in their cycles. Solar PV [photovoltaic] resources are most available in summer while this is in many climates a time of relative drought with respect to hydro resources. Winds are often stronger in winter which is also a time of peak demand in colder climates.[321]

The use of a mix of renewable energy technologies that have different types of variability, therefore, could help to significantly reduce the amount of backup capacity that is required.[322] Solar energy has the additional advantage of being able to generate electricity where it is needed, thus lessening the need for transmission and distribution capacity. This is a particular benefit for countries in the Global South and could have a dramatic impact on the development of their energy systems.

By any measure, the total magnitude of solar energy incident upon the Earth is enormous. Estimates of available solar power potential globally amount to more than 25 times the world's total electricity consumption in 2000.[323] The National Renewable Energy Laboratory has estimated that an area less than one-third of that "used for military purposes in the United States" could supply the country's entire electricity demand.[324] More importantly, however, there is already a great deal of land that could be used for solar generation without the need to disturb new areas. This includes such areas as window awnings, the tops of roofs and parking structures, along the medians of highways, and along the existing electricity transmission corridors. To give one example, it has been estimated that the entire British electricity demand could be supplied just by integrating solar panels "into existing building structures without the need for additional land."[325]

Despite the enormous potential of solar energy, the deployment of photovoltaic solar cells to capture that energy and convert it into usable electricity remains extremely small. For example, in 2003, all solar technologies combined supplied just 0.013 percent of the electricity consumed in the United States.[326] In that same year, less than 0.75 GW of new solar capacity was produced by all PV manufactures worldwide.[327] This discrepancy between solar power's potential and its realization is due primarily to the high cost of solar cells compared to other types of generation. This remains true despite the steady decrease in solar's cost since 1978.[328] The continued development of new technologies, such as high efficiency thin film PV, however, suggests that solar power could play a more important role over the coming decades.[329]

Currently there are five thin film solar cell designs that have already been commercialized or are nearly ready for commercialization.[330] Thin film technologies have two major advantages over the monocrystalline silicon designs that dominate the market today. First, thin films use significantly less semiconductor material than current solar cells allowing them to be made at a much lower cost. For example, in 2001, the IPCC estimated that thin film solar cells with comparable efficiencies to current technologies could be manufactured for less than one-third the cost of monocrystalline cells ($1.50 per watt for future technologies compared to $4 to $5 per watt today).[331] Recent breakthroughs in thin-film solar technology have reaffirmed these projections. Estimated costs based on a pilot scale plant put the cost of new thin-film solar cells at just 0.95 euros per watt (approximately $1.15 per watt).[332] The DOE and the Electric Power Research Institute projected that by 2020, "utility-scale thin-film" solar cells could have a generation cost of just 6.2 cents per kWh. This analysis is currently being updated to take into account the recent advances in solar technology, but it already shows the potential for thin-film solar cells to reach economic competitiveness with new nuclear power, particularly when the lack of transmission and distribution costs for solar are taken into account.[333]

Briefly, the second important advantage of thin film solar cells is that the semiconductor can be deposited on flexible substrates allowing them to be used in a wider variety of applications than traditional monocrystalline designs.[334] The greater versatility of thin-film technologies will help to increase the types of areas available for future solar development, further reducing any potential need to make use of previously undisturbed lands for electricity generation.

One potential drawback to these new technologies, is that most of the proposed thin-film solar cells make use of rare metals that could result in problems of materials availability if solar power were to be greatly expanded in the medium-term. In fact, at very high levels of solar penetration, the production of PV cells could become a significant consumer of rare elements such as tellurium, indium, germanium, or ruthenium. In some cases the amount of mining waste from the production of these rare metals could become a consideration with respect to the environmental impacts of solar power.[335] However, it now appears likely that photovoltaic cells can be effectively recycled at the end of their service life without adding significantly to their overall cost.[336] In addition to recycling the semiconductor materials from older cells, the use of multiple thin film technologies as part of a broad mix of renewable energy sources would further reduce any potential concerns that might arise with respect to the use of rare elements. Finally, we note that the materials availability and mining issues with solar technologies are unlikely to ever rise to the level of those that would be encountered with uranium under either the global or steady-state growth scenarios (see Appendix A).

Finally, it appears likely that sustainable biomass could also play some role in the medium-term to support the mix of renewable energy resources. Biomass has the advantage of having no intermittency, and thus of being directly capable of supplying baseload power. The burning of the biomass, however, does emit carbon dioxide into the atmosphere. However, the plant matter consumed had removed carbon dioxide from the atmosphere during its growth, thus reducing the net emissions assuming the biomass was grown with minimal amounts of fertilizers or pesticides and that it was transported only a short distance before being burned. Additional care must be taken to ensure that the biomass was grown and harvested in such as way that it does not degrade the quality of land over time and that it does not damage fragile ecosystems nearby. In 1997, the DOE and the Electric Power Research Institute projected that the cost of electricity from biomass will fall to between 6.1 and 7.0 cents per kWh by 2010 and to 5.4 to 5.8 cents per kWh by 2020.[337] This would make biomass economically competitive with other carbon mitigation options under consideration. More recently, the DOE's Energy Information Administration projected even lower costs for biomass, with electricity from "[o]pen-loop biomass" projected to fall to just 5.1 cents per kWh by 2010.[338] Finally, we note that the use of sustainable biomass may play become important due to its potential connection to coal gasification and carbon sequestration (Sections 2.4.3 and 2.4.4). If biomass is mixed with coal in an integrated gasification combined cycle (IGCC) plant, and the resulting carbon dioxide is captured for storage in a geologic formation, the net greenhouse gas emissions of such as system could actually be made negative.[339] Such a possibility could be a significant advantage in helping to reach the deep reductions in greenhouse gas emissions that are likely to be required between now and the middle part of this century.

Section 2.4.3 – Coal Gasification

As discussed in Section 2.4.1, the use of natural gas as a transition fuel will require significant investments in LNG infrastructure over the coming years and the costs will remain somewhat variable due to seasonal fluctuations in demand for heating and the increased reliance on foreign supplies. On the other hand, the domestic supply of coal in countries like the U.S., Germany, China, and India are well known and very large which make their price highly stable. For example, the University of Chicago study concluded that "[c]oal supplies worldwide are expected to be sufficiently elastic that even a doubling of demand is not expected to increase price appreciably."[340] The main economic uncertainty of coal lies in its large carbon emissions compared to fuels. It is therefore likely that coal would remain a major energy source during the transition period of this half-century if it could be economically utilized in such as way that its greenhouse gas and other emissions could be significantly reduced. The two most likely ways to achieve this goal are to switch from pulverized coal fired plants to integrated gasification combined cycle (IGCC) plants and to directly capture the emitted CO_2 and isolate it in geologic formations (see Section 2.4.4).

In an integrated gasification combined cycle plant, coal and steam are reacted at high temperature and pressure to produce a mixture of gases known as "syngas." The exact composition of the syngas produced is determined by the type of coal employed and the precise reaction conditions being used, but it is typically made up by such things as carbon monoxide, carbon dioxide, hydrogen, and methane. Most impurities in the coal will not gasify and are removed from the reactor as slag which can be resold for use in making asphalt and other products. Other impurities will react into forms that are fairly straightforward to extract from the syngas such as hydrogen sulfides and ammonia. This results in far lower emissions of particulates, sulfur dioxide, nitrogen oxides, and mercury compared to traditional coal burning plants.[341] Experience in the U.S. has shown that even "high sulfur Illinois Basin bituminous coal" can be successfully burned in an IGCC plant without negatively impact the plant's performance or leading to high levels of emissions.[342]

Once the syngas is cleaned, it is fed into a combined cycle generating unit similar to those used for natural gas fired plants. This system can achieve higher efficiencies than traditional pulverized coal plants (45 percent for today's IGCC technology versus 33 to 38 percent for pulverized coal) resulting in lower carbon dioxide emissions. For current technology, it is estimated that an IGCC plant would emit between 13.5 to 27 percent less CO_2 than a pulverized coal plant for the same level of generation. The efficiency of future IGCC plants is expected to rise to as much as 60 percent between 2020 and 2030, which is in line with the recent experience with natural gas fired plants. If realized, such an IGCC plant would emit 40 percent less CO_2 than a pulverized coal fired plant.[343]

A potential additional advantage of developing coal gasification during the transition period of 2020 to 2050 is that it is well suited to the co-production of electricity and hydrogen which could make it an important component in the transition to the hydrogen economy.[344] The Department of Energy currently has several active programs aimed at supporting the development of commercially viable IGCC plants that would also be capable of producing hydrogen.[345] In its analysis of the hydrogen economy, the National Research Council of the U.S. National Academy of Sciences and the National Academy of Engineering concluded that

> Coal is a viable option for making hydrogen in very large, centralized plants when the demand for hydrogen becomes large enough to support an associated very large distribution system. The United States has enough coal to make all the hydrogen that the economy will need for more than 200 years, a substantial coal infrastructure already exists, commercial technologies for converting coal to hydrogen are available from licensors, the cost of hydrogen from coal is among the lowest available, and technology improvements are identified to reach future DOE cost targets.[346]

In fact, the authors concluded that, using current technologies, the cost of hydrogen from large centralized IGCC plants with carbon capture and storage would be just $2.19 per kilogram.[347] Adjusting for the relative efficiency with which hydrogen could be used in fuel cells compared to the use of gasoline in internal combustion engines, a hydrogen cost of $2.19 per kilogram would be equivalent to a gasoline price of approximately $1.32 per gallon, which is well below today's gas prices even when federal and state taxes are taken into account.[348] Thus, it appears likely that the development of dual use IGCC plants could offer an economically viable option for the large scale production of hydrogen during the transition to a more sustainable long-term energy system.

While not yet fully commercialized, the viability of IGCC technologies have been demonstrated in a number of countries. The first power plant of significant capacity to use coal gasification was the 170 MW plant in Luenen, Germany, that came online in the late 1970s. Other large plants have been constructed in the Netherlands, the Czech Republic, Spain, Germany, and the United States.[349] In the U.S., for example, the Tampa Electric Company currently operates a 250 MW integrated gasification combined cycle coal plant in Florida. This plant, built at a cost of approximately $2,000 per kW, has been in operation since 1996 and supplies baseload electricity to the surrounding communities. Current industry estimates put the capital cost of similar future plants between $1,300 and $1,650 per kW with further reductions possible.[350] Supporting the move to IGCC, in October 2004, General Electric and the Bechtel Corporation announced their intent to form an alliance that would offer commercial integrated coal gasification plants in the United States.[351] The DOE has also announced plans to provide financial assistance to utilities in Minnesota and Florida to build two new advanced IGCC plants. These first-of-a-kind demonstration projects, part of the Bush Administration's Clean Coal Power Initiative, are expected to cost between $1,950 and $2,220 per kW.[352] There is ongoing interest in developing this technology outside the U.S. as well. For example, two projects in Britain have already sought planning consent to begin construction of new IGCC power plants.[353]

Neither the MIT study nor the University of Chicago study presented their own analyses for the projected cost of IGCC plants and chose instead to focus only on pulverized coal technology. In order to examine the likely cost of electricity from dedicated IGCC plants that could be fielded by 2020, we reviewed estimates from a variety of sources, as we did for our discussion of the economics of wind power.[354] These estimates, summarized in Table 2.15, are in reasonably good agreement with each other for the costs of pulverized coal and IGCC plants. The wider variation in their estimates for the electricity prices from natural gas is due mainly to the assumptions they made about the long term cost of fuel while their variation in the estimated cost of nuclear power is due mainly to the assumptions they made about the capital cost, construction time, and effective interest rates.

Table 2.15: Summary of estimated generation costs in the U.S. for integrated combined cycle coal generation (IGCC) plants that could be fielded by at least 2020 (all prices in cents per kWh). The range of costs for natural gas, pulverized coal, and nuclear power are shown only for comparison.[355]

	Natural Gas	Pulverized Coal	IGCC	Nuclear
EIA (2005)	5.04	n.a.	4.8	5.95
Deutsche Bank (2003)[a]	3.0 to 3.9	3.8 to 4.3	3.2 to 4.3	5.1 to 6.3
Bechtel (2002)	4.33	4.08	4.76	4.83
IPCC (2001)	2.9 to 3.4	3.3 to 3.7	3.2 to 3.9	5.0 to 6.0
OECD (1998)[a]	3.0 to 3.2	3.6 to 4.1	3.6 to 4.2	4.5 to 5.3
U Chicago (2004)	3.5 to 4.5	3.3 to 4.1	n.a.	5.3 to 7.1
MIT (2003)	3.8 to 5.6	4.2	n.a.	6.7

(a) As quoted in the University of Chicago study.

Given the consistency between the estimates in Table 2.15 for pulverized coal with those from the MIT and University of Chicago studies, we can conclude that the range of 3.2 to 4.8 cents per kWh is likely to be a reasonable estimate for the cost of electricity from dedicated IGCC plants in the United States over the medium-term. A somewhat wider range would have been found (3.2 to 6.1 cents per kWh) if projections from these studies for countries in Western Europe, Japan, and China had been included, but the estimates in Table 2.15 are those that are most relevant to the U.S. energy system.[356] Thus, IGCC plants appear likely to be economically advantageous compared to new nuclear power plants that could be built within the next ten to fifteen years.

While these coal gasification plants would have somewhat lower carbon emissions than traditional pulverized coal fired plants due to their higher efficiencies, and thus could make some contribution to emissions reductions by themselves, the main benefit from the development of economically competitive IGCC plants would be their suitability for use with carbon sequestration technologies as discussed in the following section. This combination could be made for plants that produced electricity alone as well as those that would produce both electricity and hydrogen.

Section 2.4.4 – Carbon Capture and Storage

Beyond the higher thermal efficiency and lower emissions, one of the main driving forces behind the interest in coal gasification is that it is particularly well suited to carbon capture and storage technologies. When oxygen, rather than air, is used in the gasification unit, carbon dioxide is produced in a concentrated stream that is easier and less expensive to separate and capture than trying to scrub it from the exhaust post-combustion.[357] In the U.S., the DOE Carbon Sequestration Program announced the creation of the Integrated Sequestration and Hydrogen Research Initiative in 2003. This effort is seeking to develop both coal gasification and carbon sequestration technologies. The plan, known as FutureGen, is to design, build, and operate a 275 MW gasified coal plant for the co-production of electricity and hydrogen which will also include carbon capture and sequestration technologies. Current estimates are that the engineering and construction costs for this first-of-a-kind power plant and its associated sequestration equipment would total $2,760 per kW, which is well above the overnight capital cost of even a new nuclear plant. The overall goal of the FutureGen program is to demonstrate technologies that could be economically commercialized by 2020 at significantly lower costs.[358] The FutureGen Industrial Alliance signed an agreement with the DOE in late 2005, and expects to select a site for the construction of the plant by the end of 2007.[359]

While carbon capture and storage technologies have not been commercialized in the electricity sector, they have been used in other industries for decades, and are widely recognized as a potentially important option for reducing carbon emissions during the transition to a more sustainable energy system.[360] For example, the scrubbing of CO_2 from the exhaust gas of a few pulverized coal plants is already occurring to produce carbon dioxide for use in the food industry.[361] In addition, carbon dioxide injection into oil fields to enhance the recovery of petroleum has been used in the U.S. since at least 1972. While most of the CO_2 used for these enhanced oil recovery (EOR) projects is derived from natural sources, about one-fourth is purchased from industrial sources, primarily natural gas processing plants. One such industrial carbon dioxide source of particular interest in the current context is the Great Plains Synfuels coal gasification plant in North Dakota. This facility, which uses a technology similar to the gasifier in IGCC plants, separates CO_2 from the syngas it generates and then pipes that carbon dioxide to southeastern Saskatchewan, Canada, where it is injected into the Weyburn oil field. Approximately one million tons of CO_2 per year has been injected into this reservoir since September 2000 to increase oil production.[362] Overall, about 43 million tons per year of carbon dioxide is currently being injected into oil fields each year in 65 enhanced oil recovery programs in the United States alone.[363]

A related source of experience with sequestration has been built up with acid gas injection from natural gas production. Acid gas, primarily a mixture of hydrogen sulfide (H_2S) and carbon dioxide, has been successfully separated from natural gas in processing plants and re-injected into depleted gas fields as well as into nearby saline aquifers. However, the amounts of gas involved in these efforts are generally far smaller than those for enhanced oil recovery projects. The separation and injection of acid gas began in Canada in 1989 as a result of tightened sulfur emission limits and has proven to be an economic choice at a number of projects in Canada as well as in the United States.[364] All told 44 such projects are currently injecting acid gas into geologic formations in Western Canada alone.[365] The successful experience with these systems is noteworthy given the greater corrosiveness of acid gas compared to carbon dioxide.

One of the most important demonstrations of carbon sequestration technology available today is in the Sleipner gas fields in the North Sea. Motivated by the imposition of a tax amounting to approximately \$140 per ton of carbon in 1996, the Norwegian company Statoil began injecting the CO_2 it separates from the natural gas it extracts into a sandstone formation under the sea floor. This sandstone formation, known as the Utsira formation, has a layer on top which is impermeable to CO_2, and is believed to be capable of retaining the injected carbon over geologic timescales. Statoil has been injecting one million tons of CO_2 every year into the formation, and plans to continue doing so for the next 20 years. Just one percent of the Utsira reservoir is capable of storing three years worth of emissions from all European power plants.[366] The fate of the injected CO_2 in this formation has been studied since 1998 by an international research effort and the initial results indicate that the CO_2 will remain successfully trapped in the aquifer and gradually dissolve into the brine over the next several thousand years.[367]

The ongoing operations at the Sleipner gas fields are not an isolated example. A similar CO_2 sequestration program began in April 2004 at the In Salah natural gas fields in Algeria. At this site, Sonatrach, BP, and Statoil plan to store up to 1.2 million metric tons of CO_2 per year in a deep sandstone reservoir near the gas field. Two additional projects for re-injecting carbon dioxide from the production of natural gas, are currently being planned by Statoil in the Barents Sea and by Chevron at Barrow Island off the western coast of Australia.[368]

If developed successfully in the near to medium-term, carbon sequestration would likely be capable of making a significant contribution to reductions in global greenhouse gas emissions. Beyond the Utsira formation, a survey of the North Sea area alone revealed aquifers with a combined capacity equal to 800 times the current annual emissions of all European power plants.[369] Estimates for the global storage capacities in different types of repositories from the International Energy Agency's Greenhouse Gas R&D Programme are shown in Table 2.16.

Table 2.16: Estimated carbon dioxide storage capacity of geologic formations worldwide and a comparison of these potential capacities to the emissions from the global electricity sector in the year 2000.[370]

Type of Formation[a]	Gigatons of Carbon Dioxide Storage	Years of Storage at 2000 Emissions Levels[b]
Depleted Oil Fields	125	16
Unmineable Coal Seams	20 to 148	2.6 to 19
Depleted Gas Fields	800	100
Deep Saline Aquifers	400 to 10,000	52 to 1,300
Total	1,345 to 11,073	170 to 1,400

(a) Only geologic formations are included in this table. The direct sequestration of CO_2 in the deep ocean has been proposed, but is not considered here in light of the potentially serious impacts from ocean acidification on marine ecosystems (see Section 1.2).
(b) The estimated carbon dioxide emissions from the global electricity sector in 2000 were 7,770 million metric tons of CO_2.[371] The estimated years of storage have been rounded to two significant figures.

The values in Table 2.16 compare well with other estimates that have been made for the global CO_2 storage capacity. For example, the IPCC estimates that 675 to 900 gigatons of CO_2 could be sequestered worldwide in known oil and natural gas fields and that up to 900 to 1,200 gigatons of CO_2 could be stored if the projected capacity of undiscovered reserves is included. In addition, the IPCC estimates that deep saline aquifers have a potential capacity of 200 to 56,000 gigatons of carbon dioxide. In reviewing these estimates, the IPCC concluded that it is "very likely that global storage capacity in deep saline formations is at least 1000 GtCO₂."[372] Finally, while estimates of the total storage capacity of bituminous coal range from 60 to 200 gigatons of CO_2, the IPCC notes that "[t]echnical and economic considerations" suggest that only roughly 7 gigatons of CO_2 is likely to be stored in these un-mined coal seams.[373]

Therefore, it can be concluded that, if fully utilized, the total storage capacity available in geologic formations would likely be sufficient to hold the equivalent of all CO_2 emissions from all global electricity generation for at least one to two hundred years. While only a fraction of the available resources could be exploited due to both technical and economic limitations, it is clear that the potential for carbon sequestration to play a role in reducing greenhouse gas emissions is quite significant. As a result, a number of government and industrial efforts are currently underway in Europe, Canada, the U.S., and Japan aimed at improving our understanding of the potential for geologic sequestration to play a role in combating climate change over the coming decades.[374]

Equally important as the total magnitude of the available repositories is the fact that they are geographically accessible to fossil fuel power plants. It has been estimated that 65 percent of carbon dioxide captured from power plants in the U.S. could be sequestered in geologic formations "without the need for long pipelines," while it has been further estimated that "all power plants in the United States are located within 500 km of possible sequestration sites."[375] The U.S. already has a good deal of experience with transporting CO_2 via pipelines as a result of enhanced oil recovery projects. Approximately 22 million tons of carbon dioxide is piped through a 3,980 kilometer (2,470 mile) system each year for injection into oilfields in the Permian Basin, while the North Dakota coal gasification plant transports its CO_2 over a 330 kilometer (205 mile) pipeline to Canada for injection into the Weyburn oilfield.[376] Further experience has been gained with the transport and injection of acid gas separated during natural gas production. In the future, access to CO_2 pipelines and suitable sequestration sites could be used as a factor in siting IGCC or natural gas fired power plants.

One concern with sequestration efforts is that carbon dioxide, while not explosive like natural gas, is nevertheless an acidic gas that is also a powerful asphyxiant. Care will have to be taken in siting the pipelines and repositories to minimize the risks from accidental releases. It is estimated that rare events such as the failure of a well closure could result in the release of 1,600 to 960,000 tons of CO_2. While this is a small fraction of the total amount that could be sequestered at each site, and it is believed that the risks from such accidents can be properly managed, it is important to consider the potential health and environmental effects of such events and to locate the repositories accordingly.[377] The importance of considering such events is illustrated by natural carbon dioxide disasters that have occurred in the past. For example, in 1986, Lake Nyos in Cameroon unexpectedly released a massive amount of CO_2 that had bubbled up naturally through the lake bottom over time. The dense, heavy cloud of CO_2 vented from the lake filled the nearby valleys, and resulted in the death of approximately 1,800 people.[378]

The costs of carbon sequestration come from three main areas; (1) the cost of separating and capturing the CO_2 at the power plant, (2) the cost of transporting the CO_2 from the plant to the repository, and (3) the cost of injecting the CO_2 into the geologic formation. Typical carbon dioxide capture rates from fossil fuel plants are generally between 87 and 88 percent. In addition to adding to the capital cost of the plants, the capture of CO_2 also reduces the efficiency of the plants, adding to the fuel costs. The costs for pipeline transport will vary in proportion to the distance traveled, due to the additional construction and material costs along with maintenance and monitoring costs due to moisture entering the pipeline and forming corrosive carbonic acid.[379] Finally, the injection costs will depend upon the type of geologic formation being used. Injection into operating oil fields will be the lowest cost option since the petroleum industry actually pays companies about $40 to $60 per ton of carbon for CO_2 delivered to an operational site. The next lowest cost option would typically be injection into oil and gas fields where the geology had already been extensively studied. These types of deposits have an added advantage over the formations like the one being used at the Sleipner gas fields in the North Sea in that they are known to have contained oil and natural gas over geological timescales increasing the confidence in their ability to retain the carbon dioxide.[380] While we do not yet have cost estimates based on experience in the electric power sector, a number of projections have been made for the likely cost of sequestration efforts (see Table 2.17).

Table 2.17: Estimated cost for the addition of carbon capture and storage to natural gas, pulverized coal, and integrated coal gasification plants (all figures in cents per kWh). These estimates include the cost of capturing, transporting, and sequestering the carbon as well as the additional fuel costs that result from the decrease in plant efficiency.[381]

Type of Generation	IPCC (2001)	Bechtel (2002)	U Chicago (2004)[a]	IPCC (2005)
Natural Gas	1.50	1.37	1.72 to 1.88	1.2 to 2.9
IGCC	2.50	2.15	1.99 to 2.33	1.0 to 3.2
Pulverized Coal	3.00	4.00	3.78 to 4.16	1.9 to 4.7

(a) Estimate for 500 kilometer transport of CO_2 via a pipeline. As noted, all existing plants are believed to be located within this distance of a suitable carbon dioxide repository.

The values in Table 2.17 are in good agreement with other estimates that have been made for the cost of sequestering carbon. Using IEA Greenhouse Gas R&D Programme's assumptions, it was estimated that carbon sequestration would add between 2.0 and 2.1 cents per kWh to the cost of electricity from a 500 MW IGCC power plant.[382] In the U.K., the Department of Trade and Industry estimated that the costs for capturing, transporting, and storing CO_2 would add between 1.0-2.3 pence per kWh (1.6 to 3.68 cents per kWh) to the cost of electricity from fossil fuel plants. The DTI also noted that "international research has identified an appreciable potential for these costs to be reduced through innovation."[383]

Finally, adding the estimated costs for carbon capture and storage capabilities to the cost of generating electricity from new fossil fuel power plants, we can compare the cost of this transition strategy with that of building new nuclear power plants (see Table 2.18).

Table 2.18: Total estimated generation costs for natural gas, pulverized coal, and IGCC plants with carbon capture and storage capabilities (all figures in cents per kWh).[384]

Type of Generation	Estimated Generation Costs without Sequestration	Estimated Generation Costs with Sequestration
Natural Gas[a]	4.1 to 5.6	5.3 to 8.5
IGCC[b]	3.2 to 4.8	4.2 to 8.0
Pulverized Coal[b]	3.3 to 4.3	5.2 to 9.0
Nuclear Power	6.0 to 7.0	6.0 to 7.0

(a) Range of generation costs correspond to the medium and high gas price scenarios in the MIT study.
(b) Range of generation costs correspond to the estimates presented in Table 2.15.

From Table 2.18 we see that, while there remains a fair amount of uncertainty with the total cost, the use of carbon sequestration is likely to be economically competitive with new nuclear power. In fact, the middle of the projected cost range for natural gas or pulverized coal fired plants is 6.9 to 7.1 cents per kWh, while the middle of the cost range for IGCC plants is just 6.1 cents per kWh. The economic comparison would improve if potential cost reductions foreseen in carbon separation and storage were realized, and if the revenue from enhanced oil, gas, or coal bed methane recovery was fully taken into account. In addition to being similar in cost, the available coal and natural gas resources, as well as the available CO_2 storage capacity in geologic formations, are sufficient to allow very significant reductions in emissions to be achieved through the use of carbon sequestration in the transition period of 2020 to 2050. Finally, natural gas, IGCC, and nuclear power are all directly suitable for supplying baseload electric power without any significant changes to the electricity transmission and distribution system. In summary, it appears very likely that carbon capture and storage efforts can be an economically viable component the transition from our current fossil fuel based energy system to more equitable and sustainable long-term possibilities.

Section 2.5 - Conclusions

Nuclear power is a "mature" technology that has been commercialized for more than 50 years. Currently, 103 nuclear plants are operating in the U.S. alone, and a total of 438 reactors are currently in existence around the world.[385] Over the last half a century, nuclear power has been the recipient of more government subsidies in the United States than any other source of electricity. Despite this support, however, by the mid-1980s the nuclear power industry had failed so completely in the U.S. that it led a Forbes Magazine cover story to call nuclear power "the largest managerial disaster in business history, a disaster on a monumental scale."[386] The large cost overruns and ballooning lead-times for construction made nuclear power an economically unattractive option, and no new reactors have been ordered in the U.S. in more than a quarter of century. Despite a number of significant improvements that have been made since the 1980s, new nuclear power is likely to remain an expensive option in the future.

Projections from studies conducted at MIT and the University of Chicago put the likely cost of electricity from new nuclear power plants between six and seven cents per kWh. While a number of potential cost reductions were considered by the authors of these two reports, it is unlikely that plants not heavily subsidized by the federal government would be able to achieve any further economic improvements beyond those already considered in the studies' base case estimates. This is particularly true in light of the fact that any improvements would have to be maintained under the very demanding timetables set by the global or steady-state growth scenario. To meet the level of nuclear growth envisioned by the authors of the MIT report, more than one reactor would have to come online somewhere in the world every 15 days for four decades. Meeting the more aggressive steady-state growth scenario would put an even greater strain on the nuclear industry, with one reactor having to come online every six days between 2010 and 2050.

At six to seven cents per kWh, the cost of electricity from new nuclear power is above the range of projected costs for competing coal and natural gas fired plants. While the recent spot market price for natural gas has been both higher and lower than the "high" fuel price considered in the MIT or U Chicago studies, the long-term price of gas can be expected to remain within a competitive range if policies on efficiency, conservation, and an increased reliance on imported liquefied natural gas are pursued. Thus, without policies directly aimed at reducing carbon emissions from the electricity sector, nuclear power is very unlikely to be an economically competitive choice for new base load generation. This can be seen quite clearly in the continued hesitance of financial institutions such as Standard & Poor's to actively support the construction of even those first few nuclear plants that would be heavily subsidized by the federal government.

The economic comparison is more complicated, however, when nuclear power is viewed in relation to other potential strategies for reducing carbon dioxide emissions from the electricity sector. It appears increasingly likely that reductions on the order of 60 to 80 percent will be required by 2050 in order to avoid the more serious potential consequences of global climate change. As such, aggressive policies will be needed in the coming decades to curb and then reverse the growth of CO_2 emissions from all sectors of the energy system. Adding to the complexity of this already very difficult problem is the fact that these reductions will have to occur at a time of increasing electricity demand throughout the Global South. Of particular note is the projected increase in electricity consumption in the world's two most populous countries, India and China. Within this context, the range of future nuclear costs appear more competitive. As originally noted by Dr. Arjun Makhijani, President of the Institute for Energy and Environmental Research, when projected over the near to medium-term, the costs of many of the available alternatives all tend to fall roughly within or just below the range of six to seven cents per kWh.

In order to achieve the large reductions in greenhouse gas emissions that are needed, a tiered approach will be required that integrates options that are available for immediate use as well as those that are not yet fully commercialized, but can be brought online within then next ten to fifteen years. The most important near-term options include efforts to increase efficiency in the generation and use of energy and the large scale expansion of wind power at favorable sites. Improvements in energy efficiency and a reduction in demand through conservation have the potential for significant benefits throughout the Global North. Unlike programs focused on increasing supply, demand side options can result in reductions in greenhouse gas emissions with low or even negative costs while simultaneously providing new jobs and opening new avenues of economic growth. Combined with efforts to reduce demand, the use of renewable energy, particularly wind power, offers the most economical alternative for supplying the required near-term incremental growth in generating capacity. At approximately four to six cents per kWh, wind power at favorable sites is already economically competitive with natural gas and new nuclear power. With the proper priorities in investment on transmission and distribution infrastructure and changes to the ways in which the electricity sector is regulated, wind power could rapidly make a significant contribution to reductions in greenhouse gas emissions. Without large scale changes to the existing grid, wind power could already expand in the very near-term to make up 15 to 20 percent of the electricity supply as compared to less than one-half of one percent today. This could be done without negatively impacting the stability or reliability of the current transmission grid.

Over the medium-term (2020 to 2050), additional strategies will be needed. Some of the economically viable options available, such as the further expansion of wind to penetrations well beyond 20 percent and the increased use of other renewable resources like thin-film solar cells, advanced hydropower, and some types of sustainable biomass, have few environmental or security impacts compared to those of our present energy system. As a result these options should be pursued to the maximum extent practicable. However, other options with more significant health and environmental tradeoffs are also likely to be needed during the next several decades in order to achieve climate stabilization. In this vein, two of the most important transition strategies are likely to be the increased import of liquefied natural gas and the use of coal gasification with carbon sequestration. Some of the most troubling aspects of these technologies, such as mountain top removal mining for coal, would be lessened by reducing the demand for coal through increases in efficiency and the expansion of alternative energy sources. In addition, the use of coal gasification technologies would greatly reduce the emissions of mercury, particulates, and sulfur and nitrogen oxides for new coal fired plants. Despite these improvements, however, the use of fossil fuels would continue to have many very serious drawbacks. Their negative impacts notwithstanding, when compared against the potentially catastrophic damage that could result from global climate change and against the uniquely dangerous problems of nuclear power such as the potential for nuclear weapons proliferation (Chapter Three), the risks of catastrophic reactor accidents (Chapter Four), and the difficulty of safely managing long-lived radioactive waste (Chapter Five), the use of liquefied natural gas and integrated coal gasification technologies with carbon sequestration appear to be preferable options for use in supporting improvements in energy efficiency, conservation, and the expanded use of renewable resources during the period of transition from where we are today to a more equitable and sustainable energy system in the future. Finally, it is important to note that the development of viable transition technologies in the Global North could also help countries like China and India, which both have large reserves of coal, to rapidly increase their electricity supplies while leapfrogging over older, dirtier technologies and avoiding large amounts of greenhouse gas emissions.

The growing threat posed by climate change will require hard choices to be made in the future, and the precise mix of energy options that will be most effective at achieving a deep reduction of emissions in the U.S. and around the world cannot yet be foreseen. However, trading the vulnerabilities of global warming for the vulnerabilities of nuclear power is not a sound energy policy when a clear set of robust and economically viable alternatives are available that pose far less severe environmental and security risks.

Chapter Three: Megawatts and Mushroom Clouds

> The somewhat frayed nonproliferation regime will require serious reexamination and strengthening to face the challenge of the global growth scenario, recognizing that fuel-cycle-associated proliferation would greatly reduce the attraction of expanded nuclear power as an option for addressing global energy and environmental challenges.[387]
>
> - *The Future of Nuclear Power* (2003)

> The development of atomic energy for peaceful purposes and the development of atomic energy for bombs are in much of their course interchangeable and interdependent. From this it follows that although nations may agree not to use in bombs the atomic energy developed within their borders the only assurance that a conversion to destructive purposes would not be made would be the pledged word and the good faith of the nation itself. This fact puts an enormous pressure upon national good faith. Indeed it creates suspicion on the part of other nations that their neighbors' pledged word will not be kept. This danger is accentuated by the unusual characteristics of atomic bombs, namely their devastating effect as a surprise weapon, that is, a weapon secretly developed and used without warning. Fear of such surprise violation of pledged word will surely break down any confidence in the pledged word of rival countries developing atomic energy if the treaty obligations and good faith of the nations are the only assurances upon which to rely.[388]
>
> - Acheson - Lilienthal Report (1946)

While concern over catastrophic accidents and long-term waste management are perhaps better known, the largest single vulnerability associated with an expansion of nuclear power is likely to be its potential connection to the proliferation of nuclear weapons. This is due both to the impact of proliferation on world security as well as to the terrible destruction that accompany the use of nuclear weapons. The bombs that destroyed Hiroshima and Nagasaki sixty years ago were responsible for an estimated 170,000 to 200,000 immediate deaths. The global economic consequences that would follow a nuclear attack on cities like New York, Tokyo, New Delhi are difficult to predict, but would almost certainly be catastrophic. For example, the U.N. High-level Panel on Threats, Challenges and Change concluded that the total economic impact following the detonation of even a simple nuclear weapon in a major city would be "at least one trillion dollars."[389]

Beyond the impacts of their use as a weapon of war, the National Cancer Institute estimated that the atmospheric testing of nuclear weapons at the Nevada Test Site caused between 11,300 and 212,000 cases of thyroid cancer in the United States alone.[390] These cancers were caused by only one of the radionuclides released in the tests, iodine-131, which the government knew, in the 1950s, was concentrating in milk and resulting in potentially high doses to children's thyroids. Companies like Eastman-Kodak were warned of fallout patterns so that they could protect their film stocks, but no such warnings were given to farmers or families so that they could protect their children.[391] No comparable official estimates for the impact of U.S. testing on neighboring countries which also received fallout, such as Mexico and Canada, has been made by the NCI or any other U.S. government agency.

The designs of the Hiroshima and Nagasaki bombs are much less complicated than those of modern nuclear weapons, and are well within the technological capacity of many countries. Given access to the internet and scientific literature, graduate students working alone have proven capable of developing workable nuclear weapons designs. By far the most difficult step in the actual construction of any nuclear weapon is the acquisition of a suitably large quantity of fissile material.[392] Controlling access to fissile materials has formed the basis of non-proliferation efforts to date, and it is exactly the acquisition of this fissile material that can be facilitated by the existence of nuclear power and its related fuel cycle infrastructure. Specifically, the enrichment of uranium and the separation of plutonium from spent fuel are the areas of greatest proliferation concern. While neither uranium enrichment nor spent fuel reprocessing is fundamentally necessary for the utilization of nuclear power, the former is integral to the types of reactors that have been most widely pursued to this point, as well as to those most likely to be built over the coming decades (i.e. light-water reactors).

As noted in Chapter One, the potential connection between the nuclear power fuel cycle and nuclear weapons was evident from the earliest days of the atomic age. The control of this technology proved difficult, however, and with research reactors and nuclear power plants spreading around the world, President Kennedy predicted that by 1975 there would be 15 to 25 nuclear weapons states. Due in large part to the 1970 Nuclear Non-Proliferation Treaty (NPT), which entered its 35th year in 2005, the number of known or suspected nuclear weapons states has been limited to nine.[393] The central bargain of the NPT was that no member state that did not already have nuclear weapons would pursue them and the five that did possess them would negotiate in good faith towards their eventual elimination while simultaneously helping the other member states to develop civilian nuclear technology. The NPT was indefinitely extended in 1995, and currently ranks behind only the United Nations Charter in the number of signatories. Only four countries in the world, India, Pakistan, Israel, and North Korea, currently remain outside the treaty regime and all four are known or believed to have produced nuclear weapons.

Some of the recent positive developments regarding nuclear weapons proliferation include: (1) South Africa's abandonment of its nuclear weapons program in the early 1990s and its dismantlement of the small number of bombs it had manufactured; (2) the choice by Ukraine and Kazakhstan to return the nuclear warheads on their territories to Russia and join the NPT as non-nuclear weapons states following the dissolution of the Soviet Union; and (3) Libya's December 2003 announcement that it was abandoning its clandestine nuclear weapons program and opening its nuclear facilities to international inspection. However, on the negative side, the U.S. and Russia continue to retain stockpiles of several thousand warheads each with a combined explosive yield beyond human comprehension. Many of these warheads are retained on high alert, ready for launch on just minutes notice. The U.S. is currently studying concepts for new nuclear weapons and has plans to develop new delivery vehicles that would extend the integration of nuclear weapons into the military. France, China, the United Kingdom, and presumably Israel maintain stockpiles on the order of a few hundred warheads each, while both India and Pakistan demonstrated their nuclear capabilities by performing underground tests in 1998. North Korea withdrew from the NPT in January 2003 and has publicly claimed to have manufactured nuclear weapons. Finally, the failure of the 2005 NPT Review Conference at the United Nations to reach any agreement whatsoever is a major challenge to the future of the treaty regime and has once again forced to the surface the long-standing issue of how to balance the non-proliferation and disarmament obligations of the treaty. The continued maintenance of large stockpiles of nuclear weapons by the states that already have them makes it much more difficult to prevent proliferation. A world of nuclear haves and have-nots is unstable, and we are beginning to see the increasing strain caused by the efforts of the nuclear weapons states to sustain this unbalanced situation.

In the current context, India, Israel, Pakistan, and North Korea are noteworthy in that they are all currently outside the NPT regime and all made use of technology and equipment ostensibly meant for use in civilian programs to facilitate their acquisition of nuclear weapons. This was also the case for the now dismantled South African and Iraqi programs. In reviewing this history, however, it is important to separate the motivations of the countries from the intended use of the technologies they pursued. India and Israel acquired supposedly civilian research reactors and reprocessing know-how and technology from Europe, Canada, and the United States under the Atoms for Peace initiative initially for civilian purposes. Iraq also pursued the research reactor path until the Osirak reactor was destroyed in 1981 by an Israeli attack. In Pakistan and Iraq secret efforts to enrich uranium were also pursued. The Pakistani effort made use of plans for commercial enrichment technologies stolen from a European company and they developed a successful program built on a mixture of foreign and domestic suppliers. Iraq, on the other hand, pursued a variety of mainly indigenous enrichment technologies with very limited success. The South African enrichment program was unique in that it was publicly acknowledged by the government as part of an effort to support both a civilian research reactor and nuclear power plant. Finally, North Korea provides perhaps the strongest link between nuclear power and nuclear weapons. Unlike the Israeli and early Iraqi programs that focused on research reactors that were not intended to produce electricity, the 5 MW-electric nuclear plant at Yongbyon, North Korea was a research reactor that also supplied electricity to the surrounding community. In addition to supplying electricity, however, this facility also supported a program to separate plutonium which has now reportedly been used in nuclear weapons.

Even though it has not proved to be the preferred route to date, there is little debate over the potential for commercial nuclear programs to play a role in enabling future nuclear proliferation. As noted by J. Robert Oppenheimer as early as 1946:

> We know very well what we would do if we signed such a convention [to abolish nuclear weapons]: we would not make atomic weapons, at least not to start with, but we would build enormous plants, and we would call them power plants -- maybe they would produce power: we would design these plants in such a way that they could be converted with the maximum ease and the minimum time delay to the production of atomic weapons, saying, this is just in case somebody two-times us; we would stockpile uranium; we would keep as many of our developments secret as possible; we would locate our plants, not where they would do the most good for the production of power, but where they would do the most good for protection against enemy attack.[394]

Today, this connection between the infrastructure of nuclear power and the potential for nuclear weapons production is the central concern driving the U.S. and European efforts to prevent Iran from developing the capability to enrich uranium. As summarized by William Sutcliffe, a Senior Physicist at the Lawrence Livermore National Laboratory and specialist on issues of nonproliferation and the nuclear fuel cycle, "it is almost certain that expertise and infrastructure intended for the development of nuclear power have supported the development of a nation's (e.g. India) capability to use nuclear weapons."[395] The U.N. High-level Panel on Threats, Challenges, and Change noted the proliferation threat posed by countries that while "acting within the letter but perhaps not the spirit" of the NPT may seek to "acquire all the materials and expertise needed for weapons programmes with the option of withdrawing from the Treaty at the point when they are ready to proceed with weaponization."[396] Finally, even Mohamed ElBaradei, the Director General of the International Atomic Energy Agency, speaks of the "latent nuclear deterrent" value inherent in the possession of commercial fuel cycle technologies given the ability of these facilities to rapidly produce weapons usable fissile material should the operator so choose.[397]

In this chapter we will address the specific proliferation concerns relating to uranium enrichment and spent fuel reprocessing technologies in light of the fact that as many as nine states continue to possess nuclear weapons and that five of those nuclear armed states make up the permanent members of the U.N. Security Council. We will then briefly address the impacts of the current U.S. practice of manufacturing tritium for use in its nuclear weapons in commercial nuclear power plants. While not related to the acquisition of fissile materials, this policy has important implications for non-proliferation efforts. Finally, we will discuss why the proposals that have been put forth for how to try and manage these risks, while simultaneously enabling an expanded role for nuclear power, are very unlikely be successful.

Section 3.1 – Uranium Enrichment

Uranium-235 is the only naturally occurring radionuclide that has been used to fuel both nuclear reactors and nuclear weapons. In nature, however, the percentage of U-235 in uranium ore is too small for use in either nuclear weapons or in the most common types of nuclear reactors currently employed. Typical uranium ore contains just 0.711 percent U-235 while the remaining material is made up of the non-fissile isotopes U-238 (99.284 percent) and trace amounts of U-234 (0.005 percent). While there are types of nuclear reactors, such as the Canadian CANDU, that are capable of using natural uranium as a fuel, the light-water reactor designs that dominate the world's installed capacity require uranium enriched to between 3 to 5 percent U-235.[398] Uranium enriched to this level is referred to as "low enriched uranium" or LEU to distinguish it from material suitable for use in nuclear weapons which typically contains 90 percent or more U-235, and is referred to as "highly enriched uranium" or HEU.

The two most common types of light-water reactors are the Boiling Water Reactors (BWRs) which use the radioactive cooling water to turn the turbines directly, and the Pressurized Water Reactors (PWR) which transfer the heat from the radioactive cooling water to a secondary system that then drives the turbine. PWRs are typically more complex than BWRs given the need for additional heat exchange equipment, but they have the advantage of preventing the turbine from becoming radioactively contaminated during normal operation. Currently, PWRs make up about two-thirds of the nuclear capacity in the United States while BWRs make up the remaining third. Globally, PWRs make up nearly 65 percent of the total installed capacity, while BWRs make up 22 percent. The remaining global capacity is made up of other designs such as heavy-water and gas-cooled reactors.[399]

For light-water reactors, enrichment forms a vital step in the front end of the nuclear fuel cycle. The two most common enrichment technologies that have been pursued on an industrial scale to date are gaseous diffusion and gas centrifuges, although other techniques, such as electromagnetic, laser, aerodynamic, and chemical isotope separation, have all been developed as well.[400] Given that the reactors most likely to contribute to the global growth scenario, including any potential contribution from high temperature gas-cooled reactors (HTGR), would all require enriched uranium fuel, the expanded use of enrichment will be a necessary element of any expansion of nuclear power. The resulting spread of uranium enrichment services around the world would raise significant concerns regarding the potential proliferation of nuclear weapons.

All five of the acknowledged nuclear weapons states under the NPT have operated uranium enrichment plants for the production of reactor fuel and all five have operated enrichment plants for the production of nuclear weapons. In the United States for example, the same enrichment plants have been used to produce both LEU for commercial purposes and HEU for use in the U.S. stockpile.[401] The fissile material in the "Little Boy" bomb that was dropped on Hiroshima consisted of approximately 60 to 65 kilograms of highly-enriched uranium produced at the Oak Ridge, Tennessee facility. Fifty-four years later, the nuclear weapons tested by Pakistan in 1998 used uranium enriched in a clandestine military facility that was based on commercial technology from the European enrichment company Urenco. The recent controversy over Iran's nuclear power program, and in particular its attempt to manufacture gas centrifuges for enriching uranium, as well as revelations of the international sale of advanced enrichment technology by a supposedly private ring centering around A.Q. Khan, the "father" of the Pakistani nuclear bomb, have served to bring a renewed attention to the threats posed by uranium enrichment technology.

There are three related concerns regarding the spread of uranium enrichment technologies that must be considered. The first concern relates to the diversion of weapons usable material from known facilities that are ostensibly intended for civilian purposes, i.e., plants supposedly built to supply fuel for research or commercial power plants. The South African weapons program is an example of this type of proliferation. The second concern relates to the construction and operation of a dedicated, clandestine facility for strictly military purposes using technology developed for commercial applications. The Pakistani program illustrates the dangers of this second proliferation route. Third, there is the concern that stockpiles of low-enriched uranium and the existence of commercial enrichment facilities could allow rapid weaponization in the future should the country so choose. To date, no country has yet followed this third route to the acquisition of nuclear weapons, but the potential "latent nuclear deterrent" of enrichment capabilities is clearly apparent in the ongoing negotiations surrounding the Iranian enrichment program.

While low-enriched uranium is not itself usable in nuclear weapons, the connection between the production of LEU and the production of bombs is heightened by the fact that a majority of the effort goes into the early stages of the enrichment process. For example, roughly two-thirds of the energy and effort required to produce HEU goes into enriching natural uranium with 0.711 percent U-235 to fuel grade low-enriched uranium with 3.6 percent U-235, while only about one-third goes into the further enrichment of that LEU to produce highly enriched uranium with 90 percent U-235. In terms of the masses required, the advantages of using LEU as a feed material rather than natural uranium are even larger. In order to produce one kilogram of highly-enriched uranium, it would require 176 to 219 kilograms of natural uranium, but just 26 to 27 kilograms of LEU.[402] Figure 3.1 shows more generally how both the mass of the feed material and the amount of enrichment services needed to produce one kilogram of HEU (90 percent U-235) varies with the enrichment of the feed stock.

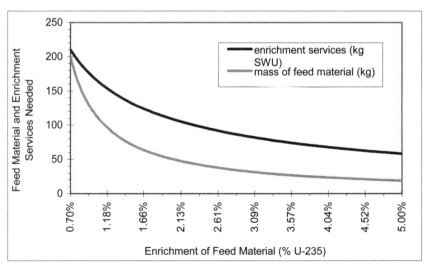

Figure 3.1: The mass of feed material (in kilograms of uranium) and the amount of enrichment services (in kilogram separative work units) required to produce one kilogram of highly enriched uranium (90 percent U-235). The lower bound on the horizontal axis represents natural uranium while the upper bound represents the typical limit of low-enriched uranium that would be used in light-water reactors.

From Figure 3.1 it is clear that possessing a stockpile of low-enriched uranium significantly lowers the barriers both in time and energy for the production of weapons usable HEU. The rapid decrease in the required feed material and enrichment services occurs at enrichments below 3 percent, and therefore any typical LEU for use in light-water reactors could serve as a useful stockpile for HEU production. In addition to facilitating the more rapid conversion of commercial facilities to military production, there is also a concern that LEU from a declared enrichment facility may be diverted for use as feed material in an undeclared enrichment facility. Such a diversion would significantly reduce the number of centrifuge stages needed in the clandestine plant as well as lowering its power consumption and other operational needs, making it even more difficult for the international community to discover.

The early technologies for uranium enrichment are fairly difficult to hide since they require very large facilities and a great deal of electrical power. In fact, at the time it was built in the 1940s, the gaseous diffusion uranium enrichment plant at Oak Ridge, Tennessee, was the largest industrial building in the world and its location was selected, in part, to take advantage of the large amounts of hydroelectric power available from the Tennessee Valley Authority. The successful development of sophisticated gas centrifuge technology in the Soviet Union, and its subsequent application by European governments and corporations has dramatically reduced the size and energy usage of enrichment plants as well as making it more difficult to determine if a given plant is configured to produce low-enriched uranium for reactor fuel or highly-enriched uranium for weapons. Modern centrifuge plants require 40 to 50 times less electricity than gaseous diffusion plants to achieve the same level of enrichment services. In addition, the centrifuge facilities do not generate as much waste heat as gaseous diffusion plants. They therefore use far smaller quantities of ozone depleting coolants like Freon and are more difficult to detect remotely. To give a sense of the scale of cooling needs at older plants, in 2002 the Paducah gaseous diffusion plant alone accounted for more than 55 percent of all airborne releases of Freon from all large industrial users in the entire United States.[403]

The smaller size, electricity needs, and heat signature of gas centrifuge plants make them the preferred choice for all future commercial enrichment facilities. As noted previously, these same characteristics, make them a particular concern with regard to clandestine programs. As noted nearly three decades ago by the Congressional Office of Technology Assessment,

> Any enrichment plant can theoretically be used for the production of weapons material while simultaneously providing immunity from international nuclear fuel embargoes, but only one type of enrichment plant—the centrifuge type—increases opportunities for proliferation on the same scale as reprocessing plants.[404]

Recognizing these risks, the authors of the MIT study concluded that "[c]landestine uranium enrichment programs, as have appeared in Iraq, Iran, North Korea and elsewhere, may present a dramatically increasing threat."[405]

The 1998 Pakistani tests of nuclear weapons made from highly enriched uranium offered a dramatic highlight to the proliferation risks associated with the spread of centrifuge technology. The European enrichment conglomerate Urenco was founded by the governments of Britain, Germany, and the Netherlands following the signing of the Treaty of Almelo in March 1970. In this treaty, the three countries agreed to jointly develop centrifuge enrichment technology in order to ensure a supply of reactor fuel that would be independent of the United States. While working for Urenco in the Netherlands, Abdul Qadeer Khan reportedly stole the centrifuge plans and a list of roughly 100 European suppliers before returning to Pakistan in the mid-1970's. Using this information, in conjunction with domestic suppliers, Khan was able to identify companies willing to sell equipment necessary for the Pakistanis to develop their own enrichment capabilities and produce HEU for use in nuclear weapons.[406]

Recent revelations regarding the spread of Pakistani centrifuge technology have raised additional concerns. In October 2003, the Italian coast guard intercepted the *BBC China*, a German ship bound for Libya. On board, the Italians found completed centrifuges as well as information and equipment necessary for manufacturing additional components.[407] Over the next several months, details emerged concerning the most far reaching illicit network for the proliferation of nuclear weapons related technology ever discovered. A.Q. Khan had organized a sophisticated multinational network that sold a nuclear weapon design as well as advanced centrifuge and other uranium enrichment technology to Libya, Iran, and North Korea. Khan and his associates sold uranium hexafluoride (the gas used in centrifuge enrichment), designs for both the older Pakistani centrifuge as well as the newer, more efficient models, as well as physical components and complete centrifuges to those foreign governments. The transnational network included not only Khan in

Pakistan, but also a Malaysian manufacturing facility where centrifuge components were manufactured and a computer company in Dubai that served to facilitate shipping and financial transactions.[408] At last count, roughly two dozen companies were known or believed to have been involved. Surprisingly, less than two months after President Bush laid out the extent of Khan's proliferation network to the public and Khan himself was granted a pardon by President Musharraf and the United States announced its intention to extend "Major Non-NATO Ally" status to Pakistan.[409] The importance of these events was summarized by Dr. El-Baradei as follows

> The relative ease with which A.Q. Khan and associates were able to set up and operate a multinational illicit network demonstrates clearly the inadequacy of the present export control system. Nuclear components designed in one country could be manufactured in another, shipped through a third (which may have appeared to be a legitimate user), assembled in a fourth, and designated for eventual turnkey use in a fifth.
>
> The fact that so many companies and individuals could be involved is extremely worrying. And the fact that, in most cases, this could occur apparently without the knowledge of their own governments, clearly points to the inadequacy of national systems of oversight for sensitive equipment and technology.[410]

While the spread of gas centrifuge technology is likely to be the greatest single concern, there are important risks associated with other enrichment technologies as well. This is particularly true for countries seeking only a limited number of weapons for use as a deterrent against external aggression rather than a sizeable stockpile for strategic or offensive use. The authors of the MIT report note that "[u]neconomic technologies may in some cases be utilized for 'batch scale' enrichment sufficient to produce HEU for a small number of nuclear weapons."[411] This was the path pursued by the Iraqi government. In the 1980s, Saddam Hussein spent several billion dollars on the development of many different types of enrichment technologies, including the inefficient and expensive electromagnetic separation technique. Despite the scale of these expenditures, however, this program succeeded in producing only a very small amount of medium enriched uranium (just above 20 percent) before it was discovered in the wake of the first Gulf War.[412]

The South African program is another example of uneconomic enrichment technologies enabling the pursuit of nuclear weapons. The South Africans built an enrichment facility using aerodynamic separation which works on a principle that is similar to that of gas centrifuges, but is typically more complicated and expensive to build and operate. This facility was supposedly built to provide low enriched fuel to the unfinished French built Koeberg nuclear power plant and more highly enriched uranium for the U.S. supplied Safari-1 research reactor. In reality, the enrichment plant also supplied an estimated 400 kg of uranium enriched to greater than 80 percent for military use. A total of seven nuclear weapons were eventually constructed from this material.[413] As summarized by the Congressional Office of Technology Assessment,

> In the longer run, *laser isotope separation* techniques and *aerodynamic separation* may have serious proliferation potential as means of producing highly enriched uranium for nuclear weapons. Openly pursued by more than a dozen non-nuclear-weapon states, *laser enrichment* technologies use precisely tuned laser beams to selectively energize the uranium-235 isotope most useful for nuclear weapons and separate it from the more common uranium-238 isotope. Laser facilities would be small in size and could enrich uranium to high levels in only a few stages. They could therefore prove to be difficult to detect and control if successfully developed as part of a clandestine program.[414]

The large increase in nuclear power envisioned in the global growth scenario would require a proportional expansion of uranium enrichment capacity and would likely lead to the establishment of fuel cycle infrastructure in a number of countries as governments sought to ensure access to domestic or regional sources of fuel. In addition, the diffusion of knowledge and the increase in global trade of the specialized materials and equipment needed to build and operate gas centrifuge facilities would make it progressively more difficult to identify clandestine programs or illicit transfers of technology as nuclear power expanded around the world. Table 3.1 shows the capacity of the commercial enrichment plants currently operating in the U.S., Europe, Russia, China, Japan, and Brazil.[415] India, Iran, Pakistan, and Israel are not included since they do not currently have known commercial enrichment programs.

Table 3.1: Summary of the commercial uranium enrichment capacity in existence around the world as of 2004. Together, Russia, the United States, and France account for more than 80 percent of the current enrichment capacity. Despite the scale of present enrichment capacity, however, the existing plants would be able to meet less than half the demand for enrichment under the global growth scenario.[416]

Country	Enrichment Capacity (MTSWU per year)	Type of Enrichment Plant(s)	Percent of Total Enrichment Capacity Needed to Fuel the Global Growth Scenario[a]
Russia	15,000	four gas centrifuge plants	12.5 to 15.0%
United States	11,300	one gaseous diffusion plant[b]	9.4 to 11.3%
France	10,800	one gaseous diffusion plant	9.0 to 10.8%
United Kingdom (Urenco)	2,300	one gas centrifuge plant	1.9 to 2.3%
Netherlands (Urenco)	2,200	one gas centrifuge plant	1.8 to 2.2%
Germany (Urenco)	1,800	one gas centrifuge plant	1.5 to 1.8%
Japan	1,050	one gas centrifuge plant	0.88 to 1.0%
China	~700	one gaseous diffusion and two gas centrifuge plants[c]	0.58 to 0.70%
Brazil	120	one gas centrifuge plant	0.10 to 0.12%
Total	45,270[d]	three gaseous diffusion and eleven gas centrifuge plants	37.7 to 45.3%

(a) A typical 1000 MW light water nuclear power plant requires approximately 100 to 120 MTSWU per year in enrichment services to provide its fuel. The increased demand for uranium under the global or steady-state growth scenario would make the higher enrichment levels more likely since the amount of uranium feed material and the amount of enrichment services needed are inversely related for a fixed percentage of U-235 in the tails. The assumptions made in the MIT report are consistent with this conclusion. In that report the authors assume the equivalent of nearly 125 MTSWU per year of enrichment services will be required for each reactor.[417]
(b) The US capacity would increase to 18,700 MTSWU per year if the Portsmouth facility, which is currently in a standby condition, was added to the capacity of the active Paducah facility listed in this table.
(c) A third gas centrifuge plant is currently under construction in China that will add an estimated 500 MTSWU per year to the country's enrichment capacity.
(d) In 2005, the IAEA estimated that the global commercial enrichment capacity, excluding Brazil, was 45,100 MTSWU per year which is in excellent agreement with our present estimates.[418]

The current enrichment capacity of the 14 plants in Table 3.1 is adequate to supply the needs of existing reactors, however, in order to fuel the 1,000 gigawatts of nuclear capacity envisioned under the global growth scenario the installed enrichment capacity by 2050 would have to increase by 120 to 165 percent over the current levels. For the steady-state growth scenario with 2,500 gigawatts of nuclear capacity online in 2050, the enrichment capacity would have to expand by more than 450 to 560 percent. The increase in the total number of enrichment plants would have to be even more dramatic given that the large gaseous diffusion plants in France and the United States, which together make up nearly 50 percent of the world's current enrichment capacity, would likely be shutdown and replaced by smaller, more efficient gas centrifuge facilities.

In fact, France already plans to replace its 26 year old gas diffusion plant with a new 7,500 MTSWU gas centrifuge facility. In the United States, the U.S. Enrichment Corporation (USEC) and Louisiana Energy Services (LES), a joint venture of Urenco, Exelon, Duke Power, and Entergy, have both filed license applications with the NRC seeking to construct new gas centrifuge plants in preparation for the eventual shutdown of the fifty year old Paducah facility. These new U.S. enrichment plants are proposed to have capacities of 3,500 and 3,000 MTSWU per year respectively. Therefore, even if all three proposed facilities in the U.S. and France are built to replace the two aging gaseous diffusion plants, there would still be a net <u>decrease</u> in global enrichment capacity of nearly 18 percent. In addition, the two larger Urenco centrifuges plants located in the Netherlands and the United Kingdom are both more than 30 years old while even the youngest of the four Russian plants is more than 40 years old.[419] In light of these facts, the IAEA concluded that

> The next decade will see something very unusual in the nuclear fuel cycle: all of the world's commercial enrichment enterprises will be engaged at the same time in re-building and to a lesser extent expanding their industrial capacities.[420]

Another way to compare the level of enrichment services needed under the global growth scenario is to note that it would require 16 to 19 times more enrichment capacity than currently deployed by Urenco in Britain, Germany, and the Netherlands combined or about 15 to 18.5 times more capacity than that of the proposed USEC and LES centrifuge plants combined. A discussion of the availability of the uranium reserves that would be required to supply these large number of enrichment plants in contained in Appendix A.

To give a sense of how significant this proliferation risk could be, we note that just <u>one percent</u> of the enrichment capacity required by the global growth scenario's reference case would be enough to make between 175 and 310 nuclear weapons every year, assuming 20 to 25 kilograms of HEU per bomb. A single 250 MTSWU per year facility, like the one being pursued by the Iranians at Natanz, would have enough enrichment capacity to produce both the LEU needed to fuel a 1,000 MW reactor <u>and</u> enough HEU to build more than 20 bombs a year if it could be reconfigured rapidly enough.[421] Thus, it is clear that, by any measure, the proliferation risks for the expansion of nuclear power under the global growth scenario would be quite severe. We will address the proposals that have been put forward by nuclear power proponents to try and manage these risks in Section 3.4, but we will first turn to the back end of the fuel cycle and consider the proliferation risks associated with the separation of plutonium through the reprocessing of spent fuel.

Section 3.2 – Reprocessing and the Plutonium Economy

The world's first nuclear explosion occurred on July 16, 1945 at the Alamogordo test site in New Mexico. The fissile material used in this "gadget," as it was called, as well as in the "Fat Man" bomb dropped on Nagasaki less than a month later was plutonium that had been recovered from spent fuel reprocessed at the Hanford Engineer Works in Washington State. The first nuclear weapons tested by the Soviet Union, Britain, and France all made use of plutonium. Until recently, much of the focus on nuclear weapons proliferation has focused on the plutonium route given that (1) smaller amounts are needed compared to uranium, (2) plutonium can be produced in research reactors as well as power plants, and (3) the use of plutonium allows for more sophisticated weapons designs to be employed, which are easier to adapt to missile delivery systems.

The focus on plutonium and its relation to nuclear weapons proliferation was heightened in the wake of India's surprise test of a "peaceful" nuclear device in May 1974. In exchange for the construction of a CANDU heavy-water reactor, the Indian government agreed that the fuel supplied by Canada would not be used for the production of nuclear weapons. While technically honoring this agreement, the Indians also used the reactor to irradiate fuel produced from indigenous supplies of uranium. This was possible because the heavy-water CANDU reactors use natural uranium for fuel, and thus requires no enrichment infrastructure. The indigenous spent fuel was then reprocessed in an Indian facility that was built using knowledge gained through training programs run by the Atomic Energy Commission under the Atoms-for-Peace initiative.[422] The plutonium was then used to make India's first nuclear device.

In the earliest days of the nuclear age, it was generally assumed that uranium was a limited natural resource, and that in order to fuel the rapid expansion of nuclear power that was then envisioned, it would be necessary to make use of plutonium as an energy resource. As such, a major focus of nuclear research and development was placed on the development of so-called fast-breeder reactors. These reactors, often cooled by liquid sodium, would make use of the relatively abundant quantities of non-fissile U-238 available to create, or "breed," more plutonium than was being used as fuel. This fuel cycle would thus provide a nearly inexhaustible energy resource. However, reprocessing turned out to be a major economic failure, and breeder programs encountered numerous difficulties including serious accidents at the Fermi reactor near Detroit in 1966 and at the Monju reactor in Fukui prefecture, Japan, in 1995 as well as the permanent closure of the French Superphénix breeder reactor in 1998 after it had achieved an effective lifetime capacity factor of just 6.3 percent.[423] Only one commercial reprocessing plant has ever been operated in the U.S., although two others were completed but abandoned before processing any spent fuel. The one U.S. plant that was operated was located in West Valley, New York and was primarily run by Getty Oil. The plant reprocessed spent fuel between 1966 and 1972 when it was shut down for economic reasons leaving the state and federal governments with a billion dollar cleanup effort.[424]

The detonation of the Indian nuclear device turned a spotlight on the proliferation risks associated with the projected growth of the plutonium economy.[425] In the same way that accidents have led to a renewed focus on reactor safety and to outstanding concerns finally being addressed by the nuclear establishment, the emergence of a sixth nuclear armed state led to a renewed focus on the potential risks inherent in the U.S. policy of advocating the widespread use of plutonium. Just one week before the presidential election in 1976, then President Gerald Ford took the first steps towards turning the U.S. away from the civilian use of plutonium by issuing a statement which declared that the United States would no longer view reprocessing as an essential part of the nuclear fuel cycle. Following his election, President Carter went further and effectively ended all commercial reprocessing efforts in the United States in 1977.[426] In that same year, the Congressional Office of Technology Assessment summarized the connections between the plutonium economy and the potential spread of nuclear weapons as follows

> Reprocessing provides the strongest link between commercial nuclear power and proliferation. Possession of such a facility gives a nation access to weapons material (plutonium) by slow covert diversion which would be difficult for safeguards to detect. An overt seizure of the plant or associated plutonium stockpiles following abrogation of safeguards commitments could, if preceded by a clandestine weapons development program, result in the fabrication of nuclear explosives within days. Furthermore, such a plant reduces a nation's susceptibility to international restraints (sanctions) by enhancing fuel cycle independence. Finally, plutonium recycle is the most likely source for both black market fissile material and direct theft by terrorists.[427]

While the official ban on commercial reprocessing in the U.S. was lifted by President Reagan in the early 1980s, the high cost and the decision by Congress to pursue direct disposal of spent fuel in a geologic repository has so far effectively kept the ban in place. The Clinton administration discouraged the use of commercial reprocessing in the United States, but agreed to "maintain its existing commitments regarding the use of plutonium in civil nuclear programs in Western Europe and Japan."[428] The current Bush administration has advocated expanded research on reprocessing technologies that are claimed by the Department of Energy to be less proliferation prone. In addition, the 2001 *National Energy Policy* report prepared by a committee chaired by Vice-president Dick Cheney concluded that "the United States will continue to discourage the accumulation of separated plutonium, worldwide."[429]

Since 2005, reprocessing has gained significant prominence within the U.S. government. For example, the U.S. House of Representatives Committee on Appropriations concluded in their report on the fiscal year 2006 budget that

> ... the Committee believes that the Department [of Energy] should embark on a concerted initiative to begin recycling our spent nuclear fuel, starting with the preparation of an integrated spent fuel recycling plan for implementation in fiscal year 2007, including selection of an advanced reprocessing technology and a competitive process to select one or more sites to develop integrated spent fuel recycling facilities (i.e., reprocessing, preparation of mixed oxide fuel, vitrification of high level waste products, and temporary process storage.)[430]

The Energy Policy Act of 2005 allows allocations of up to $580 million over the next three fiscal years for research on new reprocessing and transmutation technologies under the so-called Advanced Fuel Cycle Initiative.[431] Finally, in February 2006, the Bush Administration launched the Global Nuclear Energy Partnership (GNEP), a major initiative to promote the expansion of nuclear power in the U.S. and around the world. As part of the GNEP program, the

> The U.S. and key international partners will accelerate the demonstration and deployments of new advanced recycling technologies such as UREX+ and pyroprocessing that recycle nuclear fuel in a manner that does not produce separated plutonium – a proliferation risk inherent in existing recycling technologies.[432]

While the majority of current arguments supporting reprocessing center on long-term waste management issues, questions regarding the adequacy of the available uranium supply continue to appear. As discussed in Appendix A, however, the economically recoverable uranium reserves are likely to be sufficient to supply either the global or steady-state growth scenarios if major international development efforts were undertaken in the near to medium term. Uranium prices, even under the high demand scenarios, would likely remain below $130 per kilogram which is well below the levels required to make reprocessing and the use of MOX fuel economical.

Despite the poor economics of reprocessing, by the end of 2001 there was already 262.5 tons of separated "civilian" plutonium accumulated around the world. Of this, 165.8 tons was stored in its raw form (i.e. not as fuel rods or other finished products) at reprocessing facilities in Russia, France, Japan, and Britain.[433] For comparison, the entire amount of plutonium produced at both the Hanford and Savannah River Site complexes for use in the U.S. nuclear arsenal totaled approximately 103.4 metric tons while the estimated inventory of separated plutonium in the former Soviet Union totaled approximately 150 metric tons.[434] Therefore, through the end of 2001, the stockpile of separated plutonium from commercial fuel world wide had already grown to a roughly equal amount to that produced for both the U.S. and Soviet militaries combined. Even taking into account the differences in critical mass and the additional technical difficulties associated with the use of reactor grade plutonium, the stockpile of separated "civilian" plutonium would have been enough to produce more than 32,800 nuclear weapons through the end of 2001.[435] Since 2001, the stockpiles of plutonium at reprocessing facilities have continued to grow at several metric tons per year.

In the wake of the failure of fast breeder reactor programs around the world, the advocates of plutonium turned to the use of mixed oxide fuels or MOX.[436] In MOX fuel, plutonium oxide is mixed with uranium and the resulting fuel elements are used in modified pressurized or boiling water reactors. To date, Germany, Belgium, Switzerland, and France have all made use of MOX fuel in their power reactors. In addition, in 2003 the Federation of Electric Power Companies of Japan announced their intention to continue pursuing the MOX fuel cycle as well. In 2004, Kansai Electric Power received approval to load MOX fuel into the Takahama 3 and 4 reactors by 2007. The plans to load MOX fuel at Takahama had been put on hold in 1999 following the revelation that employees at British Nuclear Fuels Limited had falsified safety qualification data for the MOX fuel it was producing. Kyushu Electric Power has also recently announced that it plans to begin using MOX fuel by 2010.[437] The use of MOX fuel is also currently the preferred option of the Bush administration for the disposition of weapons grade plutonium that has been declared surplus to military "needs." The Russian government favors the use of surplus military plutonium in fast breeder reactors, but has for now agreed to the use of MOX.[438]

The theft of MOX fuel poses a far greater proliferation risk than uranium fuel. This is due to the fact that, unlike spent uranium fuel, MOX does not emit penetrating radiation at levels high enough to prevent easy handling and the plutonium in MOX fuel can be separated by chemical means which are far simpler to master and easier to hide than those required to further enrich LEU to weapons usable levels.[439] For MOX fuel made with 7 percent reactor-grade plutonium, just 120 kg of fuel would provide enough fissile material for one bomb using the conservative value of eight kilograms per weapon.[440] At a density of 10 grams per cubic centimeter, this amount of MOX fuel could be kept in a can measuring just nine inches on a side. Another way to picture this amount of fuel is to note that it would represent just 0.0025 percent of the MOX fuel that would be manufactured each year under the MIT global growth scenario if such a fuel cycle were in use.[441]

Commercial reprocessing plants handle very large quantities of plutonium, and therefore inventory control and materials accounting has always been an important concern with respect to the possibility of clandestine diversion. The experience to date, however, has not been encouraging. For example, in 2003 the Japanese government admitted that the Tokai-mura pilot reprocessing plant could not account for some 206 kilograms of plutonium that it had processed. This was on top of the 70 kilograms of plutonium that remain unaccounted for at a separate Japanese plutonium fuel fabrication facility. In 2003 and 2004 alone, the British Thorp reprocessing plant reported that it could not account for 49 kilograms of plutonium. Each of these are commercial facilities operating in technologically sophisticated countries under IAEA safeguards.[442] At eight kilograms per bomb, the Japanese alone have "lost" enough plutonium to make more than 34 nuclear weapons. While it is not believed that this material was purposefully diverted, its location remains a mystery. (It is presently assumed that the majority of this unaccounted for Japanese plutonium remains locked up in the pipes or in the liquid high-level waste streams.) These examples illustrate the difficulty inherent in attempting to safeguard any large scale spent fuel reprocessing plant.

The number of reprocessing facilities that would be required to support the MOX fuel cycle under the global or steady-state growth scenarios would greatly exacerbate the problem of safeguards and materials accounting. There are currently only two very large-scale commercial reprocessing plants in operation anywhere in the world. These are the La Hague plant in France and the Thorp plant in Britain. In addition, there is the RT-1 reprocessing plant in Russia, the Power Reactor Fuel Reprocessing Facility and Kalpakkam Reprocessing Plant in India, and the Tokai pilot scale reprocessing plant in Japan. Japan is also currently nearing completion of the Rokkasho commercial scale facility (see Table 3.2).

Table 3.2: Summary of existing reprocessing plants for commercial spent fuel including their cost and historical rate of performance where available.[443]

Plant	Cost	Rated Capacity (MTIHM per year)	Historical Rate of Reprocessing (MTIHM per year)
La Hague (France)	$16 billion	1,700	1,200 (between 1990 and 2000)
Thorp, Sellafield (Britain)	$5.9 billion	1,200	500 (between 1994 and 2004)
Rokkasho (Japan)	$18 billion (estimated)	800	not yet in full scale operation
RT-1, Chelyabinsk (Russia)	N.A.	400	N.A.
Power Reactor Fuel Reprocessing Facility, Tarapur (India) Kalpakkam Reprocessing Plant (India)	N.A.	~200	N.A.
Tokai (Japan)	N.A.	120	40 (between 1977 and 2002)[(a)]

(a) The Tokai reprocessing plant was shutdown in March 1997 due to an explosion at the facility's bituminization plant and did not resume full scale operations until November 2000.

The authors of the MIT report concluded that, for a balanced fuel cycle using light-water reactors, the optimal use of plutonium would limit MOX to just 16 percent of the total fuel requirements. Using this assumption, the annual reprocessing requirements for the global growth scenario would amount to approximately 15,800 MTIHM per year.[444] Assuming a 75 percent capacity factor for future reprocessing plants, the reprocessing needs of the global growth scenario would require more than 17 plants the average size of La Hague, Thorp, and Rokkasho. For the steady-state growth scenario a total of 43 large commercial scale reprocessing facilities would be required.

For the global growth scenario, approximately 155.3 metric tons of plutonium would have to be separated annually in order to supply the required MOX fuel.[445] Just <u>one percent</u> of this plutonium would be sufficient to produce more than 194 nuclear weapons every year. Given the difficulties that have been experienced to date in accounting for plutonium at existing reprocessing facilities, it would be extremely difficult to detect clandestine diversion of material sufficient for the manufacture of two or three nuclear weapons per year if reprocessing technology on this scale was being undertaken.

In order to try and overcome the proliferation concerns regarding traditional reprocessing, current proposals are looking to what they claim to be more "proliferation resistant" technologies such as UREX+ or pyroprocessing.[446] In pyroprocessing, also known as electrometallurgical processing, the spent fuel is converted into a metallic form and then dissolved in a bath of molten salt. The plutonium product can then be collected on one of two electrical plates inserted into the mixture. However, unlike the PUREX reprocessing technology in use today which results in pure plutonium as a product, pyroprocessing leaves the plutonium mixed with other transuranic elements like neptunium and americium as well as with some of the lanthanide fission products like cerium and samarium.[447] In the original UREX+ process, a modified version of the PUREX process in use today, the plutonium that is separated remains mixed only with neptunium, and would therefore have been significantly more pure than that which results from pyroprocessing.[448] To try and improve the proliferation resistance of UREX+, DOE has more recently proposed a modified process that would leave additional transuranic and lanthanide fission products mixed with the plutonium. This version of UREX+ would result in material more similar to that which would result from pyroprocessing.[449] While these alternative reprocessing technologies would have some nonproliferation benefits compared to current reprocessing technologies, they would still pose significant risks if deployed on a large scale.

First, while the material resulting from pyroprocessing is very unlikely to appeal to states that have dedicated production facilities or existing stockpiles of weapons grade plutonium due to its isotopic composition, the plutonium product would potentially appeal to terrorists or threshold states pursuing clandestine weapons programs aimed at manufacturing a limited number of warheads. There do not appear to be any technical difficulties associated with making a nuclear weapon from the plutonium that results from pyroprocessing that would be significantly more complicated to overcome than those associated with using ordinary reactor grade plutonium recovered via PUREX reprocessing.[450] In addition, unlike the spent-fuel assemblies themselves, the material that would re-

121

sult from reprocessing the existing stockpile of light-water reactor fuel would not be sufficiently radioactive to protect against diversion or theft. As summarized in a recent review

> The radiation doses from transuranics without the lanthanides are more than three orders of magnitude lower than the IAEA's threshold for self-protection.
>
> Inclusion of either of two lanthanide fission products, ^{144}Ce [cerium-144] and ^{154}Eu [europium-154], could increase the dose rate above the self-protection threshold. However, ^{144}Ce has a half-life of only 0.8 years and has already decayed away in all but the most recently discharged spent LWR fuel. ^{154}Eu has a half-life of nine years but is not recycled with the transuranics in the pyroprocessing fuel cycle. It therefore appears that keeping plutonium from aged LWR fuel mixed with other transuranics and with lanthanide fission products other than ^{154}Eu would not make it significantly more self-protecting.[451]

Further, pyroprocessing plants could be modified quite rapidly in such a way that they would produce plutonium that contained significantly smaller quantities of the transuranic and lanthanide contaminants than originally intended.[452] Thus, a state that sought to use such a facility to support a limited nuclear weapons program could potentially overcome much of the nonproliferation advantages of pyroprocessing by simply modifying the plant prior to the commencement of military production.

The second major proliferation concern surrounding pyroprocessing is that it would likely be more difficult to safeguard than traditional reprocessing schemes. Pyroprocessing facilities can be made smaller than PUREX facilities, making them more difficult to detect remotely. Thus, pyroprocessing has the potential to exacerbate proliferation from plutonium in much the same way that the smaller, more efficient centrifuge enrichment plants exacerbated the proliferation risk from uranium enrichment. In addition, it would be even more difficult to accurately monitor the amount of plutonium being processed by a pyroprocessing plant than it is at current PUREX plants due to the inhomogeneous nature of the molten salt mixture.[453] This would make it even more difficult to verify that clandestine diversion for nuclear weapons development was not taking place under a globalization of reprocessing facilities.

Finally, the large stockpiles of already separated plutonium creates additional concerns. For example, a government could simply choose to take possession of plutonium stockpiles built up at commercial facilities and use them directly to manufacture nuclear weapons. Ichiro Ozawa, the head of the Liberal Party in Japan, highlighted this potential dual nature of plutonium from commercial fuel in April 2002. During a public discussion, Ozawa commented that if Japan felt threatened by China's military modernization program, that Japan could quickly and easily build a nuclear arsenal of several thousand warheads using plutonium from its commercial reactors. In October 1999, Shingo Nishimura was forced to resign as vice-defense minister due to the public outcry over his suggestion that Japan might consider arming itself with nuclear weapons. Ozawa's statement, however, was followed not by his resignation, but by Yasuo Fukuda, the Chief Cabinet Secretary, stating that times were changing such that even revising the constitution's ban on these weapons was being discussed, and that, "depending on the world situation," Japan might eventually need to acquire a nuclear capability.[454] The total amount of separated plutonium stored in Japan as of 2001 (5.6 tons) could make roughly 700 bombs while the total amount of plutonium owned by Japan (38 tons, most of which is stored at reprocessing plants in France and Britain) could make up to 4,750 bombs.[455]

The expense and concerns over nuclear weapons proliferation led the authors of the MIT study to recommend the use of the once-through fuel cycle in which the spent fuel is stored until it is ready for geologic disposal. In particular, the authors note that

> The result of our detailed analysis of the relative merits of these representative fuel cycles with respect to key evaluation criteria can be summarized as follows: *The once through cycle has advantages in cost, proliferation, and fuel cycle safety*, and is disadvantageous only in respect to long-term waste disposal; the two closed cycles have clear advantages only in long-term aspects of waste disposal, and disadvantages in cost, short-term waste issues, proliferation risk, and fuel cycle safety.[456]

Based on these findings, the MIT study went on to conclude that

> For the next decades, government and industry in the U.S. and elsewhere should give priority to the deployment of the once-through fuel cycle, rather than the development of more expensive closed fuel cycle technology involving reprocessing and new advanced thermal or fast reactor technologies.[457]

While we do not believe further effort should be placed on the development of any aspects of the front end of the nuclear fuel cycle, we share in the conclusion that the use of plutonium in commercial reactors is expensive and poses unacceptable risks of nuclear weapons proliferation. We would add that existing stockpiles of both military and commercial plutonium should rapidly be put under international supervision to reduce its susceptibility to diversion and should be converted into a form suitable for permanent disposal such as mixing it with high-level waste prior to vitrification.[458] The discovery of a discrepancy of at least 300 kilograms between the estimates for the amount of plutonium in the waste at the Los Alamos National Laboratory highlights the importance of immediately undertaking an international effort to accurately account for all civilian and military plutonium stocks.[459]

Section 3.3 – Tritium Production

The acquisition of plutonium or highly enriched uranium remains the primary barrier to the production of nuclear weapons, and is thus the area of greatest concern with respect to proliferation. However, once the nuclear threshold is crossed, the production of tritium (a radioactive isotope of hydrogen with a half-life of just over 12 years) can greatly enhance the "effectiveness" of a country's nuclear stockpile. As summarized in the DOE's 1991 *Draft Environmental Impact Statement for the Siting, Construction, and Operation of New Production Reactor Capacity*,

> The use of tritium in nuclear weapons makes it possible to build smaller, yet more powerful weapons and also makes it possible to reduce the amount of plutonium in each weapon.[460]

A reduction in warhead size and weight is an important step in being able to fit the weapons onto missiles, and thus greatly increase the threat they represent. The use of tritium in warhead designs is also the first step towards making the leap from pure fission weapons to thermonuclear weapons (i.e. hydrogen bombs) which can be hundreds of times more powerful. Currently all of the weapons in the U.S. arsenal make use of tritium as a vital component.

Nearly all of the tritium that has been used in the U.S. stockpile was produced in the reactors at the Savannah River Site (SRS) in South Carolina. Since the shutdown of the K reactor at SRS in 1993, the U.S. has been recycling tritium from retired warheads to maintain its active stockpile. Claiming a need to restart production due to the fact that tritium decays at roughly 5.5 percent per year, Energy Secretary Bill Richardson announced on May 6, 1999, that the TVA's Watts Bar and Sequoyah nuclear power plants in eastern Tennessee would be modified by the DOE for the production of tritium to be used in the U.S. arsenal. In December

1999 the TVA Board of Directors approved the agreement and in September 2002 the NRC approved the license amendment for Watts Bar and Sequoyah Units 1 and 2. On October 20, 2003, the Watts Bar reactor officially commenced full scale production of tritium for the nuclear stockpile. The lithium target rods were irradiated for 18 months before being returned to the DOE for reprocessing at the Savannah River Site.[461]

These events mark a major departure from the long-held U.S. position on distinguishing between civilian power plants and military production reactors. As summarized by the DOE itself in 1991,

> The production of nuclear material for defense purposes by commercial power reactors licensed by the U.S. Nuclear Regulatory Commission would be contrary to the longstanding national policy to separate commercial nuclear power generation from the nuclear weapons program.[462]

In fact, the U.S. government is now having to actively subvert the spirit of export controls in order to allow their plans for tritium production to go forward. The DOE has had to arrange for completely domestic supplies of uranium to fuel these three reactors since the major foreign suppliers like Canada and Australia are prohibited from supplying uranium for any use related to nuclear weapons production. In addition, the DOE has had to arrange for the acquisition of components such as integrated circuit boards from nations with less stringent export controls than major supplier states like Japan, since the Japanese prohibit its equipment from being used in the manufacture of nuclear weapons. The DOE has even agreed to pay the TVA an additional $30 million over ten years to compensate them for the added inconvenience these kinds of arrangements will require.[463]

One argument that has been put forward by proponents of the current tritium production strategy is that, because the Tennessee Valley Authority is wholly owned by the federal government, that the implications for non-proliferation efforts are less severe than if some other commercial reactors had been chosen for the manufacture tritium. For example, Ernest Moniz, one of the co-chairs of the committee that wrote the MIT nuclear power study and an Undersecretary in the DOE at the time the new tritium production strategy was being decided upon, told the *Los Angeles Times* that because the TVA reactors were technically the property of the government that "the blurring is less."[464] There are, however, two important reasons that this logic is flawed. The first is that the Watts Bar and Sequoyah plants are not meaningfully different than any other commercial power reactors in either technical details or how they are regulated by the NRC. They are commercial reactors that were built by civilians with the sole intention of supplying electricity to the consumer market and they have not served any military function whatsoever before this time.[465] Second, and more important, is the dangerous message that this type of reasoning sends to other countries that have state owned utilities about the use of those facilities for the production of nuclear weapons related materials.

The continued pursuit of this tritium policy by the United States has the potential to make it more likely that other countries may choose to follow suit, especially if the number of light-water reactors were to expand significantly as envisioned by the global or steady-state growth scenarios. This concern is heightened by the potential for the diffusion of tritium production know-how from the commercial TVA facilities.[466] Finally, the U.S. backtracking on its commitment to a clear and unambiguous separation of commercial nuclear power plants from military production reactors may make it more difficult to prevent other countries in the future from turning their civilian nuclear infrastructure towards the production of fissile materials for nuclear weapons. This general blurring of the line between what is "civilian" and what is "military" is likely to be the most significant impact of the current U.S. tritium policy.

Section 3.4 – Strengthening Non-Proliferation Efforts

The authors of the MIT study themselves concluded that the expansion of fuel cycle infrastructure and the limitations of the current safeguards regime "raise significant questions about the wisdom of a global growth scenario that envisions a major increase in the scale and geographical distribution of nuclear power."[467] We share this conclusion and note that it would be an even more acute concern in the case of a larger increase in nuclear power such as envisioned under the steady-state growth scenario. The MIT report goes on to note that while much of the proposed nuclear expansion would occur in countries like the United States and China which already have dedicated nuclear weapons complexes, other countries like Japan, South Korea, and Taiwan would also see significant expansion of nuclear power and that these countries might eventually seek nuclear weapons in response to emerging pressures from either China or a nuclear armed North Korea. Finally, the authors also acknowledge that the significant growth of nuclear power in India and, to a lesser extent, Pakistan might provide a potential pathway for facilitating the expansion of these countries' arsenals as well.[468] In addition, the 2003 Pentagon study examining the possible implications of rapid climate change (see Section 1.2) also included Israel, Germany, Iran, and Egypt among the countries that might seek to develop or expand their nuclear weapons capability in response to future national security pressures.[469] The withdrawal of North Korea from the NPT and its apparent open pursuit of nuclear weapons also makes it a concern for future proliferation should they maintain access to nuclear power technologies. Table 3.3 summarizes the current nuclear capacity of these ten countries, as well as their projected capacity in 2050 under the MIT high and low nuclear growth scenarios.

Table 3.3: The current and projected nuclear power capacity under the MIT global growth scenario for ten countries that are considered to be of potential concern for future nuclear weapons proliferation.[470]

Country	Effective Capacity in 2000 (GW)[a]	Effective Capacity in 2050 (MIT low nuclear growth)[a]	Effective Capacity in 2050 (MIT high nuclear growth)[a]
Egypt	0	5	10
Germany	17	33	49
India	2	87	175
Iran	0	11	22
Israel	0	1	2
Japan	31	61	91
North Korea	0	3	5
Pakistan	0	10	20
South Korea	11	26	37
Taiwan	4	11	16
Total	65	248	427

(a) The "effective capacity" reported by the MIT study is equal to the capacity required to supply the projected level of electricity production assuming a 100 percent capacity factor for the nuclear plants. For the 85 percent capacity factor assumed in the economic analysis of Chapter Two, the actual installed capacity in these countries would be greater than that given in this table by 18 percent.

In just the ten countries listed in Table 3.3, nuclear power would expand by 280 to 560 percent through 2050. More than 80 percent of that growth would occur in Germany, India, Japan, and South Korea. In addition, there are other countries that, while unlikely to pose a serious risk of proliferation, cannot be completely ignored given their history. Countries such as Argentina, Brazil, Libya, and South Africa have all had known or suspected nuclear weapons programs in the past and have all had some level of access to fuel cycle technologies. In fact, Apartheid South Africa actually built and then dismantled a total of seven nuclear bombs using enriched uranium produced under the cover of a supposedly civilian nuclear program.[471] All four of these countries, however, have officially renounced nuclear weapons and are currently non-nuclear weapons states under the NPT. Under the low and high nuclear growth scenarios, these four countries would be projected to have a total of 30 to 60 GW of nuclear capacity in 2050 compared to just 2 GW today.[472]

Such a large expansion of nuclear power would likely be accompanied by the spread of enrichment and, potentially, reprocessing facilities as well. Even programs short of full scale commercial facilities could raise important proliferation concerns. As noted by the authors of the MIT study,

> The rapid global spread of industrial capacity (such as chemicals, robotic manufacturing) and of new technologies (such as advanced materials, computer-based design and simulation tools, medical isotope separation) will increasingly facilitate proliferation in developing countries that have nuclear weapons ambitions. A fuel cycle infrastructure makes easier both the activity itself and the disguising of this activity. Indeed, even an extensive nuclear fuel cycle RD&D [research, development, and demonstration] program and associated facilities could open up significant proliferation pathways well before commercial deployment of new technologies.[473]

Considering the complex history of the countries listed in Table 3.3 both domestically, and in their relations to their neighbors (for example Egypt and Israel, Israel and Iran, India and Pakistan, India and China, Japan and China, China and Taiwan, North Korea and South Korea, North Korea and Japan, and so on), the added tensions and uncertainties that expanded fuel cycle infrastructure could add to future conflicts is an important vulnerability of nuclear power.

Unfortunately, as the withdrawal of North Korea from the Non-Proliferation Treaty and its apparent resumption of nuclear weapons production have shown quite clearly, even when inspections and treaty obligations are able to prevent current weapons activity, they are not always a guarantee for the future. A further example of how situations may change dramatically over time is the case of Iran. In the mid-1970s the United States was allied with the Shah of Iran and sought to encourage not only the development of nuclear power reactors in that country, but also commercial scale fuel cycle facilities as well. In fact, under the Ford Administration, the U.S. had began negotiations with the Iranian government concerning the potential development of a binational or multinational reprocessing center in Iran as well as the development of domestic fuel fabrication capabilities. One specific advantage that the U.S. then saw in aiding the development of such a facility was that access to an Iranian reprocessing plant might help prevent Pakistan from seeking to develop its own fuel cycle capabilities.[474] What the implications for the region might have been had Iran been in possession of nuclear power plants and a commercial scale reprocessing center at the time of the Islamic revolution is difficult to say, but the Israeli attack on the Iraqi Osirak reactor in 1981 raises a number of serious questions. In summarizing the potential security implications of nuclear development, Mohamed ElBaradei noted that, when faced with such an uncertain security position,

> ...a country might choose to hedge its options by developing a civilian nuclear fuel cycle – legally permissible under the NPT – not only because of its civilian use but also because of the "latent nuclear deterrent" value that such a programme could have, both intrinsically and in terms of the signal it sends to neighboring and other countries. The unspoken security posture could be summarized as follows: "We have no nuclear weapons programme today, because we do not see the need for one. But we should be prepared to launch one, should our security perception change. And for this, we should have the required capacity to produce the fissile material, as well as the other technologies that enable us to produce a weapon in a matter of months." Obviously, the narrow margin of security this situation affords is worrisome.[475]

One notable exception to the consensus regarding the risks associated with nuclear power and the spread of nuclear weapons is the study conducted at the University of Chicago discussed in Chapter Two. The authors of this study conclude that "[t]here is a lack of agreement about whether or not the availability of current reprocessing and enrichment technology under current regulatory mechanisms are increasing nonproliferation risk." They go on to claim, however, that issues of nuclear weapons proliferation are outside the scope of their analysis because they believe that "[t]he future economic viability of nuclear power... does not depend on their resolution."[476] Given the clear evidence of the proliferation risks associated with uranium enrichment and reprocessing technologies outlined in Sections 3.1 and 3.2, as well as the fact that the existing safeguards regime has been able to limit, but not prevent the spread of nuclear weapons, it seems quite hard to justify the conclusions of the U Chicago study.

Given the concerns regarding the potential spread of nuclear weapons accompanying the spread of nuclear power, the authors of the MIT study concluded that

> ...the current non-proliferation regime must be strengthened by both technical and institutional measures with particular attention to the connection between fuel cycle technology and safeguardability. Indeed, if the nonproliferation regime is not strengthened, the option of significant global expansion of nuclear power may be impossible, as various governments react to real or potential threat[s] of nuclear weapons proliferation facilitated by fuel cycle development.[477]

The impacts from such proliferation are difficult to quantify, but they could be quite severe. The spread of nuclear weapons and their further introduction into regional conflicts would make their use increasingly likely. The continued desire of the U.S. nuclear weapons establishment to pursue new and more specialized battlefield nuclear weapons adds to this potential. In addition, the spread of nuclear weapons technology and fissile materials could make their acquisition by non-state actors more possible. This latter fact is a particular concern given the difficulties in deterring non-state actors and the greater likelihood for the surprise use of a nuclear weapon against a large population center. Beyond the terrible human consequences that would accompany the use of such a weapon, the direct and indirect financial impacts would also be staggering if a major center like New York, Tokyo, London, or New Delhi were attacked with a nuclear weapon.

Unfortunately, despite a recognition of the important problems confronting nuclear power with respect to proliferation, the aggressive proposals put forward by the authors of the MIT report and others to try and overcome these concerns are very unlikely to be successful.

Section 3.4.1 – Enhanced Inspections under the IAEA

One of the oldest proposals for how to allow the expansion of nuclear power and the spread of research reactors without simultaneously increasing the risk of nuclear weapons proliferation is to rely on a system of international safeguards and inspections. These inspections would seek to verify that countries are both living up to their commitments not to use civilian infrastructure for military purposes as well as to verify that countries are not operating undeclared clandestine facilities. Despite the attractiveness of this option in principle, however, it was realized from the very earliest days of the nuclear age that it was unlikely for any viable inspection regime to be complete or intrusive enough to guarantee that a country wasn't cheating.

As early as November 15, 1945, the leaders of the United States, Canada, and Great Britain issued a joint declaration in which they noted that "[t]he military exploitation of atomic energy depends, in large part, upon the same methods and processes as would be required for industrial use." In light of this connection, President Truman and the Prime Ministers of Britain and Canada stated their belief that

> No system of safeguards that can be devised will of itself provide an effective guarantee against production of atomic weapons by a nation bent on aggression.[478]

Similar conclusions were shared by General Leslie Groves, the head of the Manhattan Project during World War II. In his testimony before Congress on November 28, 1945, General Groves summarized the difficulties he saw in relying on inspections as follows:

> Senator Tydings: So, if we do have an atomic-energy operated world, all the inspections will be pretty much dissipated -- the value, rather -- because once the development of atomic energy is assured to different nations and the means for producing it is set up, it is a very short step from there, both in time and in mechanics and intellect and everything else that enters into it to change that into making a bomb with it?
>
> General Groves: That is correct. If that came to pass and I had anything to say about the inspections, I would want an inspector of my own in every plant that this material was being used in for the production of energy and I would also want somebody in there watching that man to make certain he was still my man.[479]

These conclusions regarding the efficacy of inspections were further supported by the findings of a major study commissioned in 1946 by then Under-Secretary of State Dean Acheson on the future of nuclear power and its connection to nuclear weapons proliferation. The Committee on Atomic Energy, as it was called, was chaired by David Lilienthal, then the Chairman of the Tennessee Valley Authority and later the first Chairman of the Atomic Energy Commission. The committee also included the presidents of the New Jersey Bell Telephone Company, Monsanto, and General Electric, as well as Robert Oppenheimer who had headed the bomb design work at Los Alamos Laboratory during the Manhattan Project. In their final report, the authors concluded that "[t]he development of atomic energy for peaceful purposes and the development of atomic energy for bombs are in much of their course interchangeable and interdependent."[480] In light of this conclusion, the committee considered the potential effectiveness of various efforts to try and enable the spread of nuclear power without simultaneously enabling the spread of nuclear weapons. While they concluded that inspections would be a necessary piece of any system of control, their conclusions with respect to the value of inspections alone are worth quoting at length:

> We have concluded unanimously that there is no prospect of security against atomic warfare in a system of international agreements to outlaw such weapons controlled only by a system which relies on inspection and similar police-like methods. The reasons supporting this conclusion are not merely technical, but primarily the inseparable political, social, and organizational problems involved in enforcing agreements between nations each free to develop atomic energy but only pledged not to use it for bombs. National rivalries in the development of atomic energy readily convertible to destructive purposes are the heart of the difficulty. So long as intrinsically dangerous activities may be carried on by nations, rivalries are inevitable and fears are engendered that place so great a pressure upon a system of international enforcement by police methods that no degree of ingenuity or technical competence could possibly hope to cope with them.... We are convinced that if the production of fissionable materials by national governments (or by private organizations under their control) is permitted, systems of inspection cannot by themselves be made "effective safeguards.... to protect complying states against the hazards of violations and evasions."[481]

Despite theses strong warnings, over the next several decades the spread of nuclear technology was encouraged under President Eisenhower's Atoms for Peace initiative relying mainly on national efforts to control the diffusion of fuel cycle technology and safeguards built around international inspections. The International Atomic Energy Agency, created in 1957, was tasked with enforcing the nuclear safeguard agreements and with performing the required inspections. The agency's powers were significantly expanded in 1970 when the Nuclear Non-Proliferation Treaty came into force. As predicted, the efforts by the IAEA and individual countries were able to reduce but not eliminate the proliferation of nuclear weapons. The 1974 nuclear test conducted by India and the revelations in 1986 by Mordechai Vanunu of the extent of Israel's clandestine nuclear program stand as examples of the limitations of these strategies when they were not universally applied. (Recall that both India and Israel are outside the NPT regime, and therefore their research reactors and reprocessing plants were not subject to IAEA safeguards). Since the early 1990s, three important discoveries have further highlighted the limitations of international inspections even when they are applied to NPT member states.

Following the 1991 Gulf War, Iraq agreed to highly intrusive international inspections as a condition of the ceasefire agreement. These inspections revealed that Iraq had developed an extensive clandestine nuclear weapons program at undeclared sites that had gone undiscovered by the IAEA through its normal safeguards agreement with Iraq. This secret Iraqi program had included large investments aimed at developing a variety of uranium enrichment technologies. Despite the aggressive nature of this program, however, the Iraqis were never able to enrich more than a small quantity of medium enriched uranium. The second important event was the revelation in 1992-93 that North Korea had extensively violated its safeguards agreements and had been secretly pursuing the acquisition of plutonium for nuclear weapons. In the wake of these two failures by the IAEA to rapidly uncover fairly large and well developed clandestine nuclear weapons programs, negotiations were undertaken to significantly expand the Agency's inspection powers. This effort led to the creation of the Model Additional Protocol which was adopted by the IAEA's Board of Governors on May 15, 1997. When ratified by a member state, the Additional Protocol adds the following important requirements to the country's cooperation with the IAEA:

(1) The member state is required to provide an expanded declaration of nuclear related activities, even if they do not currently involve nuclear material. This expanded declaration must also include information on such things as uranium mines (regardless of whether they are currently in use) and disclose all trade in materials or equipment which are controlled by the Nuclear Suppliers Group.
(2) The members state is required to provide access to IAEA inspectors at all facilities specified by the agency as well as at all facilities included in the expanded declaration in the event of any questions or inconsistencies identified by the agency.

(3) The member state is required to provide IAEA inspectors visas upon request within one month which will remain valid for one year.

(4) The member state is required to accept the use of environmental monitoring by IAEA inspectors as part of their analysis, including at sites not specified by the member state in its declaration.[482]

As with other aspects of the NPT regime, the Additional Protocol is not implemented in the same way in nuclear weapons states as it is in non-nuclear weapons states. For example, non-nuclear weapons states must adopt all provisions of the Model Additional Protocol while nuclear weapons states can pick and choose as they see fit. In addition, while the IAEA is obligated to inspect all declared facilities in non-weapons states, they have the right, but not the obligation, to inspect them in nuclear weapons states. Of the 250 non-military nuclear facilities made available for inspection by the U.S. to the IAEA under its voluntary agreement (i.e. commercial power plants, research reactors, commercial fuel fabrication plants, etc.) the Agency has only inspected 17 to date.[483]

The discovery of an extensive uranium enrichment program in Iran that had gone undetected by the IAEA for as much as 20 years further highlighted concerns over the ineffectiveness of older safeguards agreements. These revelations began following the September 2002 announcement by Iran that it planned to install as much as 6,000 MW of nuclear capacity and that it planned to simultaneously pursue the associated fuel cycle facilities to support these plants. Following a February 2003 visit by Mohamed ElBaradei, Iran acknowledged that they had conducted a wide range of nuclear related activities that were not previously declared to the IAEA. These undeclared activities included: (1) the importation of approximately 1.8 metric tons of natural uranium in various chemical forms (some of which was used for isotope production experiments while some was converted into uranium metal), (2) the nearly complete construction of a pilot scale enrichment plant which would hold 1,000 centrifuges, (3) the commencement of construction on a production scale enrichment facility which would hold 50,000 centrifuges, and (4) plans to construct a 40 MW-thermal heavy water research reactor, a heavy water manufacturing plant, and a related fuel fabrication facility.[484] These initial admissions were followed by subsequent evidence that Iran was still not being fully honest with the IAEA.

In the original exchanges with the IAEA, the Iranians claimed that their work on centrifuge technology had begun in 1997 and was based on information from open sources and their own domestic development work using inert gas for testing. Environmental samples taken by inspectors in mid-2003, however, revealed the presence of trace amounts of highly enriched uranium at the centrifuge facility. In the wake of this discovery, the Iranians then admitted that the decision to start their centrifuge program had actually begun in 1985 and that around 1987 they had obtained drawings for centrifuges from a foreign intermediary, now believed to have been a part of the A.Q. Khan network. Prior to 1997 they received centrifuge components from overseas as well. It is believed that the HEU detected by the inspectors was a result of contamination brought into the country on the centrifuge components that had been acquired overseas.[485] This is consistent with the conclusion that the components came from Pakistan through Khan's network in light of Pakistan's known HEU production for nuclear weapons. As a result of the ongoing questions regarding the Iranian program, Iran was officially declared to be in noncompliance with its safeguards agreement by the IAEA Board of Governors in September 2005 and was reported to the United Nations Security Council in February 2006.[486]

The fact that such an extensive program had once again gone undetected by the IAEA for so long despite the application of safeguards, led to a renewed pressure for universal application of the Additional Protocol. At the beginning of 2003, only 28 countries had ratified Additional Protocols with the IAEA. Significantly, of the five acknowledged nuclear weapons states, only China had ratified their agreement with Agency, and they had not done so until March of 2002.[487] Under pressure, Iran signed an Additional Protocol with the IAEA on December 18, 2003, and agreed to abide by its conditions prior to ratification.[488] However, in February 2006 Iran ended its voluntary cooperation with the IAEA following the Agency's reporting of Iran to the U.N. Security Council.[489]

In light of the experience to date regarding the efficacy of safeguards agreements, the authors of the MIT study proposed the following far reaching reworking of the basic framework of the Nonproliferation Treaty:

> We suggest a new approach that retains this [NPT] framework and is based on technical assessment of risk, but politically non-discriminatory. This approach centers on classifying states as "privileged" of nuclear reactors or as "fuel cycle states." Declared "privileged states" would operate nuclear reactors according to their internal economic decisions about nuclear power versus alternatives, with international support

> for reactor construction, operational training and technical assistance, lifetime fresh fuel, and removal of spent fuel. Privileged states would not be eligible for fuel cycle assistance (enrichment, fuel fabrication, reprocessing). Thus "privileged" states would be low risk for proliferation and would gain several benefits: absence of intrusive safeguards and inspections, relief from expensive fuel cycle infrastructure development costs, and in particular elimination of nuclear spent fuel/waste management challenges....
>
> On the other hand, the "fuel cycle states" would be subject to a new level of safeguards and security requirements, along the line of those recommended above. Both groups of states would be subject to the Additional Protocol with respect to undeclared facilities.[490]

While this type of framework would allow the IAEA expanded inspection powers in the countries that pursue enrichment or reprocessing capabilities, they would remain focused mainly on preventing the misuse of declared commercial facilities. The problems of uncovering clandestine efforts would not be significantly reduced by this increased attention on known plants. In addition, facilities in "fuel cycle states" that are also acknowledged nuclear weapons states would likely continue to be routinely ignored by the IAEA while those in countries like Iran would receive intense scrutiny. Such a system would perpetuate the existing double standard of nuclear haves and have-nots, and would likely continue to be viewed as unfair and unacceptable.

The slow progress to date in simply achieving universal ratification of the Additional Protocol highlights the difficulties encountered in efforts to expand the IAEA's inspection powers. Despite the increased pressure brought about by the revelations of the Iranian enrichment program and the endorsement of universal application by the United Nations General Assembly, the NPT Review Conference, and the IAEA General Conference, as of June 2005 only 67 countries had ratified their Additional Protocol agreements with the IAEA, while another 30 countries had signed, but not ratified, such agreements.[491] In addition, Taiwan has a nongovernmental agreement for inspections with the Agency and Libya has agreed to apply the terms of their Additional Protocols while they are awaiting entry into force. Of the five nuclear weapons states, the Additional Protocols with China, France, and the United Kingdom have now gone into force, but the agreements with Russia and the United States have still not been ratified.[492]

A further challenge to strengthened inspection regimes is the issue of commercial interest and the protection of proprietary technology. Brazil has the sixth largest known recoverable uranium deposits in the world. At an estimated 197,000 tons, they are nearly 90 percent larger than those in the United States.[493] Brazil is currently operating nuclear power plants that require enriched uranium for fuel and, while they supply their own uranium, the enrichment services are carried out in Europe which adds to the total cost. In order to provide a domestic source of enrichment, as well as to potentially allow it to break into the international enriched uranium market, the Brazilians are currently preparing to commence the operation of a gas centrifuge enrichment plant near Rio de Janeiro with a capacity of approximately 120 MTSWU per year. The planned startup of this facility, however, was delayed by the refusal of the Brazilian government to allow full and unfettered inspections by the IAEA. Claiming a need to protect proprietary information, Brazil offered to allow inspectors to conduct environmental monitoring as well as to monitor the uranium inputs and outputs, but they did not want to permit visual inspections inside the plant.[494] In late 2004, the Brazilian government and the IAEA reportedly reached an agreement on the nature of allowable inspections and the plant is now expected to commence commercial operation.

The concern over the potential abuse of the inspections process to allow the theft of corporate secrets is not new. In Mohamed Shaker's 1980 review of the negotiations surrounding the Non-Proliferation Treaty he notes that the non-nuclear weapons states had expressed fears that "international inspection might turn into industrial espionage."[495] In fact, the United States claims certain exemptions for its commercial facilities under the Additional Protocol it signed in 1998, but has yet to ratify. In her January 2004 testimony before Congress, Susan Burk, acting Assistant Secretary of State for Nonproliferation, sought to reassure senators that if the IAEA chose to inspect a U.S. facility that "so-called 'managed access' techniques can be used to protect sensitive proprietary or commercially sensitive information from disclosure." The U.S. has also reserved the right of final approval over all members of IAEA inspection teams.[496] In addition, Burk noted that

> The national security exclusion [which is part of the U.S. Additional Protocol agreement], therefore, gives the United States an extraordinary legal means to protect and prevent the transfer of information to the IAEA and exclude inspectors' access in the United States whenever required for the protection of activities of direct national security significance to the

United States or of information or locations associated with such activities.[497]

As with other types of double standards, the acceptance of special treatment for the commercial facilities in the nuclear weapons states weakens the inspection regime as a whole over the long run by undermining its legitimacy. It will grow progressively more difficult to argue that other countries should accept safeguards and intrusive inspections that the U.S. has rejected for use in its own facilities.

Overall, the difficulties facing inspectors today are not far removed from those facing inspectors in the 1950s. In fact, with the spread of smaller, more energy efficient gas centrifuge plants and the general increase in overall technological sophistication of countries, the difficulties facing the IAEA are likely increasing rather than decreasing. Even with the advent of the Additional Protocol, the continued reliance on international inspections is not likely to be adequate to prevent the proliferation of nuclear weapons should nuclear power expand dramatically. Referring to such activities as the mining of uranium and thorium, the enrichment of uranium, the operation of nuclear reactors, and the operation of reprocessing plants the Acheson - Lilienthal committee concluded that

> If nations or their citizens carry on intrinsically dangerous activities it seems to us that the chances for safeguarding the future are hopeless.[498]

The inability of all but the most highly intrusive, expensive, and time consuming inspection regimes to safeguard against proliferation in a world where the existing nuclear weapons states seek to retain their weapons indefinitely has led to a variety of proposals for restricting access to and control over fuel cycle technology, even for commercial purposes. These proposals are discussed in the following section.

Section 3.4.2 – Restricting Access to Fuel Cycle Technologies

Proposals to restrict access to fuel cycle technology are nearly as old as the technology itself. As discussed above, the Acheson - Lilienthal committee concluded in 1946 that no inspections regime alone would be adequate to prevent nuclear weapons proliferation if countries were allowed to develop and operate their own reactors and related facilities such as enrichment and reprocessing plants. The committee's conclusion, therefore, was "that only if the dangerous aspects of atomic energy are taken out of national hands and placed in international hands is there any reasonable prospect of devising safeguards against the use of atomic energy for bombs."[499] They envisioned the creation of an "Atomic Development Authority" that would "bring under its complete control world supplies of uranium and thorium" as well as build, own, and operate all fuel fabrication plants, nuclear reactors, and reprocessing plants.[500]

A modification of the Acheson - Lilienthal proposal was presented to the United Nations on June 14, 1946, by Bernard Baruch, President Truman's special representative to the U.N. Energy Commission. In the Baruch Plan, an Atomic Development Authority (ADA) would be set up under the auspices of the United Nations and would control all nuclear research and development, would own of all facilities potentially related to nuclear weapons production, would be responsible for licensing and inspecting all other nuclear related activities carried out by individual countries, and would be responsible for overseeing the eventual elimination of all existing nuclear weapons.[501] The plan failed for a number of reasons. Chief among these reasons were: (1) the desire of the U.S. to retain the weapons it had built and to continue the development of its nuclear capabilities until the establishment of the U.N. agency was complete, (2) the proposal's elimination of the Security Council veto on atomic matters which was strongly opposed by the Soviet Union, and (3) the desire by the Soviet Union to break the U.S. monopoly and develop its own nuclear arsenal.[502]

In contrast to these internationalist proposals was the "Atoms for Peace" initiative put forth by the Eisenhower Administration in 1953. This plan not only allowed individual countries to control both research and power reactors, but actively sought to promote the expansion of these technologies around the world. The non-proliferation focus of the Atoms for Peace initiative centered on controlling the spread of fuel cycle technology by promising that the United States would provide a secure supply of enriched uranium to countries seeking to build light-water reactors. However, access to the fuel cycle was technically allowed and Article IV of the 1970 Nuclear Non-Proliferation Treaty enshrined in international law the bargain that, in exchange for a promise to not seek the bomb, non-nuclear weapons states had the right to acquire nuclear power and all related "peaceful" technologies.

The promise of the United States to provide enrichment services, notwithstanding, large commercial fuel cycle facilities eventually began to proliferate over time. The Netherlands, Germany, and the United Kingdom formed the uranium enrichment corporation Urenco in 1970 and developed advanced centrifuge enrichment plants in each of these three countries. France joined with Spain, Italy, Belgium, and later Iran to form the enrichment company Eurodiff which built a large gaseous diffusion plant in southern France. Finally, Japan developed its own reprocessing and fuel cycle facilities, while South Africa developed an indigenous uranium enrichment plant and commercial scale enrichment or reprocessing technologies were offered for sale to countries like Pakistan, South Korea, and Brazil.[503]

The concerns over nuclear weapons proliferation highlighted by the 1974 Indian nuclear test and the U.S. renunciation of commercial reprocessing in 1976-77, led to a revival of the idea of international control in the 1970s and 80s. Among the more important of these efforts were: "the IAEA study on Regional Nuclear Fuel Cycle Centres (1975-77); the International Nuclear Fuel Cycle Evaluation programme (1977-80); the Expert Group on International Plutonium Storage (1978-82); and the IAEA Committee on Assurances of Supply (1980–1987)."[504] However, each of these efforts failed to make any significant progress towards facilitating the establishment of fuel cycle facilities under regional or international control.

Today commercial enrichment and reprocessing services remain in the hands of individual countries and private companies. However, the recent revelations about the proliferation network of A.Q. Khan and the concerns that have been raised by the U.S. and Europe over Iran's efforts to develop uranium enrichment capabilities, have given renewed life to proposals seeking to limit access to fuel cycle technologies. Two basic variations of these proposals have been put forward recently by those seeking a revival of nuclear power. The first proposal harks back to the Eisenhower "Atoms for Peace" initiative with the countries that already possess these technologies providing fuel to the rest of the world, while the second variation draws upon the Acheson - Lilienthal and Baruch plans and advocates for international control of the nuclear fuel cycle.

In 2004, President Bush proposed that, in order to combat nuclear weapons proliferation under the guise of civilian nuclear programs,

> The world's leading nuclear exporters should ensure that states have reliable access at reasonable cost to fuel for civilian reactors, so long as those states renounce enrichment and reprocessing. Enrichment and reprocessing are not necessary for nations seeking to harness nuclear energy for peaceful purposes.[505]

He went on the propose that no country be allowed to import any nuclear technology unless they have ratified the Additional Protocol and that

> The 40 nations of the Nuclear Suppliers Group should refuse to sell enrichment and reprocessing equipment and technologies to any state that does not already possess full-scale, functioning enrichment and reprocessing plants.[506]

A similar proposal for restricting access to fuel cycle technologies was included as an integral part of the Bush Administration's Global Nuclear Energy Partnership as well.[507]

These proposals for a total freeze on the development of fuel cycle facilities in new countries represent a fundamental shift in the interpretation of the Nonproliferation Treaty's guarantee of assistance to any country that has not been found to be in violation of its safeguard agreements. Specifically, Article IV of the NPT states that:

> 1. Nothing in this Treaty shall be interpreted as affecting the inalienable right of all the Parties to the Treaty to develop research, production and use of nuclear energy for peaceful purposes without discrimination and in conformity with articles I and II of this Treaty.
>
> 2. All the Parties to the Treaty undertake to facilitate, and have the right to participate in, the fullest possible exchange of equipment, materials and scientific and technological information for the peaceful uses of nuclear energy. Parties to the Treaty in a position to do so shall also cooperate in contributing alone or together with other States or international organizations to the further development of the applications of nuclear energy for peaceful purposes, especially in the territories of non-nuclear-weapon States Party to the Treaty, with due consideration for the needs of the developing areas of the world.[508]

Despite this very unambiguous wording, the U.S. now argues that "[t]he plain language of Article IV creates no 'right' to any particular nuclear activities or facilities, nor does it require the transfer of any particular technology." The U.S. further argues that the obligations to not acquire or aid others in acquiring nuclear weapons under Articles I and II of the NPT allows states to withhold access to fuel cycle technology when there are questions about its possible use in military programs regardless of whether or not the IAEA has found the country to be in compliance with its safeguards.[509]

Insight into why the Bush administration's proposal to create a permanent monopoly on uranium enrichment is not likely to be acceptable to the global community can be gained by considering the history of how similar promises made under Eisenhower's "Atoms for Peace" initiative were broken by the United States.[510] When the Nixon administration came to power in 1969, it sought to privatize the enrichment of reactor fuel, which had until then been conducted by the Atomic Energy Commission.[511] This was part of the administration's larger efforts to privatize activities that had previous been government functions.

At the time when President Nixon entered office, the United States was still the world's largest provider of enriched uranium ,and was the only reliable supplier for many countries. In order to justify privatization, the Nixon Administration sought to disrupt the smooth functioning of the AEC's enrichment enterprise. In January 1973, the contract requirements for enrichment services were quietly changed to require customers to make very long term projections of their enrichment needs and to impose larger financial penalties for failing to accept a scheduled order. The AEC also raised the cost of enrichment services by 42 percent over the 1966 price, built in future biannual price increases, and decoupled the rate charged to international customers from that charged to domestic utilities allowing for the possibility of discriminatory pricing in the future.[512]

Under this artificially created pressure, utilities began signing long-term contracts with the AEC for enrichment services in order to avoid being left short of fuel. By June 1974, the AEC enrichment capacity through the year 1982 had already been oversubscribed. On July 2, 1974 the signing of new enrichment contracts was suspended and 98 pending applications were declined.[513] Despite complaints from the State Department that provoking this enrichment crisis was damaging U.S. nonproliferation efforts, the AEC refused to undo the change in its contracting procedures. By the time the State Department was able to force concessions from the Commission, the damage to the U.S. reputation as a stable and reliable supplier of enrichment had already been done. The European enrichment consortiums Urenco and Eurodiff attracted a number of new contracts in the wake of these events, and countries like South Korea, Pakistan, and Brazil sought their own fuel cycle capabilities citing the bottleneck in U.S. enrichment services as justification.[514]

The current global environment make it even less likely that countries would be willing to trust a predominantly U.S., Russian, and European monopoly with control of its energy security. For example, India is expected to make up a large percentage of any future nuclear growth scenario, and they currently have no large scale commercial enrichment capacity. It is not reasonable to believe that they would allow any other country or group of countries to unilaterally control the fuel for 15 to 30 percent of its electricity production over the next 45 years as envisioned under the global growth scenario.[515] The potential for political or economic disturbances in Russia to disrupt that country's ability to provide a counterbalance to western supplies further weakens the acceptability of the Bush administration's proposal. For example, in early January 2006, a conflict between Russia and Ukraine over natural gas prices led Russia to reduce the amount of gas it was shipping through the Ukrainian pipelines. This in turn caused disruptions in supply to many European countries highlighting concerns over relying on Russia as a major energy supplier.[516]

Finally, as the Congressional Office of Technology Assessment concluded a decade ago

> Reserving enrichment and reprocessing for the nuclear-weapon states would so badly aggravate the existing discriminatory nature of the international nonproliferation regime that this option must be considered politically untenable.[517]

The inclusion of a handful of other European countries, Japan, and possibly Brazil to such a monopoly (see Table 3.1) would do little to make it more politically acceptable today. Iran's continuing refusal to give up its commercial uranium enrichment program can be seen, in part, in this context. It is this unacceptability of national monopolies that have led to a revival of proposals for international control.

The highest profile proposal for international control over the fuel cycle has come from Mohamed ElBaradei. This is a particularly significant fact given that the stated goal of the IAEA is to help facilitate the development of nuclear power around the world. Speaking a few months after President Bush announced his proposal for freezing access to enrichment and reprocessing facilities, the IAEA Director General proposed that

> We should consider limitations on the production of new nuclear material through reprocessing and enrichment, possibly by agreeing to restrict these operations to being exclusively under multinational controls. These limitations would need to be accompanied by proper rules of transparency and, above all, by international guarantees of supply to legitimate would-be users.[518]

In early 2005, Dr. ElBaradei elaborated on his proposal as follows:

> The first step [to improving world security]: put a five-year hold on additional facilities for uranium enrichment and plutonium separation. There is no compelling reason to build more of these facilities; the nuclear industry has more than enough capacity to fuel its power plants and research centres. To make this holding period acceptable for everyone, commit the countries that already have the facilities to guarantee an economic supply of nuclear fuel for bona fide uses. Then use the hiatus to develop better long-term options for managing the technologies (for example, in regional centres under multinational control).[519]

Dr. ElBaradei's plan therefore incorporates elements of President Bush's proposal in the near term, but attempts to overcome its political unacceptability by explicitly linking the national monopolies to the longer term creation of an international monopoly. Similar proposals put forth by others have also included the option for international storage of spent fuel as a means of preventing clandestine reprocessing.[520]

Although they raise different concerns, multinational plans such as that proposed by Dr. ElBaradei are likely to be little more acceptable than the national monopolies of President Bush's proposal. The group of experts brought together by the IAEA to examine the feasibility of the Director General's proposal summarized the failure of efforts undertaken in the 1970s and 80s to establish multinational control over the fuel cycle as follows:

> All of these initiatives failed for a variety of political, technical and economic reasons, but mainly because parties could not agree on the non-proliferation commitments and conditions that would entitle States to participate in the multilateral activities. Moreover, differences of views prevailed between those countries and/or regions that did not plan to reprocess or recycle plutonium and those that favoured it (the latter group being concerned, in particular, about the availability of fuel supplies and the possibility of the interruption of supplies by suppliers).[521]

In addition, the IAEA report also conclude that, while so-called Multilateral Nuclear Approaches (MNAs) would likely have some advantages from a nonproliferation standpoint,

> However, the case to be made in favour of MNAs is not entirely straightforward. States with differing levels of technol-

> ogy, different degrees of institutionalisation, economic development and resources and competing political considerations may not all reach the same conclusions as to the benefits, convenience and desirability of MNAs. Some might argue that multilateral approaches point to the loss or limitation of State sovereignty and independent ownership and control of a key technology sector, leaving unfairly the commercial benefits of these technologies to just a few countries. Others might argue that multilateral approaches could lead to further dissemination of, or loss of control over, sensitive nuclear technologies, and result in higher proliferation risks.[522]

There are additional concerns, such as economic considerations and what would be done with the military facilities in the five nuclear weapons states, that add to the difficulty of negotiating multilateral approaches. At current market prices, the enrichment services necessary to fuel 1,000 gigawatts of nuclear capacity would amount to roughly $10 to $14.4 billion in annual sales.[523] Current enrichment services are provided primarily by private or quasi-private companies. In light of these kinds of considerations, the Congressional Office of Technology Assessment concluded that

> Instituting an international nuclear material control regime would involve the internationalization of enrichment, reprocessing, and, possibly, fuel fabrication facilities.... As such, an international control regime would involve drastic changes to the way the uranium and plutonium markets now operate, affecting the ownership and operation of many billions of dollars worth of existing facilities. Dramatic changes would be required to the international legal regime, along with extensive treaty negotiations.

> It would be very difficult to create such a regime. Non-nuclear-weapon states would likely object strongly to a regime that reinforced the discriminatory aspects of the NPT by denying them the ability to operate nuclear fuel-cycle facilities by themselves, while permitting the nuclear weapon states to do so in their military programs. Given the magnitude of the changes such a policy would require, it would likely be possible only with sustained effort over many years, if at all.[524]

The conflicts between the United States and its NATO allies France, Belgium, and Germany in the lead up to the 2003 invasion of Iraq illustrate the difficulties that can be encountered in trying to implement multilateral security programs even among countries with close economic, political, and military relationships who share many similar overall security concerns and international goals.

A further potential difficulty to implementing such programs was created by the issuance of the *Joint Statement Between President George W. Bush and Prime Minister Manmohan Singh* on July 18, 2005. In this statement, President Bush announced his intention to request that Congress amend federal laws and to have the U.S. and its allies amend "international regimes" in order to allow "full civil nuclear energy cooperation and trade with India" including the sale of uranium fuel for the Tarapur reactor. In addition, the statement announced the U.S. intention to "consult" with other countries on allowing India to join in ITER [International Fusion Energy Research project] and the Generation IV International Forum. In exchange the Indian Prime Minister agreed to a "phased" separation of India's civilian and military nuclear infrastructure, as well as placing the declared civilian infrastructure under voluntary IAEA safeguards, continuing the test moratorium, working with the U.S. on advancing negotiations of a Fissile Material Cut-off Treaty, and "refraining from transfer of enrichment and reprocessing technologies to states that do not have them" in line with the Bush Administration's proposal.[525]

In addition to the specific steps outline above, the joint statement included a particularly important pair of statements that shed additional light on the significance of this major change in U.S. policy. The U.S. part of the joint statement noted that

> President Bush conveyed his appreciation to the Prime Minister over India's strong commitment to preventing WMD proliferation and stated that as a responsible state with advanced nuclear technology, India should acquire the same benefits and advantages as other such states.[526]

The definition of "other such states" was not clarified further by the U.S., however, the Indian half of the joint statement noted that

> The Prime Minister [Manmohan Singh] conveyed that for his part, India would reciprocally agree that it would be ready to assume the same responsibilities and practices and acquire the same benefits and advantages as other leading countries with advanced nuclear technology, **such as the United States**.[527]

Thus, while not explicitly declaring it, by issuing a joint statement which includes a specific reference to India accepting the "same responsibilities and practices" on nuclear issues as those of the United States, the Bush

Administration has taken a significant first step towards what is likely to be viewed as a de facto recognition of India as a nuclear weapons state akin to those under the NPT. The similarity of India's acceptance of safeguards on self-declared "civilian" facilities while prohibiting access to its "military" facilities to the practices of the nuclear weapons states further strengthens this implication of the joint statement.

Implementing this agreement between the U.S. and Indian governments would require the Nuclear Suppliers Group to fundamentally change its rules against exporting sensitive nuclear technology to countries that have not ratified the Non-Proliferation Treaty. By allowing India, a country whose nuclear weapons program goes back well before 1974 and who has tested nuclear weapons as recently as 1998, to be given full access to advanced nuclear power technology would severely erode the primary bargain of the NPT; namely that such technology could only be acquired in exchange for a pledge not to develop or acquire nuclear weapons. If the President's proposals are carried out, it would create a significant new double standard that would make it increasingly difficult to stop NPT members like Iran from acquiring fuel cycle technology while India would retain its nuclear arsenal outside the treaty and enjoy unrestricted access to nuclear technology and materials.

It is very unlikely that either national or international monopolies on the fuel cycle would be sustainable, as the deadlock at the 2005 NPT Review Conference and the ongoing standoff between the U.S., the EU and Iran demonstrate. However, even if one of these proposals was somehow brought into force, it would still not be capable of totally eliminating the proliferation risks associated with a large expansion of nuclear power. As noted by William Sutcliffe, a Senior Physicist at the Lawrence Livermore National Laboratory, "[t]here is little that can be done to prevent the diffusion of technology and therefore its potential for misuse."[528] Mohamed ElBaradei concluded that

> The technical barriers to mastering the essential steps of uranium enrichment – and to designing weapons – have eroded over time, which inevitably leads to the conclusion that the control of technology, in and of itself, is not an adequate barrier against further proliferation.[529]

As noted above, the extent of the A.Q. Khan proliferation network and fact that it was reportedly able to transfer advanced centrifuge technology and components to Iran, Libya, and North Korea before it was identified and finally stopped highlights the limitations of export controls and current efforts to prevent clandestine proliferation. Such concerns have given rise to an increased focus on punishing proliferation after it has been discovered as will be discussed in the next section.

Section 3.4.3 – Increased Consequences for Suspected Proliferators

The proposals put forth for strengthening the inspection regimes and restricting access to fuel cycle technologies are both proactive efforts that seek to detect and prevent nuclear weapons programs from developing. However, for such efforts to have any significant effect on limiting proliferation, the proponents of nuclear power acknowledge that mechanisms would have to be in place to punish countries that violate their international agreements as well. As summarized by Bernard Baruch in his June 1946 speech before the United Nations on the international control of atomic energy

> We must provide the mechanism to assure that atomic energy is used for peaceful purposes and preclude its use in war. To that end, we must provide immediate, swift, and sure punishment of those who violate the agreements that are reached by the nations. Penalization is essential if peace is to be more than a feverish interlude between wars.[530]

The imposition of enforcement mechanisms, however, creates a number of concerns over the viability of these non-proliferation efforts. These concerns are particularly acute for proposals that would be strong enough to facilitate a widespread expansion of nuclear power without resulting in a spread of nuclear weapons as well.

Outside of naval blockades and other direct military action, the most aggressive of the proposed enforcement mechanisms is the imposition of United Nations sanctions. The NPT already allows the U.N. Security Council to impose a penalty on countries that are found to be in violation of their safeguards agreements with the IAEA. However, to date the United Nations has been reluctant to exercise this power, and this has prompted calls for greater action by some proponents of nuclear power's revival. The authors of the MIT study recommended that the "[t]he U.N. Security Council should develop guidelines for multilateral sanctions in the event of serious violations of safeguards agreements." In addition, they recommend that "more stringent restrictions" be placed on countries that remain outside the non-proliferation framework.[531] Another example can be found in a recent statement by Mohamed ElBaradei. In a June 2004 speech, Dr. ElBaradei noted that

> For 12 years, the Democratic People's Republic of Korea (DPRK) has been in non-compliance with its NPT obligations. In January 2003, the DPRK capped its non-compliance by declaring its withdrawal from the NPT. Naturally, the Agency reported the situation to the United Nations Security Council. But now, more than a year later, the Security Council has not even reacted. This lack of response, this inaction, may be setting the worst precedent of all, if it conveys the message that acquiring a nuclear deterrent, by whatever means, will neutralize any compliance mechanism and guarantee preferred treatment.[532]

A final example of these calls for stronger action by the Security Council was the Bush administration's efforts to see Iran sanctioned by the United Nations. This effort led to Iran being reported to the Security Council in February 2006.[533]

There are three main reasons why a more stringent sanctions regime is unlikely to be effective at successfully deterring the production of nuclear weapons and should not be supported. The first reason is that economic sanctions can actually work to strengthen a government's internal position while primarily hurting ordinary people. Cuba and Iraq are two examples of this effect. Despite an intensive economic embargo and a long history of covert military and political operations sponsored by the U.S. against it, the government of Fidel Castro has outlasted nine U.S. presidents and remains in power today.[534] In Iraq, the dictatorship of Saddam Hussein remained in power during more than a decade of intense sanctions coupled with ongoing U.S. and British bombing campaigns in parts of Iraq. As a result, the United Nations Children's Fund estimated that, had Iraq not been under sanctions and instead been allowed to continue investing in its social sector, between 1990 and 1998 a staggering half a million children who died before reaching their fifth birthday would have otherwise survived.[535] This human toll should be more than enough reason to resist the use of these kinds of coercive sanctions.

The second reason that sanctions are likely to be an ineffective strategy in the current context is that many countries of potential proliferation concern are either permanent members of the Security Council themselves (i.e. the five nuclear weapons states under the NPT) or are strategically important to at least one, and often several, of the permanent council members. Given the absolute power of the veto in the United Nations, such countries or their close allies are therefore very unlikely to be targets of international sanctions. For example, the U.S. use of its veto power to systematically shield Israel from condemnation by the Security Council despite Israel's possession of an undeclared nuclear arsenal and the extension by the United States of "Major Non-NATO Ally" status to Pakistan shortly after the A.Q. Khan proliferation network became public provide relevant examples.[536] The importance of the veto in limiting the ability of the U.N. to hold certain countries accountable was recognized from the very start of the nuclear age. In his June 1946 speech on the international control of atomic energy and atomic weapons, Bernard Baruch, the U.S. negotiator, concluded

> It would be a deception, to which I am unwilling to lend myself, were I not to say to you and to our peoples that the matter of punishment lies at the very heart of our present security system. It might as well be admitted, here and now, that the subject goes straight to the veto power contained in the Charter of the United Nations so far as it relates to the field of atomic energy.[537]

The Baruch proposal to eliminate the veto power in matters relating to violations of non-proliferation agreements was strongly objected to by the Soviet Union, and became one of the central disagreements that caused the negotiations for international control of atomic energy to fail.[538] No proposal to limit or eliminate the Security Council veto today would fare any better.

The third reason that sanctions are unlikely to be effective is that the acquisition of nuclear weapons by a country creates great pressures on other countries to pursue diplomatic negotiations and against further economic and political isolation. The threat posed by a nuclear armed state is such that engagement and dialogue are likely to have significant security advantages for its neighbors or rivals which might be hampered by the imposition of harsh economic sanctions. This has so far proved true in the case of North Korea. Since the latest crisis began, both China and South Korea have actually expanded trade with North Korea and have shown reluctance to join in any discussion of sanctions or other more confrontational strategies. On the other hand, Japan has reduced its economic contacts with the North. As implied by the June 2004 comments from Dr. ElBaradei quoted above, it would be hard for potential proliferators to compare the invasion and occupation of Iraq and the hard line taken by the United States towards the Iranian commercial enrichment program to the treatment of North Korea's open pursuit of nuclear weapons without drawing the conclusion that the possession of a nuclear deterrent is a potential pathway to enhanced national security when facing threats from far more powerful countries. This is particularly important in the case of countries that are facing potential threats from states that are already armed with nuclear weapons and have declared policies that would allow their use against non-nuclear armed states.[539]

Given that strengthened sanctions regimes have potentially devastating human impacts and that they are likely to be of limited value in preventing proliferation, a variety of new strategies have been proposed to augment export controls and to try and prevent the illicit transfer of nuclear weapons related technologies. A central element of these so-called "counter proliferation" efforts has been the establishment of the Proliferation Security Initiative (PSI). The PSI is a group of countries that, without explicit United Nations authorization, have agreed to work together to intercept shipments of nuclear, chemical, and biological weapons as well as missile technology and materials or equipment related to any of these types of programs.[540] The first of the so-called "Interdiction Principles" agreed to by the PSI member states that these countries will

> Undertake effective measures, either alone or in concert with other states, for interdicting the transfer or transport of WMD, their delivery systems, and related materials to and from states and non-state actors of proliferation concern. "States or non-state actors of proliferation concern" generally refers to those countries or entities that the PSI participants involved establish should be subject to interdiction activities because they are engaged in proliferation through: (1) efforts to develop or acquire chemical, biological, or nuclear weapons and associated delivery systems; or (2) transfers (either selling, receiving, or facilitating) of WMD, their delivery systems, or related materials.[541]

The interdiction of the *BBC China* by the Italian coast guard in October 2003, and the seizure of uranium enrichment technology and related equipment bound for Libya has been widely heralded as a major success for the PSI. However, certain aspects of this initiative are of dubious legality and raise serious concerns with respect to their long-term impact on the Law of the Sea.[542] In particular, the Law of the Sea, a loose collection customary law and international treaties, outlaws such activities as piracy, slave trading, trafficking in illegal drugs, and unauthorized broadcasting of television or radio, but includes no prohibition against the transport of nuclear weapons, missiles, or their related technology. In fact, the United States, France, Britain, China, and Russia have all argued against the imposition of any such prohibition because of their desire to continue transporting their nuclear weapons and missiles over the world's oceans.

Under the PSI's interdiction principle quoted above, the decision as to which states are classified as a "proliferation concern," and thus possible targets for interdiction, are made entirely by the PSI countries themselves with no review or oversight authority. For example, missile sales by the United States are allowed to go unchallenged while a missile shipment from North Korea to Yemen has already been intercepted, although not ultimately prevented.[543] As Christopher Clary, a research associate at the U.S. Naval Postgraduate School, concluded in 2004, "[a]t present, the line between interdiction and piracy is thin and blurry."[544]

In order to try and overcome the questionable legality of this interdiction policy, the U.S. has entered into a series of bilateral agreements with major shipping countries. In February 2004, the United States signed an agreement that gave it the explicit legal authority to board any Liberian vessel in search of nuclear, chemical, or biological weapons, their delivery vehicles, or related materials. This was followed in May by the signing of a similar ship boarding agreement with Panama and in August by an agreement with the Marshall Islands. By August 2005 three additional ship boarding agreements had been signed with Belize, Croatia, and Cyprus. Panama and Liberia have the largest and second largest ship registries in the world respectively. When added to the ships registered to PSI member states, the six ship boarding agreements give the U.S. a legal right to search more than 50 percent of the world's commercial shipping fleet.[545]

The United States has called on NPT member states to endorse "cooperation to interdict illegal transfers of nuclear material and equipment that is fully consistent with domestic legal authorities and international law and relevant frameworks, such as the Proliferation Security Initiative."[546] However, both China and Russia initially refused to join the PSI due to concerns over its unilateral nature and lingering questions over its legality.[547] The clear double standard it represent makes it unlikely that interdiction strategies like the PSI will be able to achieve the universality required to prevent proliferation in the event of a greatly expanded nuclear power infrastructure. If interdiction was used routinely, the impact on international law in general, and on the Law of the Sea in particular, could be serious as this small number of countries, representing just 15 percent of the world's population, would be essentially seeking to unilaterally re-write the rules governing ocean transport by themselves.[548]

Finally, and most troubling, are the proposals for an expanded use of "preemptive" military strikes against countries suspected of violating their safeguards and pursuing clandestine nuclear programs. In 1981, the Israeli air force invaded Iraq and used U.S. made planes to bomb the French built Osirak reactor about 15 miles east of Baghdad. This attack was prompted by the Israeli concern over the reactor's ability to support a nuclear weapons program. A French technician working at the site was killed in the attack. The bombing was condemned by all the members of United Nations Security Council, including the United States. The Security Council unanimously passed a resolution that strongly condemned the attack by Israel which it said was "in clear violation of the United Nations charter and the norms of international conduct" and was "a serious threat to the entire International Atomic Energy Agency safeguards regime."[549]

Despite this rare public disavowal of Israeli actions in the United Nations, after the 1991 Gulf War, then Secretary of Defense Dick Cheney sent an aerial photograph of the wrecked Osirak reactor to the commander of the Israeli Air-Force, on which Cheney reportedly wrote:

> For Gen. David Ivry, with thanks and appreciation for the outstanding job he did on the Iraqi nuclear program in 1981 -- which made our job much easier in Desert Storm.[550]

The support for preemptive military action against suspected nuclear, chemical, and biological weapons was formally incorporated into U.S. policy in the September 2002 *National Security Strategy of the United States of America*. This policy document stated that

> The United States has long maintained the option of preemptive actions to counter a sufficient threat to our national security. The greater the threat, the greater is the risk of inaction—and the more compelling the case for taking anticipatory action to defend ourselves, even if uncertainty remains as to the time and place of the enemy's attack. To forestall or prevent such hostile acts by our adversaries, the United States will, if necessary, act preemptively.
>
> The United States will not use force in all cases to preempt emerging threats, nor should nations use preemption as a pretext for aggression. Yet in an age where the enemies of civilization openly and actively seek the world's most destructive technologies, the United States cannot remain idle while dangers gather.[551]

The option of preemptive strikes against nuclear, chemical, or biological weapons was also included in the December 2002 *National Strategy to Combat Weapons of Mass Destruction*. This document warned that

> The United States will continue to make clear that it reserves the right to respond with overwhelming force – including through resort to all of our options – to the use of WMD against the United States, our forces abroad, and friends and allies.
>
> ...
>
> Because deterrence may not succeed, and because of the potentially devastating consequences of WMD use against our forces and civilian population, U.S. military forces and appropriate civilian agencies must have the capability to defend against WMD-armed adversaries, including in appropriate cases through preemptive measures. This requires capabilities to detect and destroy an adversary's WMD assets before these weapons are used.[552]

Finally, the Bush Administration's position on preemptive military action was reiterated in the March 2006 *National Security Strategy of the United States of America*. This update of the 2002 policy stated that

> If necessary, however, under long-standing principles of self defense, we do not rule out the use of force before attacks occur, even if uncertainty remains as to the time and place of the enemy's attack. When the consequences of an attack with WMD are potentially so devastating, we cannot afford to stand idly by as grave dangers materialize. This is the principle and logic of preemption. The place of preemption in our national security strategy remains the same.[553]

The March 2003 invasion and occupation of Iraq by the U.S. and Britain premised on the supposed continuation of Iraqi programs to develop illicit weapons is the most stark and deeply troubling example of these doctrines in action. Adding to such concerns is the speculation that Israel or the United States may be prepared to attack Iran's nuclear facilities if it continues to make progress towards developing uranium enrichment capabilities.[554] Finally, Richard Perle, a key administration advisor and one of the architects of the war on Iraq, told reporters regarding the North Korean nuclear program

> But I don't think anyone can exclude the kind of surgical strike we saw in 1981... We should always be prepared to go it alone, if necessary.[555]

As the Director General of the IAEA noted, there is only a "narrow margin of security" that accompanies the development of civilian fuel cycle technology in a country given its ability to support a future nuclear weapons program.[556] As we have discussed above, it is unlikely that access to fuel cycle technology could be limited in any meaningful or sustainable way if the use of nuclear power was to greatly expand over the coming decades. The embracing of "preemptive" military strikes by powerful states like the U.S. and Israel (which have both demonstrated their willingness to carry out such a policy), further erodes this margin of security. Such military actions present a serious challenge to the future growth of nuclear power around the globe if proliferation were to be held in check by increased regional tensions and the risk of conventional war.

Section 3.4.4 – Disarmament and Nonproliferation

Finally, the two most important issues that limit the effectiveness of efforts to prevent nuclear weapons proliferation while expanding nuclear power is the inability to adequately address the problem of countries outside the NPT regime and the inability to address the unwillingness of the five acknowledged nuclear weapons states to live up to their NPT obligation to bring to a conclusion negotiations on disarmament.[557] The institutionalization of a system in which some states are allowed to possess nuclear weapons while dictating intrusive inspections and restricting what activities other states may pursue is not sustainable over the long-term. This situation is aggravated by the fact that the only country to actually use nuclear weapons in a time of war is one of the central countries claiming that others are not responsible enough to be trusted with these weapons. The devastating impact of nuclear bombs and their potential use as a weapon of surprise or terror adds further complications to a world of nuclear "haves" and "have-nots." As summarized by Mohamed ElBaradei in his remarks to the 2005 NPT Review Conference

> As long as some countries place a strategic reliance on nuclear weapons as a deterrent, other countries will emulate them. We cannot delude ourselves into thinking otherwise.[558]

First, any attempt to reduce future proliferation concerns will have to address the four countries currently outside the NPT regime. The existence of nuclear weapons in India, Pakistan, Israel, and presumably North Korea creates serious pressures on their neighbors. In this context it is interesting to note that despite discussion of three of these four programs, the MIT study does not contain a single mention of the Israeli nuclear weapons program despite its size, technical sophistication, and implications for the long-term stability of the Middle East. Estimates following the release of information by Mordechai Vanunu in 1986 put the Israeli stockpile nearly on par with those of France or the United Kingdom. The Egyptian government opposed the indefinite extension of the NPT due to the security implications of Israel and other countries remaining outside the regime.[559] Mohamed ElBaradei has stated that "the Israeli military nuclear programme is a cause of great concern in the Middle East and in the world as a whole."[560] Finally, the Congressional Office of Technology Assessment concluded that

> Even if Israeli weapons of mass destruction are not themselves deemed to threaten the United States or U.S. interests, however, their implicit acceptance complicates nonproliferation policy.[561]

It has been 10 years since the goal for a nuclear weapons free zone in the Middle East was adopted by the NPT member states. While the United States government officially supports the creation of a Middle East nuclear weapons free zone, it has yet to put any meaningful pressure on Israel. In a constructive sign, during the visit of Mohamed ElBaradei to Israel in 2004, then Prime Minister Sharon and Minister of Foreign Affairs Silvan Shalom told the Director General that, within the larger context of negotiations with the Palestinians, that the Israelis were prepared to begin a discussion on the eventual establishment of a nuclear weapons free zone.[562] However, the treatment by the Israeli government of Mordechai Vanunu, the man that exposed the scale of Israel's secret nuclear weapons program to the world, highlights the long road that such a process would likely represent.

Mordechai Vanunu was released from prison in April 2004 after spending 18 years in jail following his conviction for espionage and treason. He reportedly spent more than eleven of those years in solitary confinement. Upon his release he was barred by the Israeli government "from leaving the country, approaching its borders," or of "talking to foreigners without permission."[563] These prohibitions were to last at least one year. from the time of his release Less than seven months later, Vanunu was rearrested by the Israelis and placed under house arrest for speaking to reporters and the European Social Forum without authorization. This arrest was made despite the fact that no new secret information was revealed.[564]

Second, successful nonproliferation efforts must necessarily address disarmament in the five acknowledged nuclear weapons states. This has been recognized to be the case from the earliest days of the nuclear age. For example, India's official justification for refusing to join the NPT regime was the treaty's lack of clearly required steps for disarmament as well as its lack of legally binding assurances that nuclear weapons would not be used against non-nuclear states.[565] This connection between disarmament and non-proliferation efforts remains strong today. In an op-ed published in the New York Times in February 2004, Mohamed El-Baradei wrote

> Of course a fundamental part of the nonproliferation bargain is the commitment of the five nuclear states recognized under the nonproliferation treaty -- Britain, China, France, Russia and the United States -- to move toward disarmament.... A clear roadmap for nuclear disarmament should be established -- starting with a major reduction in the 30,000 nuclear warheads still in existence, and bringing into force the long-awaited Comprehensive Test Ban Treaty.[566]

The authors of the MIT report, however, do not address the problems of disarmament in the context of their nonproliferation proposals. This is a noteworthy omission given that the study's two co-chairs (Ernest Moniz and John Deutch) had both been high ranking officials in the Department of Energy, which oversees the U.S. nuclear weapons establishment. In fact, Ernest Moniz has referred to the maintenance of a so-called safe and reliable nuclear stockpile as "a supreme national interest for this country" while John Deutch has co-authored an op-ed with Henry Kissinger and Brent Scowcroft opposing ratification of the Comprehensive Test Ban Treaty. In this op-ed the authors stated that they were "no fans" of the CTBT and that "for the foreseeable future, the United States must continue to rely on nuclear weapons to help deter certain kinds of attacks on this country and its friends and allies."[567]

While a complete discussion of the role of nuclear weapons in U.S. and NATO military strategy is well beyond the scope of this report, a brief review will serve to illustrate the challenge that these policies present to current and future nonproliferation efforts.[568] Prior to 1999, the United States had been pursuing policies that were slowly coming more directly into line with its disarmament obligations. These policies included the negotiation of the Strategic Arms Reduction Treaties (START) with Russia, the 1992 nuclear test moratorium, the signing of the Comprehensive Test Ban Treaty (CTBT), and the sharp reduction in tactical nuclear forces deployed in Europe. Since the failure of the U.S. Senate to ratify the CTBT, however, the U.S. has been steadily turning away from its disarmament commitments.

The withdrawal from the Anti-Ballistic Missile Treaty in December 2001, the open hostility of the Bush Administration to the CTBT, the issuance of the 2001 Nuclear Posture Review, the announcement of the NATO doctrine for continued reliance on nuclear weapons, the commencement of tritium production in commercial power reactors, and the rejection of the START framework in exchange for the unverifiable Moscow Treaty that sets no intermediary goals and is set to expire upon its entry into force in 2012, all point to the conclusion that the U.S. intends to expand rather than shrink its reliance on nuclear weapons. For example, the Nuclear Posture Review states that

> Nuclear weapons play a critical role in the defense capabilities of the United States, its allies and friends. They provide credible military options to deter a wide range of threats, including WMD and large-scale conventional military force. These nuclear capabilities possess unique properties that give the United States options to hold at risk classes of targets [that are] important to achieve strategic and political objectives.[569]

The Nuclear Posture Review also reveals that the U.S. is planning to develop a new land-based ICBM for deployment by 2020, a new submarine launched missile for deployment by 2030, and a new intercontinental heavy bomber for deployment by 2040.[570] The 2006 National Security Strategy notes that "[t]he new strategic environment requires new approaches to deterrence and defense" and goes on to declare that "[s]afe, credible, and reliable nuclear forces continue to play a critical role."[571] Finally, Table 3.4 shows the recent budget requests for DOE's National Nuclear Security Agency to pursue new nuclear weapons designs and capabilities.

Table 3.4: Budget requests by the Bush Administration and Congressional allocations for new nuclear weapons related program between fiscal year 2004 and 2006.[572]

Program	FY04 Budget (in millions)		FY05 Budget (in millions)		FY06 Budget (in millions)	
	Request	Allocation	Request	Allocation	Request	Allocation
Enhanced Test Readiness	24.9	24.74	30.0	26.78	25.0	---
Modern Pit Facility	22.8	11.55	29.8	6.95	7.7	---
Robust Nuclear Earth Penetrator	15.0	7.41	27.6	0	8.5[(a)]	---
"Advanced" Concepts[(b)]	6.0	6.0	9.0	0	N.A.	---
Reliable Replacement Warhead[(b)]	N.A.	N.A.	N.A.	8.93	9.35	---
Total	68.7	49.7	96.4	42.7	50.6	---

(a) This request included $4 million for the NNSA and $4.5 million for other Department of Defense efforts related to this program.
(b) The Advanced Concepts included the so-called "low-yield" nuclear weapons as well as weapons designed to destroy chemical or biological agents. The Reliable Replacement Warhead program has replaced the more expansive Advanced Concepts efforts.

The continued reliance on nuclear weapons in the military and political spheres is not just an issue in the United States. For example, the Russian Military Doctrine which was approved in April 2000 states that

> The Russian Federation regards nuclear weapons as a means of deterrence of an aggression, of ensuring the military security of the Russian Federation and its allies, and of maintaining international stability and peace.[573]

In addition, the 2001 *NATO Handbook* notes that

> NATO's nuclear forces contribute to European peace and stability by underscoring the irrationality of a major war in the Euro-Atlantic region. They make the risks of aggression against NATO incalculable and unacceptable in a way that conventional forces alone cannot. They also create uncertainty for any country that might contemplate seeking political or military advantage through the threat or use of Nuclear, Biological or Chemical (NBC) weapons against the Alliance. By promoting European stability, helping to discourage threats re-

lating to the use of weapons of mass destruction, and contributing to deterrence against such use, NATO's nuclear posture serves the interests not only of the Allies, but also of its Partner countries and of Europe as a whole.[574]

Finally, in January 2006, French President Jacques Chirac stated that, "in the face of the concerns of the present and the uncertainties of the future, nuclear deterrence remains the fundamental guarantee of our security."[575]

Attempting to institutionalize a world of nuclear "haves" and "have-nots" would make it extremely difficult to prevent the spread of nuclear weapons. As Mohamed ElBaradei summarized

> We must abandon the unworkable notion that it is morally reprehensible for some countries to pursue weapons of mass destruction yet morally acceptable for others to rely on them for security -- indeed to continue to refine their capacities and postulate plans for their use.[576]

Without concrete, verifiable programs to irreversibly eliminate the tens of thousands of existing nuclear weapons, no nonproliferation strategy is likely to be successful over the long-term no matter how strong it would otherwise be. This is likely to prove to be one of the most difficult obstacles to overcome in any attempt to revive the nuclear power industry.

Section 3.5 - Conclusions

As summarized nearly sixty years ago by the Committee on Atomic Energy, "[t]he development of atomic energy for peaceful purposes and the development of atomic energy for bombs are in much of their course interchangeable and interdependent."[577] This overlap between the nuclear fuel cycle and the infrastructure required to produce nuclear weapons makes nuclear power unique among all sources of electricity. It is this connection to the potential proliferation of nuclear weapons that is likely to be the largest single vulnerability associated with an expansion of nuclear power around the world.

For light-water reactors, enrichment forms a vital step in the nuclear fuel cycle. In order to fuel the number of nuclear plants under the global or steady-state growth scenarios, large increases in enrichment capacity will be needed. For example, fueling the global growth scenario alone would require roughly 18 times more capacity than is currently deployed by the European enrichment corporation Urenco in all of Britain, Germany, and the Netherlands combined. The needs of the steady-state growth scenario would be even greater (see Figure 3.2).

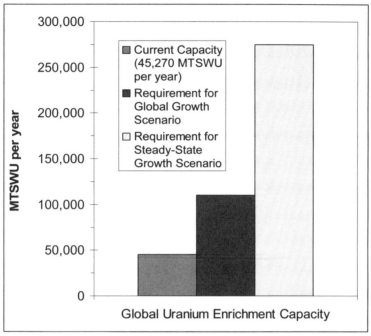

Figure 3.2: Comparison of the current global uranium enrichment capacity to the levels that would be required to fuel the global and steady-state growth scenarios operating on the once-through fuel cycle. For simplicity, these figures assume that an average of 110 MTSWU per year is required to fuel a single 1,000 MW reactor on average. Larger levels of enrichment services may be required in the future if higher fuel burnup becomes common or if the price of natural uranium rises significantly (see Appendix A).

While the 14 enrichment plants in operation today are capable of fueling the existing reactors, all of these facilities would likely be replaced over the next few decades if nuclear power was aggressively expanded. The planned closure of the two large gaseous diffusion plants in the U.S. and France, which together account for nearly half of the world's enrichment capacity, and their replacement by smaller gas centrifuge plants would be of particular importance to the global enrichment market.

The development of such large numbers of enrichment plants would create a significant risk of nuclear weapons proliferation as gas centrifuge technology spread around the world. For example, just one percent of the enrichment capacity required by the global growth scenario would be enough to supply the highly-enriched uranium for approximately 175 to 310 nuclear weapons every year. If the spent fuel was reprocessed to allow the separated plutonium to be fabricated into MOX fuel, the requirements for enrichment would be decreased somewhat, but would still remain significantly above present levels. For example, the authors of the MIT study found that the optimal use of MOX would reduce the enrichment needs by only 16 percent. In addition to the continued need for large increases in enrichment capacity, the separation of weapons useable plutonium would add significantly to the proliferation risks of nuclear power if the MOX fuel cycle were to be used.

The existing commercial reprocessing capacity is well below what would be needed to support the widespread use of MOX. Even if only 16 percent of the fuel was MOX as assumed by the authors of the MIT study, in order to supply the fuel requirements of the global or steady-state growth scenarios a reprocessing capacity three and a half to nine times greater than that in existence today would be required (see Figure 3.3).

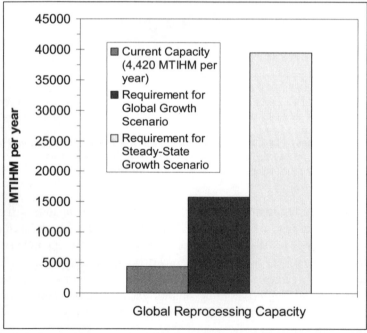

Figure 3.3: Comparison of the current reprocessing capacity to that required to fuel the global and steady-state growth scenarios. This calculation assumes that the reprocessing plants have an effective capacity factor of 75 percent. In addition, just 16 percent of the total fuel requirements are supplied by MOX, while the remaining 84 percent remains low-enriched uranium.

For the global growth scenario alone, more than 155 metric tons of plutonium would be separated annually in order to supply the required MOX fuel. Just one percent of this plutonium would be sufficient to produce more than 190 nuclear weapons every year. Given the difficulties that have been experienced to date in accounting for plutonium at existing reprocessing and fuel fabrication facilities, the use of MOX would pose a significant proliferation risk. While newer reprocessing technologies like UREX+ or pyroprocessing would have some advantages over the current PUREX process from a proliferation standpoint, the resulting plutonium would still be useable in nuclear weapons making them a serious concern. In addition, materials accounting at pyroprocessing facilities would likely be even more difficult than at current plants, making clandestine diversion more difficult to detect.

The proposals that have been put forward to try and reduce the proliferation risks posed by a revival of nuclear power are unlikely to be sustainable over the long term in light of the continued refusal of the nuclear weapons states to disarm and their continued focus on nuclear weapons as a corner stone of their national security posture. In effect, the current nonproliferation efforts seek to permanently institutionalize a discriminatory system of nuclear weapons "haves" and "have-nots." Even efforts to enhance the power of inspections by the International Atomic Energy Agency are undermined by the preferential treatment given to the five acknowledged nuclear weapons states. Despite the favorable conditions in the agreements, as of February 2006, neither Russia nor the United States had ratified their Additional Protocols with the IAEA further undermining efforts to achieve universal adherence.

Despite the clear language of the Nuclear Nonproliferation Treaty giving any member the right to posses uranium enrichment and reprocessing facilities as part of a civilian nuclear power program, proposals from the Bush Administration and the IAEA now seek to restrict access to these fuel cycle technologies. The proposals to limit the spread of enrichment and reprocessing are motivated by the inability of all but the most highly intrusive, expensive, and time consuming inspection regimes to prevent proliferation in the event of a state determined to develop nuclear weapons, as well as the possibility that states could pull out of safeguards in the future and use formerly civilian infrastructure for weapons production.

Proposals to create national or international monopolies on the nuclear fuel cycle, however, are very unlikely to be acceptable. The implication of these proposals is that certain countries, including the only country to have used nuclear weapons in a time of war, can be trusted with the fuel cycle while no one else can. This is clearly a highly discriminatory policy, and not one likely to gain significant support. The deadlock at the 2005 NPT Review Conference in New York and the continued refusal of Iran to abandon its development of an indigenous enrichment capability are clear examples of the difficulties faced by attempts to prevent countries from controlling the production of their own nuclear fuel. This difficulty would be greatly increased under the global or steady-state growth scenarios given the presumed increased reliance on nuclear power in many countries that do not current posses fuel cycle facilities.

The use of punitive sanctions or military intervention such as air strikes or the interdiction of ships on the high seas in order to enforce these proposed restrictions on fuel cycle technologies would add greatly to their unacceptability. First, such actions would further the discriminatory nature of these proposal. Specifically, the nuclear weapons states, which are also the permanent members of the Security Council, along with their allies would be shielded from any of these kinds of consequences. To illustrate this one has only to look at the difference between the U.S. approach to the Iranian uranium enrichment efforts and to the existence of a presumably significant stockpile of modern nuclear weapons in Israel. Second, it would make nuclear power appear increasingly unattractive if the proliferation of nuclear weapons was held in check by increased regional tensions, sanctions that hurt ordinary people, and a heightened risk of conventional war.

To trade the serious vulnerabilities of global warming for the environmental and security impacts that would accompany the potential proliferation of nuclear weapons is not a sound energy policy when, as discussed in Chapter Two, a clear set of robust and economically viable alternatives are available.

Chapter Four: A Culture of Safety?

> With regard to implementation of the global growth scenario during the period 2005-2055, both the historical and the PRA [Probabilistic Risk Assessment] data show an unacceptable accident frequency. The expected number of core damage accidents during the scenario with current technology would be 4. We believe that the number of accidents expected during this period should be 1 or less, which would be comparable with the safety of the current world LWR fleet. A larger number poses potential significant public health risks and, as already noted, would destroy public confidence.[578]
> - *The Future of Nuclear Power* (2003)

> We should remember that risk assessment data can be like the captured spy: if you torture it long enough, it will tell you anything you want to know.[579]
> - William Ruckelshaus, the head of the U.S. Environmental Protection Agency under Presidents Nixon and Reagan (1984)

> Of course, the whole point of the regulatory system is not simply to enable adequate **average** performance, but to ensure that every plant is operating safely all the time. And we clearly have not always achieved that objective. Indeed, a recurrent theme over the past decade is the need for improvement of safety culture at plants that are encountering serious difficulties.[580]
> - Richard Meserve, Chairman of the U.S. Nuclear Regulatory Commission (2002)

The risk associated with the proliferation of nuclear weapons discussed in the previous chapter is one of the unique vulnerabilities associated with the use of nuclear power. The potential for a catastrophic accident or a well coordinated terrorist attack such as those of September 11,

2001, to result in the release of a large amount of radiation and to contaminate a vast area is a second unique vulnerability of nuclear power. Perhaps the only other type of accidents within the energy system that could begin to approach the immediate environmental impacts of a serious accident at a nuclear plant are the failure of a large hydroelectric dam or the rupture of a supertanker carrying oil through a fragile ecosystem such as what happened with the Exxon Valdez on March 24, 1989.[581] However, the impacts on human health and the environment, the difficulty and cost involved in cleanup and decontamination efforts, and the long time scales of the impacts of the contamination all make the large-scale release of radiation from a nuclear plant a truly unique concern.

This chapter will focus on safety issues with light-water reactors given that they both dominate the existing fleet of nuclear plants, and are also the ones most likely to be built over the coming decades if nuclear power were to be expanded as part of an effort to reduce carbon emissions. It is important to note that, despite their dominance in the market, the initial choice to pursue light-water reactor designs was not based upon a detailed consideration of their safety compared to other designs. As summarized by Alvin Weinberg, the director of Oak Ridge National Laboratory from 1955 to 1973,

> Nowadays I often hear arguments about whether the decision to concentrate on the LWR was correct. I must say that at the time I did not think it was; and 40 years later, with around 400 commercial LWRs on-line or being built throughout the world, we realize more clearly than we did then, that safety must take precedence even over economics – that no reactor system can be accepted unless it is first of all safe. However, in those earliest days we almost never compared the intrinsic safety of the LWR with the intrinsic safety of its competitors. We used to say that every reactor would as a matter of course be made safe by engineering interventions. We never systematically compared the complexity and scale of the necessary interventions for, say heavy-water, graphite, or light-water reactors.[582]

One of the non-light-water reactor designs that might become commercially viable in the medium term, and thus warrants a brief discussion, is the high-temperature gas cooled reactor. In fact, the authors of the MIT report believe that "[i]t is possible with success at every turn that HTGR deployment could make up as much as one third of the global growth scenario." Given their smaller capacity per plant, however, if gas-cooled reactors were to make up one-third of the capacity under the global growth scenario then there would have to be about two or three times as many HTGRs in operation as light-water reactors by 2050.[583]

The concept for the gas cooled reactor dates back to the earliest days of the Manhattan Project. While not widely pursued in the United States, the British have built a number of gas cooled Magnox reactors and AGRs [Advanced Gas-cooled Reactors]. The choice by the U.K. to develop graphite moderated, gas-cooled reactors was driven in part by the desire to build reactors that could produce both electricity and plutonium for the British nuclear weapons program. All fourteen of the AGRs built in Britain came in over budget and took much longer to complete than expected. As a result, the U.K. has chosen to abandon this design, and the last nuclear plant completed in Britain was a light-water reactor.[584] The two commercial HTGRs that have been operated in the United States have had a poor operating history as note in the introduction to Chapter Two. The Peach Bottom 1 reactor in Pennsylvania was shut down in November 1974 after less than seven years in operation, while the Fort St. Vrain reactor in Colorado was closed in August 1989 after achieving a lifetime capacity factor of just 14.5 percent.[585]

The current concepts for high-temperature gas-cooled reactors are graphite moderated and use helium gas instead of water to cool the fuel. While this arrangement makes it unlikely that a loss-of-coolant could lead to a meltdown of the fuel, this design is susceptible to other types of serious accidents that could potentially lead to the dispersal of radioactivity. This possibility is increased by the fact that, in order to save money, gas-cooled reactors are proposed to be built without strong secondary containment structures that could serve as an additional barrier to radionuclide release. One type of accident that could occur with a gas-cooled reactor is the ejection of a control rod. As detailed by the Union of Concerned Scientists, in the event that a control rod was rapidly ejected from the pressurized reactor core of an HTGR, the sudden rapid rise in reactivity due to its loss, combined with the leakage of the helium coolant out the hole where it used to be, could lead to "a very large release of radioactivity to the environment."[586]

In addition, the high purity graphite used as the moderator in HTGRs is flammable and prone to burn if exposed to air at high temperatures. On October 10, 1957, the graphite moderator in the air cooled Windscale reactor in Britain caught fire and burned for over 24 hours before it was finally extinguished. The accident is estimated to have released between 16,200 and 27,000 curies of radioiodine to the environment as well as a number of other gaseous fission products. On October 12, the British authorities stopped the distribution of milk from 17 farms, and by October 15 they had restricted the distribution of milk from a 200 square mile area around the reactor. All told, approximately 2 million liters (528,400 gallons) of contaminated milk had to be dumped into the sea and rivers.[587]

Finally, reactor designs that employ a heat exchanger and steam driven turbines have additional safety concerns related to the leakage of water into the core. Once in the reactor, water can cause a reactivity excursion similar to the loss of a control rod. In addition the water can chemically degrade the fuel. Finally, the presence of water in the core can also lead to the evolution of combustible gases through reactions with the graphite.[588] It is worth noting that the three largest accidental releases of radioactive iodine-131 have all been from graphite moderated reactors, although they all used different types of coolants (Chernobyl in the Ukraine, water cooled; Windscale in Britain, gas cooled; and the Sodium Reactor Experiment in California, sodium cooled).[589]

In this chapter we will begin by discussing the history of accidents that have already occurred as well as the potential consequences of a worst case accident at a commercial light-water reactor. We will then turn to the question of how to determine the likelihood of such an accident and examine the limitations of current techniques such as probabilistic risk assessments. We will close with a look at what the potential risks might be if the large expansion of nuclear plants envisioned under the global or steady-state growth scenarios were to become a reality.

Section 4.1 – The Record of Safety

The primary safety concern with the operation of light-water reactors relates to their need for constant cooling to prevent the uranium fuel rods from melting. In the complete absence of cooling water, the decay heat from irradiated fuel is capable of damaging the rods in as little as five seconds and capable of completely destroying their integrity in just 20 seconds. The conditions of a real loss of coolant accident, in which only part of the fuel is uncovered would result in a slower rate of fuel damage. However, even under these conditions, core melting can still occur within minutes.[590] Following the onset of melting, the molten fuel could begin to pool at the bottom of the reactor vessel. If enough fuel melts, the combined decay heat would become sufficient to melt through the steel shell of the reactor vessel. Once outside the reactor vessel the molten core could come into contact with the concrete floor and, rather than melting straight through as suggested by the so-called China Syndrome, the fuel would more likely react violently with the concrete releasing large amounts of gas that could cause the containment building to rupture.[591]

One of the primary defenses against uncovering the core and allowing the fuel to melt is the Emergency Core Cooling System (ECCS). In the event of an accident, this system is intended to inject high-pressure water into the reactor vessel to maintain cooling. However, important questions remain over its ability to function properly under all realistic accident conditions.[592] Some of the evolutionary reactor designs discussed in the introduction to Chapter Two make use of gravity fed emergency cooling water systems and other "passive" cooling features to try to mitigate the risks associated with loss of coolant accidents. While some of these changes will likely decrease the probability of a meltdown occurring, these reactors remain fundamentally vulnerable to this type accident.

During a meltdown, a large amount of radioactivity is released from the core in the form of gaseous and particulate fission products. Many of the gaseous elements would be noble gas isotopes which do not have significant biological activity. However, radioactive iodine-131 would also be released from the core and is a particular concern given that it concentrates in milk and can thus lead to large doses to children's thyroids. In addition, a fraction of the cesium-137 and strontium-90 in the molten fuel can also be released during some accidents. These two radionuclides are of particular concern given that they are both produced in significant quantities, they are both biologically active (cesium replaces potassium in the body, while strontium replaces calcium), and they are both fairly long half-lived (cesium-137 has a half-life of 30 years and strontium-90 has a half-life of 28 years).

In order for the radionuclides to be released into the environment, the containment structure surrounding the reactor has to be breached. This can happen is a number of different ways, with the specific vulnerabilities of the containment depending strongly on its design. There are currently five different types of containment structures in use at reactors in the United States; two types use large dry containment domes that are designed to withstand very high pressures, two related designs use smaller structures that contain pools of water intended to condense the steam escaping from the core in the event of an accident and thus reduce the peak pressure, and one type that also has a smaller containment structure but uses ice is large wire baskets instead of water to condense the escaping steam. Nearly 60 percent of the reactors in the U.S. use the stronger dry containment structure which is the most robust against a variety of accident types.[593]

The primary threat to large dry containment structures is an accident scenario known as direct containment heating (DCH). Significantly, DCH was not recognized as a problem until after the Three Mile Island accident. At TMI, the meltdown was caused by a pressure release valve which stuck open rather than a large pipe breaking. If such a meltdown was not stopped in time, the molten fuel could be sprayed out of the core by the high pressure coolant remaining in the vessel. The molten fuel would subsequently burn, resulting in a rapid heating and pressurization of the containment dome that could lead to rupture. This scenario is a particular concern given that the core damage and containment failure would occur close together in time which could lead to a very large release of fission products. While the NRC officially closed the books on this safety concern in the 1990s, important uncertainties remain over the likelihood that direct containment heating could lead to the rupture of a dry containment dome during a meltdown.

Compared to large dry domes, containment designs that rely on steam suppression in combination with weaker structures have been found to be far more vulnerable to rupture under a variety of possible accident conditions. These risks include such things as, (1) the buildup and ignition of hydrogen gas from the reaction of the zirconium fuel cladding with heated steam, (2) direct melting of the containment walls due to the heat generated by a pool of molten fuel, and (3) rupture due to a pressure spike resulting from rapid steam generation following the reactor vessel being breached. In fact, an analysis by Sandia National Laboratory found that, in the event of a loss of electrical power, the ice condenser containment structure at the Sequoyah nuclear plant had a 97 percent chance of rupturing during a core melt accident.[594]

In this section, we will examine the record of safety at both new plants as well as plants that are nearing the end of their design lifetime. We will then turn to the possible safety implications of the changes that have been made to the licensing process and regulatory oversight mechanisms that would be applied to a new generation of reactors in the United States.

Section 4.1.1 – The Problems of Youth

The accident rate at nuclear plants, like that for most complex technical systems, can be expected to follow what David Lochbaum has called the "bathtub curve."[595] Specifically, the accident rate is expected to be higher during the initial shakedown phase when the plant is new. As the equipment is tested and broken in and the operators gain experience, the failure rate is expected to fall until it reaches a relatively steady rate where it remains for a majority of the plant's operation. Eventually the equipment in the plant begins to age and wear out while the operator's accumulation of experience has the potential to lead to over confidence in the performance of the plant. During this wear out stage, the accident rate will begin to rise, and continue to grow over time until the plant is finally shut down. The average failure rate over the whole lifetime of the plant is the parameter of most interest in determining the risk, and will thus not be accurately reflected by ignoring the higher values during either the shakedown or wear-out phases.

Examining the history of accidents that have occurred at civilian and military reactors to date, we see that they are all likely to have occurred on the front end of the bathtub curve. Table 4.1 summarizes the seven major accidents that have so far resulted in damage to the fuel and the release of radioactivity from the reactor core.

Table 4.1: Summary of the seven nuclear reactor accidents that have occurred to date that have resulted in damage to the reactor and the release of radioactivity. These reactors were both research and commercial reactors and employed a number of different technologies for cooling and moderation.[596]

Plant Name and Location	Type of Reactor	Date of First Criticality	Date of Accident	Description of Accident
Chalk River (NRX) Canada	Experimental heavy-water moderated, light-water cooled reactor	July 22, 1947	December 12, 1952	accidental super-criticality and partial fuel meltdown
Windscale (Pile 1) United Kingdom	Graphite moderated, gas-cooled reactor	October 1950	October 9 - 10, 1957	graphite fire in one of the plutonium production reactors
Sodium Reactor Experiment (SRE) United States	Graphite moderated, sodium cooled reactor	April 25, 1957	July 12 - 26, 1959	fuel cladding failure and extensive damage to nearly 1/3 of the reactor's fuel
Stationary Low-Power Reactor Number One (SL-1) United States	Boiling water reactor (BWR)	August 11, 1958	January 3, 1961	accidental super-criticality resulting in a steam explosion that destroyed the reactor
Fermi I United States	Sodium cooled fast breeder reactor	August 23, 1963	October 5, 1966	partial meltdown of reactor fuel due to loss of coolant circulation
Three Mile Island (Unit 2) United States	Pressurized water reactor (PWR)	March 27, 1978	March 28, 1979	partial meltdown of reactor fuel due to loss of coolant
Chernobyl (Unit 4) Ukraine	Graphite moderated, water cooled reactor	December 20, 1983	April 26, 1986	power excursion caused by a failed safety experiment resulting in a steam explosion that destroyed the reactor

All of these accidents occurred within one to seven years of the reactor's first criticality. The five most recent accidents all occurred within about one to three years of the reactor's startup. Overall, the average length of time that the seven reactors in Table 4.1 had been operating before suffering their respective accidents was less than three and a half years.

In addition to the accidents listed in Table 4.1, there have been a number of other important accidents and near misses that have occurred which, for one reason or another, did not result in the release of radiation. One of the most important of these accidents was the cable fire that occurred at the Browns Ferry 1 boiling-water reactor on March 22, 1975. Until the meltdown at Three Mile Island four years later, this fire was considered to be the most serious accident to have occurred at a commercial U.S. power plant. While checking for leaks from the containment dome with a candle, a technician accidentally ignited the insulation on some of the control cables for the reactor. The fire spread, and quickly grew out of control. It would eventually burn for seven and half hours and destroy 1,600 electrical cables including 618 related to safety systems. Despite the fact that the reactor was at full power when the fire began and the complete destruction of all of the control systems relating to emergency cooling, the operators were able to bring the reactor under control through a series of ad hoc measures. This accident occurred just 1.6 years after the reactor came online and led to a detailed reexamination of the fire safety systems at similar plants.[597] A more recent accident was the leak of 640 kilograms (1,410 pounds) of non-radioactive sodium from the secondary cooling system at the Monju fast breeder reactor in Japan on December 8, 1995. This spill occurred just 1.7 years after the reactor's first criticality.[598]

While these reactors were of many different designs and different power levels, the history of their accidents is consistent with the expectations from the bathtub curve. To date all of these accidents have likely been populating the front end of the curve. With the aging of the U.S. commercial power fleet, these accidents should serve as a cautionary tale for what to expect in the future as discussed in the following chapter.

Section 4.1.2 – The Problems of Aging

Nuclear plants were originally designed and licensed to operate for up to 40 years. However, as of January 31, 2006, 40 reactors had received twenty year license extensions, 12 other requests were under review by the NRC, and a number of additional applications were expected to be made over the coming years.[599] Without these extensions, the operating licenses for all but 20 plants would have expired by 2025.[600] Now, however, the Energy Information Administration now predicts that there will likely be no nuclear power plants retired before 2025, and that the net generating capacity of the existing fleet will actually increase slightly due to ongoing power expansion programs.[601] Table 4.2 summarizes the ages of the currently operating reactors as well as their average capacity.

Table 4.2: The age and average power level of commercial light-water reactors in the United States as of the end of 2003.[602]

Years of Commercial Operation (as of December 2003)	Number of Reactors	Average Capacity (MW)
0 – 9	1	1,121
10 – 19	36	1,145
20 – 29	46	935
30 – 35	21	689

The first thing to note is that even the youngest of the U.S. plants, the Watts Bar reactor in Tennessee, has already been in operation for more than nine years. The second thing to note is that 84 percent of the installed nuclear capacity is in plants that are between 10 and 29 years old. As these higher power reactors continue to age and are eventually run beyond their design lifetime, we can expect to see an increasing rate of accidents and near accidents due to equipment failures. Already the NRC has issued more than 100 technical reports on the degradation of such things as motors, valves, cables, pipes, switches, concrete, and tanks due to the effects of aging. Although not involving the release of radiation, four workers were killed at the Surry nuclear power plant in Virginia when a pipe that had been weakened through corrosion broke, scalding them with superheated steam.[603] In 2004, five workers at a Japanese nuclear plant were killed in a similar accident.[604]

The single most important discovery to date regarding the degradation of reactor safety due to aging was the corrosion of the reactor vessel head that occurred at the Davis-Besse plant near Toledo, Ohio. Given its importance, we will provide a detailed discussion of this event as well as its underlying causes in this section. The Davis-Besse plant is an 873 MW pressurized water reactor (PWR) that was brought online in April 1977.[605] During inspections ordered by the NRC, the operator of the plant, FirstEnergy Nuclear Operating Company (FENOC), discovered in

March 2002 that boric acid leaking from inside the core had eaten a hole approximately the size of a pineapple through the carbon steel top of the reactor vessel. The only material left to contain the superheated cooling water, exerting more than 2,180 pounds per square inch of pressure inside the reactor core, was a stainless steel liner just 0.125 inches thick. If this lining had ruptured, it would have led to a potentially serious loss of coolant accident as water would have rushed out the 4 by 5 inch hole in the vessel head and potentially damaged the nearby control rod.

The corrosion was so extensive that the entire vessel head had to be replaced before the reactor could be restarted. It was more than two years (nearly 780 days) before the plant was finally returned to 100 percent power. The total direct cost of this incident was over $640 million, including $293 million for operations, repairs, and maintenance and $348 million for the purchase of replacement electricity.[606] This is more than ten and half times the cost of a typical refueling outage and nearly nineteen times longer. In fact, the total cost incurred as a result of this corrosion amounted to nearly a third of the original overnight construction cost of the entire Davis-Besse plant.[607]

Although the extent of the corrosion at Davis-Besse was not discovered until March 2002, the history behind this problem dates back much further. As far back as the late 1970s, the NRC had become aware that boric acid corrosion was a potential safety concern at pressurized water reactors. Such corrosion was subsequently discovered at the Turkey Point Unit 4 and Salem Unit 2 reactors in the late 1980s. In 1999, the NRC actually issued a violation to Davis-Besse citing them for having an inadequate program to control boric acid corrosion after the discovery that several bolts had been corroded. In addition, although not initially linked to the concerns over large-scale boric acid corrosion, the first evidence that the nozzles which penetrate the vessel head were cracking as they aged was discovered at a French pressurized water reactor in 1991.[608]

Between November 2000 and April 2001, cracked nozzles were discovered at four pressurized water reactors in the United States. In February 2001, inspectors at the Oconee Unit 3 nuclear plant near Greenville, South Carolina, found signs that cracks had caused leaks in more than one of the vessel head nozzles. Of particular concern to the NRC was the discovery of a crack spreading around the circumference of a control rod nozzle at the Oconee Unit 3. The NRC was worried that if this type of circumferential crack went all the way through the wall of a nozzle that the nozzle could break and allow the control rod to be ejected by the high pressure water inside the reactor. Such an accident would lead not only to the loss of a control rod, but also to the simultaneous loss of reactor coolant as the superheated water rushed out the resulting hole in the vessel head. Both the NRC and the nuclear industry considered the discovery of circumferential cracking to be a "very significant safety concern."[609]

In response to this threat, the NRC issued a bulletin to all operators of pressurized water reactors on August 3, 2001 which provided guidance for determining the susceptibility of their reactors to a control rod ejection accident. The NRC bulletin contained criteria for dividing the reactors into four categories. Categories One and Two were for plants that either had known cracking in their nozzles or those that had a high likelihood of having developed cracks. Categories Three and Four were for plants that were determined to be at moderate to low risk of having developed cracks in their nozzles. All plants determined to be in Categories One or Two were ordered to perform a special inspection of their vessel head nozzles by December 31, 2001 (150 days from when the NRC issued the safety bulletin). The plants in Category One were required to perform a more sophisticated and sensitive inspection than those in Category Two. Of the 69 then operating pressurized water reactors, five were determined to be in Category One, seven were in Category Two, and the remaining 57 were in either Categories Three or Four.[610] In other words, this meant that more than one in six pressurized water reactors had either already developed cracks in their vessel head nozzles or were believed to be at a high risk of having developed such cracks. It was later discovered during a routine inspection that one plant in Category Three had also developed circumferential nozzle cracks.[611]

By the beginning of November, 10 out of the 12 Category One and Two plants had either already performed the necessary inspections, or had made arrangements with the NRC to shutdown before the end of the year in order to carry them out. For the two plants that had yet to demonstrate that they would meet the December 31 deadline, the NRC had begun preparations to issue shutdown orders. These preparations were an indication of how significant the NRC considered the potential safety implications of this cracking, given that only the Commission has only once actually issued a shutdown order for an operating plant due to safety concerns. (The Peach Bottom plant was ordered shutdown in 1987). The NRC quickly reached an agreement with one of the two holdouts, D.C. Cook 2, allowing the operator to delay their inspection by 19 days to correspond with the plants scheduled refueling outage in January

2002. This compromise, left only Davis-Besse facing a potential shutdown order.[612]

The initial response of FirstEnergy to the NRC demand was to inform the regulators that it did not intend to shutdown until March 30, 2002, when the Davis-Besse plant was next scheduled to shutdown for refueling. This proposed shutdown date was almost 90 days beyond the NRC inspection deadline. As part of their response, FirstEnergy pointed out to the NRC the "adverse financial and other consequences" of having to shutdown to inspect the vessel head nozzles and then shutdown again a few months later for refueling.[613] As noted, the cost of any shutdown can be quite significant for a nuclear reactor. When the purchase of replacement power is included, the cost can run into the hundreds of thousands of dollars per day. Incurring this cost twice in such a short time was obviously a concern for FirstEnergy.

However, the six other Babcock and Wilcox reactors similar to Davis-Besse that had already been inspected all showed evidence of cracking, with at least half of those showing clear signs of circumferential cracks. Given this fact, the Director of the NRC's Office of Nuclear Reactor Regulation concluded that

> [A]dequate protection of the public health and safety cannot be assured without successful completion of the recommended inspections [and it] is unacceptable for a facility to continue operation beyond December 31, 2001, without performing the recommended inspections.[614]

These conclusions were included in a cover letter accompanying a draft shutdown order that was sent to the NRC Executive Director for Operations on November 16, 2001. This draft order had been prepared by the technical staff and approved by both the management in the Office of Nuclear Reactor Regulation and the NRC's Office of the General Counsel. The draft order was forwarded to the NRC Commissioners five days later by the Executive Director.[615]

In order to avoid being forced to shutdown the reactor by the end of December, FirstEnergy offered a number of "compensatory measures" to the NRC. The most significant of these measures was an offer to move up the date for the refueling shutdown to February 16, 2002. This was the earliest possible date that the operators could acquire the necessary replacement fuel.[616] This new shutdown date was roughly halfway between the NRC deadline and the originally scheduled refueling outage planned for Davis-Besse. However, the technical specifications for the

Davis-Besse plant do not permit any leakage of coolant from the reactor vessel. The existence of cracks that penetrated the nozzles would therefore represent a violation of this requirement, and the NRC specifications specifically require that a plant be shutdown within 36 hours of the discovery of leakage of any coolant through the containment vessel.

According to the Director of the Office of Nuclear Reactor Regulation, however, the fact that the NRC did not know for sure that there were cracks leaking at Davis-Besse precluded the agency from legally issuing the shutdown order. This excuse is belied by the fact the fact that FirstEnergy's own analysis estimated that there were between one and nine leaking nozzles.[617] In fact, the draft shutdown order that had been submitted to the NRC Executive Director for Operations had concluded that it was "highly probable that Davis-Besse Nuclear Power Station, Unit No. 1, is also currently experiencing pressure boundary leakage and is operating in violation of its technical specifications."[618] In addition, a senior attorney from the NRC's Office of the Chief Counsel stated that his office had approved the shutdown order and that the Commission had all the legal authority it needed to issue the order.[619] Despite the vote of three staff members in support of moving forward with the shutdown order, and the fact that four out of the five safety principles outlined in the NRC's risk-informed decision making guidance would not be met by postponing the inspections, the NRC decided in late November 2001 to allow Davis-Besse to delay inspections until February 16.[620]

When the inspection was finally carried out, FirstEnergy found a total of 24 cracks in five nozzles, including nine that had penetrated all the way through the nozzle wall allowing boric acid to leak onto the vessel head. During this inspection, FirstEnergy removed roughly 900 pounds of boric acid crystals and powder from the reactor vessel head. Far more troubling than the discovery of circumferential cracks, however, was the discovery that a large section of the carbon steel that comprises the reactor vessel head had been eaten away by the boric acid that had leaked from a long axial crack in the nearby control rod nozzle.[621] A preliminary report released by the Commission in May 2004 concluded that the containment vessel at Davis-Besse could have ruptured with as little as two to 13 months of additional operation. These preliminary findings were subject to a number of uncertainties surrounding the rate of leakage and the mechanisms governing the rates of crack growth and vessel corrosion.[622] However, in light of these initial estimates, it is important to recall that the original shutdown date proposed by FirstEnergy was approximately one and a half months after the date that was agreed upon.

Both the NRC's "Lessons Learned" Task Force and the GAO concluded that corrosion of the vessel head could have been prevented if the NRC and the reactor operators had acted properly. The conditions that led to the vessel head corrosion had gone undetected for at least four years, despite knowledge of the concerns posed by boric acid leakage. At least twice, the operators of the Davis-Besse plant put off careful examination of the vessel for corrosion damage despite identification of possible boric acid leaks. These choices were driven by the operator's assumption that the corrosion rates were insignificant.[623] Following these revelations, NRC Chairman Richard Meserve concluded that "[i]n short, the inspections at Davis-Besse have revealed that the head corrosion problem was a direct result of a degraded safety culture."[624]

In their investigations, the NRC uncovered a number of indications that the operators of the Davis-Besse plant had emphasized the generation of electricity over the safety of the plant.[625] In fact, FirstEnergy's own public presentation recognized as much when it stated that

> There was a focus on production, established by management, combined with taking minimum actions to meet regulatory requirements, that resulted in the acceptance of degraded conditions.[626]

The NRC's Office of Inspector General reached a similar conclusion:

> The fact that FENOC [FirstEnergy Nuclear Operating Company] sought and staff allowed Davis-Besse to operate past December 31, 2001, without performing these inspections was driven in large part by a desire to lessen the financial impact on FENOC that would result from an early shutdown.[627]

As a result of this event, ten managers and executives at FirstEnergy left the company. However, despite these changes, more than a year and a half after the initial shutdown, an internal survey conducted by FirstEnergy found that one out of every six employees still felt that management valued cost reductions and the restart schedule more highly than resolving outstanding safety and quality issues.[628]

In April 2005, the NRC proposed imposing the largest fine it its history against the operators of Davis-Besse. The $5.45 million fine was based upon the fact that the utility provided incomplete and inaccurate information to the NRC and upon the operator's decision to restart the reactor after their inspection in 2000 without fully characterizing and stopping the boric acid leaks. In the same action, the NRC also banned Andrew

Siemaszko, the engineer at Davis-Besse responsible for the inspection of the vessel head in 2000, from working in NRC-regulated activities for five years.[629] Despite these actions, the GAO concluded that

> While NRC has not yet completed all of its planned actions, we remain concerned that NRC has no plans to address three systemic weaknesses underscored by the incident at Davis-Besse. Specifically, NRC has proposed no actions to help it better (1) identify early indications of deteriorating safety conditions at plants, (2) decide whether to shut down a plant, or (3) monitor actions taken in response to incidents at plants. Both NRC and GAO had previously identified problems in NRC programs that contributed to the Davis-Besse incident, yet these problems continued to persist. Because the nation's nuclear power plants are aging, GAO recommended that NRC take more aggressive actions to mitigate the risk of serious safety problems occurring at Davis-Besse and other nuclear power plants.[630]

One of the most important lessons to be learned from the events at Davis-Besse is that, as one NRC engineer said, the corrosion issue "was not on the radar screen" prior to the dramatic discovery in March 2002.[631] Back in 1993, the nuclear industry and the NRC had concluded that the likelihood of serious vessel head corrosion going undetected was low because nozzle leaks would be detected long before significant degradation of the steel could occur. The NRC guideline for unidentified leakage was one gallon per minute, however, the leak rate from the nozzle responsible for the vessel head corrosion at Davis-Besse was estimated at just 0.025 gallons per minute (approximately one quarter of a can of soda per minute). Detecting such small leaks is a very challenging task, and the NRC has concluded that the detection systems installed at both French and Swedish plants are not sensitive enough to identify this level of coolant loss.[632]

Despite the evidence of boric acid leaks in 2000 and the known likelihood of vessel head nozzles having cracked, the operators at Davis-Besse decided to continue operating the reactor with minimal efforts at addressing these issues. Similar examples of a "normalization of deviance" can be found throughout the history of accidents in high reliability systems.[633] In September 1977, one and a half years before the meltdown at Three Mile Island, the pilot-operated relief valve (PORV) stuck open at the Davis-Besse plant. This was the same failure that would occur at TMI in March 1979 and, significantly, the operators at Davis-Besse made the same type of mistakes in responding to the failure that would later be made during the TMI accident. Luckily, however, the Davis-Besse reactor was operating at only nine percent power at the time the valve stuck open, and thus no serious consequences occurred at that time. In fact, by the time the meltdown at TMI occurred, there had been a total of nine separate instances at similar plants where the PORV stuck open. Despite these previous warnings, no actions were undertaken to ensure that the problem was adequately resolved or that the operators were adequately trained to handle the failure should it occur. The one change that was made (i.e. the addition of an indicator that the valve had received the close command) was insufficient, and ended up contributing to the severity of the accident at Three Mile Island. Had information from the previous near misses been known by the operators of TMI, the core damage might have been avoided.[634]

The so-called "normalization of deviance" has also found to have been an important contribution to the loss of both the Challenger and Columbia Space Shuttles. As summarized by the commission investigating the loss of Columbia

> The initial Shuttle design predicted neither foam debris problems [like that which led to the destruction of the shuttle Columbia in February 2003] nor poor sealing action on the Solid Rocket Booster joints [like that which led to the destruction of the shuttle Challenger in January 1986]. To experience either on a mission was a violation of design specifications.... These engineers decided to implement a temporary fix and/or accept the risk, and fly. For both the O-rings and foam, that first decision was a turning point. It established a precedent for accepting, rather than eliminating, these technical deviations.[635]

In both of these cases, the failures being experienced were not an expected part of the shuttle's operation, but with each successful launch the concern over their importance diminished. Instead of being viewed as warning signs and subsequently resolved, the failures were not considered to be safety critical and were addressed by, at best, ad hoc fixes. It was only after a catastrophic loss of the shuttle and its crew, that the true importance of these failure modes became widely recognized.

The significance of the Davis-Besse vessel head corrosion goes beyond this one plant and this one type of aging related failure. In assessing the overall safety of nuclear plants, the NRC is not able to independently verify much of the vital information it needs, and it must therefore rely heavily on its licensees to provide the necessary data. A number of NRC inspectors, both active and those no longer with the agency, told the

GAO that, as a result, "they cannot easily distinguish a safe plant from an unsafe one."[636] As summarized by the GAO in 2004

> The underlying causes of the Davis-Besse incident underscore the potential for another incident unrelated to boric acid corrosion or cracked control rod drive mechanism nozzles to occur. This potential is reinforced by the fact that both prior NRC lessons-learned task forces and we have found similar weaknesses in many of the same NRC programs that led to the Davis-Besse incident.[637]

Between 1986 and 1997, 41 nuclear reactors were placed on the NRC's "Watch List" which was designed for plants with declining safety performance that require additional oversight from the Commission. An additional 43 reactors were considered seriously, but eventually not placed on the Watch List. In other words, roughly four out of every five reactors in the U.S. has either been on or been considered for placement on the NRC Watch List. The placement of a plant on the Watch List is a fairly serious matter. For instance, the Salem plant was not placed on the list until after the NRC had required it to remain shut down due to safety problems and the Cooper plant was never placed on the list despite also being held closed by the NRC due to safety concerns.[638] Finally, the NRC consistently assessed Davis-Besse as a "good performer" throughout the years that the vessel head degradation was occurring which was part of the reason that closer oversight of the operator was not given.[639]

With the continued aging of the U.S. and world nuclear fleets, the bathtub theory indicates that more failures of this kind are likely to emerge. The likelihood of accidents occurring is increased by the failures of the NRC to ensure that all reactor operators are enforcing an adequate culture of safety.

Section 4.1.3 – The Problems of New Reactors

Several of the regulatory changes that have been made or proposed as part of efforts to revive the nuclear power industry are likely to exacerbate some of the problems highlighted in the two previous sections, and thus have a negative impact on the safety of future plants. First, the Energy Policy Act of 1992 changed the licensing rules to allow a utility to be granted a combined construction and operating license (COL) for a new plant. The new licensing process allows a reactor to be operated without further regulatory review or possibility for public participation and comment if the NRC is satisfied that it has been built according to the agreed upon design specifications. This process will likely weaken the ability of the public to effectively raise concerns over important safety issues. The history of nuclear construction in the U.S. shows that intervenors and outside experts have had a positive impact on the safety of existing plants by raising important questions that the NRC and nuclear industry had to address more thoroughly.

As a result of the controversy over these regulatory changes, the government is offering large subsidies as an incentive for the first few utilities that are willing to test the new licensing process. This federal support places additional pressure upon the NRC to approve the licenses in order to show that the taxpayer money given to the utilities was well spent. So far, a consortium led by Dominion Resources is seeking $250 million over six years in federal subsidies for the preparation of a COL to build a reactor at the site of the existing North Anna plant in Virginia. Dominion is already receiving DOE assistance to seek an early site permit for this location. NuStart Energy Development, a consortium made up of Entergy and other nuclear utilities, is seeking $400 million in subsidies over seven years to prepare a COL. No specific site has been selected, but the consortium is currently seeking two early site permits for possible sites in Mississippi and Illinois. In both of these cases, the assistance sought would cover half of the expected cost of applying for the combined license. In addition, Duke Power (currently a member of NuStart) is considering filing its own application for a COL and a consortium led by the Tennessee Valley Authority is seeking $2 million in DOE funds to study the feasibility of building a new boiling water reactor at the site of its unfinished Bellefonte plant. However, no one has yet committed to actually starting construction on any of these proposed reactor's, even if the licenses are eventually granted.[640]

More troubling than these federal subsidies, however, is the proposal put forth by President Bush in April 2005 that the government pay for the cost of any delays in nuclear plant development that might be caused by the regulatory process. Specifically, the President directed the DOE

> ... to work on changes to existing law that will reduce uncertainty in the nuclear plant licensing process, and also provide federal risk insurance that will protect those building the first four new nuclear plants against delays that are beyond their control.[641]

This proposal, which has since been enacted in the 2005 Energy Policy Act (see Section 2.1.2), will effectively punish the NRC for carrying out its responsibility to protect the public health. The NRC will be particularly hard pressed to justify delaying the licensing of one of the nuclear plant's covered by this "insurance" given that the government, which is already running a record deficit, would end up paying the utility for the delay. Unlike most subsidies which simply transfer public money to private hands, this subsidy would have a chilling effect on the regulators and further weaken the oversight provided by the NRC.

All together, the history of the nuclear industry and the NRC, the move to combined construction and operating licenses, the federal funding of the first few applications, and the creation of "federal risk insurance" all cast serious doubt about the ability of the regulatory community to adequately ensure the safety of a new generation of plants. This doubt is increased by the fact that many of the new designs that have been proposed exist only on paper and have had no practical real world experience. In looking to the future, it is important to keep in mind the record of past safety, including the many surprises that have occurred after construction, and the fact that very few serious accident scenarios were seen as likely on an engineer's drawing prior to their occurring.

Section 4.2 – The Impacts of A Catastrophic Accident

Shortly after 1:22 am on April 26, 1986, Unit Number 4 of the Chernobyl nuclear power plant was destroyed by a steam explosion that resulted from a runaway power excursion. The graphite moderator continued to burn for 10 days after the accident while the leakage of radiation from the site continued for some time after the fires were finally extinguished.[642] Officially, much of the initial emphasis on the impacts of this accident have been placed on the 31 people who were killed by acute radiation sickness within the first few months of the accident. This focus, however, misses the far more important impacts of this disaster on the people of Russia, Belarus, Ukraine and throughout Europe. When the overall health and environmental effects are included, the Chernobyl accident likely ranks along side the methyl isocyanate release at Bhopal, India as one of the two largest industrial disasters in human history.

In the first few days after the accident, an exclusion zone 30 kilometer in radius was established around the destroyed reactor. Initially, 130,000 people were evacuated from this area and all agricultural and commercial activities were suspended. High fallout areas extended well beyond this range, however, and were detected as much 100 to 300 km from the site of the accident. Two months after the explosion, a further 113 villages were evacuated outside the original exclusion zone. An estimated 220,000 people were eventually forced to relocate as a result this accident and large areas of agricultural land had to be abandoned.

The official Soviet estimate for the release of radionuclides other than the noble gases only included the amounts that were deposited within the U.S.S.R. despite the fact that fallout is known to have occurred in every country in the northern hemisphere. In addition, the official estimates for the dose received by the people within the former Soviet Union included only the contribution of external radiation and ignored the additional exposures that occurred due to the consumption of contaminated milk and other food stuffs. For example, it is estimated that, outside of Mongolia and parts of Asia where few food crops were being grown at the time of the accident, the Soviet's focus on the external dose alone underestimated the actual exposure from cesium-137 by more than two-thirds.[643] The cesium-137 contamination in the environment will remain important for decades as the radionuclide slowly decays. In addition, there was also fallout of the gamma emitter cesium-134 as a result of the accident. While this radioisotope has a shorter half life than cesium-137 (two years for cesium-134 versus 30 years for cesium-137), it contributed to internal and external doses from the accident through the mid-1990s.

A far more important implication of the fact that the Soviet estimate focused only on external radiation, however, was that it severely underestimated the dose received by children's thyroids due to their consumption of radioactive iodine-131 in cow's and goat's milk. In areas around the Chernobyl site the sale of milk was banned for more than a year after the accident affecting as many as 20 to 25 million people. Despite the ban, however, it is known that milk was still produced and consumed in some of the nearby areas. It is also known that the impact of the Chernobyl accident on children extended far beyond the boarders of the Soviet Union. For example, levels of iodine-131 contamination well above the action level of 15,000 picocuries per liter set by the U.S. Food and Drug Administration were reported in several countries including Italy (160,000 pCi/L), Romania (78,000 pCi/L), Sweden (78,000 pCi/L), Hungary (70,000 pCi/L), Poland (54,000 pCi/L), Switzerland (50,000 pCi/L), Czechoslovakia (42,000 pCi/L), Austria (41,000 pCi/L), Germany (32,000 pCi/L), and the United Kingdom (31,000 pCi/L).[644]

Already significant increases in the incidence of thyroid cancer among people who were children at the time of the accident have occurred in many areas. It is now an accepted fact that the observed increase in these cancers in areas of Belarus and Ukraine are a direct result of the fallout from Chernobyl. Given that fallout occurred throughout the Northern Hemisphere, it is likely than millions of people were exposed to enough radiation to significantly increase their risk of developing cancer, and that many of these people will die as a result. An analysis of the impacts of this accident carried out in 1988 by researchers at the Lawrence Livermore National Laboratory, the Electric Power Research Institute, and the University of California, Davis estimated that the cumulative dose from the cesium-137 released during the accident will total as much as 93 million person-rem, with 35 percent of that dose being received by people in the former Soviet Union, 62 percent by those in Europe, and the remaining 3 percent by those in other countries throughout the Northern Hemisphere. Using their model of radiation risk, this exposure translated into as many as 17,400 excess fatal cancers occurring around the world as a result of the Chernobyl disaster. They further concluded that a reasonable upper bound estimate would be that as many as 51,000 fatal cancers might occur.[645] Both higher and lower estimates for the number of deaths expected have been made, but these results are generally consistent with the estimate made by the Environmental Protection Agency and published by the Nuclear Regulatory Commission. In 1987, the EPA concluded that roughly 14,000 fatal cancers in Europe and the former Soviet Union could be expected to occur as a result of the accident.[646] The geographical distribution of the expected cancers, however, was markedly different, with the government's analysis concluding that the majority of these deaths would occur within the former Soviet Union while the later study found that the majority of the deaths would likely occur in European countries.

A 2005 study of the effects of Chernobyl conducted by the United Nations estimated that 4,000 people "among the higher-exposed Chernobyl populations, i.e., emergency workers from 1986-1987, evacuees and residents of the most contaminated areas" will eventually die from cancer due to the accident.[647] This estimate however, ignores the far larger number of people that were exposed to lower levels of fallout from the Chernobyl accident. Our best understanding of radiation health effects is that every unit of exposure, no matter how small, leads to a proportional increase in the risk of developing cancer.[648] Thus, the actual number of cancer fatalities that will occur throughout Europe and the former Soviet Union as a result of the Chernobyl disaster is almost certain to be far higher than the estimate of 4,000 arrived at in the U.N. study.

By any estimate, however, the number of fatal cancers that are expected to occur as a result of fallout from Chernobyl is a very small percentage of the number that would be expected to occur over this same time from all other causes and would thus it would not be detectable by epidemiological means. In addition, the detection of the health impacts in areas surrounding the reactor site is made particularly difficult given the breakdown in medical services, which has been particularly acute since the collapse of the Soviet Union. This breakdown has increased the number of deaths from other causes and makes it more likely that victims of Chernobyl may not be recognized as such. However, just because the impact may not be statistically detectable against a larger background of other cancers, doesn't mean that it is not there. The deaths of a few hundred thousand people from cancers unrelated to Chernobyl does not change the fact that thousands of people can be expected to die from cancers that are a result of Chernobyl.

While light-water reactors are not susceptible to the kind of accident that occurred at Chernobyl, they have their own safety concerns as noted in the previous section. In addition, the violence of the explosion at Chernobyl made its impacts unique from those of a meltdown or cooling pool fire. By injecting a portion of the radionuclides into the upper atmosphere, the fallout was very widely distributed resulting in smaller areas around the plant receiving the extremely intense levels of contamination that could occur with an accident at a light-water reactor. An additional consideration is that the population surrounding some light-water reactors may be larger for some locations than the population surrounding Chernobyl. Within 30 kilometers of the Chernobyl site, there were approximately 130,000 people, while the largest city within that range had a population of 45,000. In the United States there are a number of reactors with far larger surrounding populations. For example, the cities of New York, New York (population of 8.01 million in 2000), Philadelphia, Pennsylvania (population 1.52 million), Charlotte, North Carolina (population 541,000), New Orleans, Louisiana (population 484,700), Omaha, Nebraska (population 390,000), Minneapolis, Minnesota (population 383,000), Miami, Florida (population 362,000), and Toledo, Ohio (population 314,000) are each within approximately 30 to 40 km of operating nuclear reactors. The addition of the populations surrounding these city centers would add to the number of people within the area at risk.[649]

In this section will examine the human, environmental, and financial impacts that could result from a worst case nuclear accident at a light-water reactor in the United States. We will also touch upon the potential impacts that such an accident could have on the energy system as a whole due to the loss of public confidence in nuclear power that would be likely to accompany another catastrophic accident.

Section 4.2.1 – Human Consequences of an Accident

Quantitative estimates for the impact of a worst case nuclear accident are very complicated and subject to numerous uncertainties due to the large number of variables involved and the sensitivity of the results to such things as the fraction of radionuclides that would be released from the core, the timing of the failure of the containment structure, the meteorological conditions prevailing throughout the accident, and the precise population distribution in the surrounding areas and what their level of shielding is. Despite the changes that have occurred in the population, the last major government investigation of the impacts from worst case nuclear accidents was completed more than twenty years ago.[650] Following efforts by the Union of Concerned Scientists, the results from this 1981 Sandia National Laboratory report were eventually released to the public.[651] Tables 4.3 and 4.4 show the ten nuclear plants with the largest number of estimated peak early fatalities and peak number of latent cancer fatalities that could result from a worst case accident (i.e. a maximum "credible" accident occurring during the worst possible meteorological conditions).

Table 4.3: Peak Early Fatalities Resulting from a Nuclear Reactor Accident as Estimated by Sandia National Laboratory in the 1981 CRAC-2 Study.[652]

Reactor	Nearest Major Population Center	Early Fatalities[a]
Salem 1 & 2	Wilmington, DE	100,000
Waterford 3	New Orleans, LA	96,000
Limerick 1 & 2	Philadelphia, PA	74,000
Peach Bottom 2 & 3	Lancaster, PA	72,000
Susquehanna 1 & 2	Berwick, PA	67,000
Indian Point 2 & 3	New York, NY	50,000/46,000
Catawba 1 & 2	Rock Hill, NC	42,000
Three Mile Island 1	Harrisburg, PA	42,000
Dresden 2 & 3	Morris, IL	42,000
Surry 1 & 2	Newport News, VA	31,000

(a) An early fatality is defined as a death resulting from radiation exposure that occurs within the first nine years.

Table 4.4: Peak Number of Cancer Deaths Resulting from a Nuclear Reactor Accident as Estimated by Sandia National Laboratory the 1981 CRAC-2 Study[653]

Reactor	Nearest Major Population Center	Cancer Deaths
Salem 1 & 2	Wilmington, DE	40,000
Millstone 3 & 2	New London, CT	38,000/33,000
Peach Bottom 2 & 3	Lancaster, PA	37,000
Limerick 1 & 2	Philadelphia, PA	34,000
North Anna 1 & 2	Richmond, VA	29,000
Susquehanna 1 & 2	Berwick, PA	28,000
Three Mile Island 1	Harrisburg, PA	26,000
McGuire 1 & 2	Charlotte, NC	26,000
Beaver Valley 1 & 2	McCandless, PA	24,000
Pilgrim 1	Plymouth, MA	23,000

Strikingly, the maximum total number of people that could be expected to die as a result of a worst case accident at the Salem nuclear power station would be comparable to the number of people killed in either the Hiroshima or Nagasaki bombings. This plant is not unique in the magnitude of its consequences. The total number of fatalities from a worst case accident at the Peach Bottom, Limerick, or Susquehanna plants could all begin to approach this level of destruction. Given that the U.S. population has been growing increasingly concentrated over time, the peak accident consequences for some areas could be even larger today. In addition, the scientific understanding of radiation risks has also increased significantly since the early 1980s which would also tend to increase the expected consequences of a major accident at a nuclear plant.

The latest assessment of cancer risks from radiation exposure conducted by the National Research Council was published in 2005. This report reaffirmed that there is no safe level of radiation exposure, and that every increment of exposure leads to a proportional increase in the risk of cancer.[654] While the fatal cancer risks have remained virtually unchanged due to improvements in available treatment, the estimated cancer incidence per unit of exposure recommended by the BEIR Committee is approximately 35 percent higher than the value recommended by the EPA in 1999. The BEIR Committee also reaffirmed the fact that women and children have a higher risk for developing cancer compared to adult males.[655] In addition, the BEIR Committee also concluded that it is likely that other health effects besides cancer are associated with high levels of radiation exposure. In particular they noted that

> Health endpoints other than cancer have been linked with radiation exposure in the LSS cohort [the Life Span Study of Hiroshima and Nagasaki survivors]. Of particular note, a dose-response relationship with mortality from non-neoplastic disease mortality was demonstrated in 1992 and in subsequent analyses in 1999 and 2003 have strengthened the evidence of this association. Statistically significant associations were seen for the categories of heart disease, stroke, and diseases of the digestive, respiratory and hematopoietic [blood cell forming] systems.[656]

The epidemiological data available is not yet sufficient to determine if these non-cancer health effects will occur at all doses similar to the risks for cancer or whether they would only appear at relatively high doses (greater than 50 rem). However, there would likely be sufficient exposures to some people following a worst case accident at a nuclear plant to require the inclusion of these additional health impacts in future impact assessments. For exposures on the order of 100 rem, the BEIR Committee estimated that the life-time risk from non-cancer diseases would be "similar to those for solid cancer for those exposed as adults, and about half those for solid cancer for those exposed as children."[657] Finally, adding to concerns over the long-term health impacts of radiation exposures, the BEIR Committee concluded that, while not yet proven conclusively in human studies, "there are extensive data on radiation-induced transmissible mutations in mice and other organisms" and that "[t]here is therefore no reason to believe that humans would be immune to this sort of [inheritable genetic] harm."[658]

Additional health impacts could occur in the event of a worst case accident due to the dislocation of large numbers of people from the areas that would be heavily contaminated with longer-lived radionuclides such cesium-137 (half-life 30 years) and strontium-90 (half-life 28 years). Even in areas where these contaminants could eventually by cleaned up to an acceptably low level, the decontamination efforts would require a significant amount of time to complete. This would, therefore, still necessitate the evacuation of the local populations during that time as well as the potential destruction of locally produced food if it was found to be contaminated at dangerous levels. Providing adequate shelter and care, including mental health services, would be important to avoid adding to the long-term impacts of the accidents. A study of the exposed populations at Chernobyl found a high incidence of distress and behavioral disorders six years after the accident. A later study of evacuees, conducted 11 years after the accident, found that women who were pregnant or had young children at the time of Chernobyl "experienced substantially poorer health," including a higher incidence of depression, post-traumatic stress disorder, and other psychological problems, than control populations.[659] The problems associated with displaced populations would be particularly severe in densely populated countries like India and China which are assumed to make up a large part of the expansion of nuclear power under the proposed growth scenarios. However, the aftermath of Hurricane Katrina, which struck the Gulf Coast in 2005, revealed the difficulties that can be encountered with caring for refugee populations even in the United States.

The likelihood of such a serious accident occurring at any particular nuclear plant is very small given the safety features built into the design of nuclear plants and the rare occurrence of the worst case meteorological conditions. Under typical weather patterns, the impacts from a serious reactor accident would be much smaller than the peak values discussed above, however they would remain quite severe. In addition, it is important to keep in mind the lesson learned from the release of toxic methyl isocyanate gas at the Union Carbide plant in Bhopal, India, that killed nearly 4,000 people and injured 200,000; namely that sometimes a worst case accident can, in fact, occur.[660] With the large number of reactors envisioned under the global or steady-state growth scenarios, the probability of an accident occurring will grow proportionately. In addition, a successful large scale terrorist attack on the scale of those carried out on September 11, 2001, could lead to a major accident with a subsequent release of radiation. While the likelihood of this kind of attack occurring is also small, more reactors mean more targets, and we should not forget that the probability of the towers of the World Trade Center collapsing due to the impact of civilian aircraft was also considered to be small before they fell. Already at least once since September 11 the Federal Aviation Administration has issued an order temporarily banning all general aviation flying within 10 nautical miles (11.5 miles) of 86 nuclear power and nuclear weapons production sites due to the threat of terrorist actions.[661]

Finally, it is important to note that even if an accident or terrorist attack damages the reactor core, but does not result in the release of large quantities of radioactivity, there would still be important social and psychological impacts. For example, the Presidential Commission set up to investigate the accident at Three Mile Island, noted the negative impacts on the surrounding communities that resulted from the "severe mental stress" experienced during the accident. This stress was particularly high for those living within about five miles of the facility, as well as for those people living further away who had young children at the time.[662] Follow-on studies have found "long-term behavioral disturbances in mothers of young children" who were living in the area during the accident.[663] As noted in Section 4.2.4, the possibility that public opinion could turn sharply against the widespread use of nuclear power in the event of an accident is a serious vulnerability with plans that envision a heavy reliance on this energy source.

Section 4.2.2 – Economic Consequences of an Accident

In addition to the potential loss of life and the social disruption caused by the displacement of entire communities, the direct and indirect financial costs arising from a catastrophic nuclear accident could also be quite significant. The economic consequences of nuclear accidents fall into two categories. The first category includes the costs to the utility from the loss of the reactor and potentially of other co-located facilities while the second category covers the costs to offsite populations from the health effects of radiation releases as well as the costs associated with the contamination of buildings, land, food crops, water resources, and other types of property.

Costs under the first category will be incurred regardless of whether or not there are dangerous offsite releases of radiation to the environment. The major expense of such an accident to the utility will be the loss of the reactor itself, which represents a large capital investment, as well as the cost of decontamination efforts at the plant. As summarized by Peter Bradford, a former commissioner of the NRC,

> The abiding lesson that Three Mile Island taught Wall Street was that a group of N.R.C.-licensed reactor operators, as good as any others, could turn a $2 billion asset into a $1 billion cleanup job in about 90 minutes.[664]

Added to the financial losses accompanying the destruction of a reactor are the costs of purchasing replacement power to make up for the loss of generating capacity. The NRC estimates that a routine shutdown typically costs a utility between $249,000 and $310,000 per day including the purchase of replacement electricity.[665] Events from the last few years, however, have demonstrated that these numbers may underestimate the cost of unexpected or extended shutdowns. When the steam generator tubes ruptured at Indian Point 2 in February 2000, spilling radioactive cooling water, the plant's operator announced that customers would have to pay approximately $600,000 per day in order to offset the cost to the utility of purchasing replacement power.[666] It was not until December 2000, roughly 10 months after the accident, that Indian Point 2 was finally brought back online. Another example is that during the two years that the Davis-Besse plant was shutdown following the discovery that leaking boric acid had eaten through the carbon steal vessel head (see Section 4.1.2), the cost to the utility of replacement power alone averaged more than $446,000 per day.[667]

The potential financial consequences to the utilities are increased by the construction of multiple reactors at a single site. During construction and licensing, the co-location of facilities reduces costs and simplifies the development process for the utilities and reactor vendors. However, in the event of an accident that does breach containment and release radiation to the environment, it is possible that other facilities on the site may become highly contaminated and have to be shut down for extended periods during cleanup. This decontamination would be expensive and the costs of purchasing replacement power would also be higher in the event that two or three co-located reactors were affected simultaneously.

While the costs to utilities from an accident can be quite significant and run as much as a few billion dollars, the financial impact to offsite populations could potentially be far larger. In addition to determining the health impacts from a worst case accident, the 1981 analysis conducted by Sandia National Laboratory also evaluated the potential financial costs. While it is very difficult to accurately place a dollar figure on the human and environmental impacts that would follow a large release of radioactivity, Table 4.5 summarizes Sandia's estimates for the largest costs that could have been incurred for an accident occurring in the early 1980s.

Table 4.5: Top 20 Peak Costs of a Worst-Case Nuclear Reactor Accident as estimated by Sandia National Laboratory in the 1981 CRAC-2 Study. (escalated to year 2000 dollars)[668]

Reactor	Cost of Accident in Billions of Dollars[a]
Indian Point 2	656.9
Indian Point 3	573.2
Limerick 1	445.6
Limerick 2	412.1
San Onofre 2	389.1
San Onofre 3	380.8
Millstone 3	364.0
Seabrook 1	343.1
Diablo Canyon 1	330.5
Diablo Canyon 2	324.3
Salem 2	313.8
Susquehanna 1	299.2
Susquehanna 2	286.6
Fermi 2	284.5
Millstone 2	282.4
Salem 1	282.4
Nine Mile Point 2	280.3
Waterford 3	274.1
Braidwood 1	265.7
Braidwood 2	255.2

(a) The costs of the accident include "estimates of lost wages, relocation expenses, decontamination costs, lost property and the cost of interdiction for property and farmland."[669]

The first thing to note from the estimates in Table 4.5 is that a worst case accident at any of these plants could result in financial losses more than 20 times greater than the Price-Anderson Act's liability limit of $10.9 billion as set forth in the 2005 Energy Policy Act.[670] Luckily, as with the health impacts, the financial consequences of an accident that occurred during typical meteorological conditions would be greatly reduced. For example, the Sandia report estimated that the average cost of a catastrophic accident at Indian Point 3 would be $24 billion (in 2000 dollars) compared to a peak cost of $573 billion (in 2000 dollars).[671] Even this lower amount, however, would still be more than twice the liability limit set by the Price-Anderson Act.

As noted in the discussion of health impacts, there are a number of factors that would tend to increase the Sandia estimates made in the early 1980s. These include the changes in population size and distribution, the additional reactors that have come online since 1981, and the increase in land and property values in many cities. In fact, just adjusting the older Sandia estimate for the cost of a worst case accident at Indian Point 3 to put it in 2004 dollars, without any other adjustments, would alone add more than $48 billion to the cost. There are additional cost uncertainties introduced by the way in which the "value" of a human life is determined in these models. Significant work on such questions has been done in connection with the September 11 compensation programs and would need to be taken into account in future assessments. Finally, the estimated decontamination costs would also likely increase if the cleanup standards were set so as to protect women and children, who are known to be at higher risk from radiation compared to men.

Secondly, we can use the estimates from Table 4.5 to compare the financial impacts of a single, worst case accident to the total projected capital cost of building all 300 nuclear plants envisioned under the MIT global growth scenario for construction in the United States. The cost that would have to be absorbed by society in the event of such an accident (i.e. the cost after taking into account the $10.9 billion in private insurance payments that would be made under the current version of the Price-Anderson Act), would be between 40 and nearly 110 percent of the total cost to build all 300 reactors ($600 billion under the MIT reference case economic model). While the likelihood of such an accident occurring is quite small, the scale of its potential costs to society are enormous. Nuclear power should be made to internalize the risks associated with these potential costs by removing the Price-Anderson Act liability limits and allowing the generation cost of nuclear power to reflect more accurately the economic risk posed by its potential for catastrophic accidents.

Finally, there would be additional financial costs resulting from an accident due to the psychological impacts on the public. The potential for a severe accident or successful large scale terrorist attack, particularly one that caused thousands of deaths and billions of dollars in financial losses, to spark widespread public pressure for a rapid phase-out of nuclear power could be highly disruptive to the energy system and the economy as a whole. This disruption would be particularly expensive if the choice had already been made to make the world increasingly dependent upon nuclear power. The cost of such a disruption and of the ad hoc strategies that might have to be applied, particularly if there were simultaneous efforts to continue deep reductions in carbon emissions, is unknown, but it could potentially be more significant than the direct consequences of the accident.

Section 4.2.3 – The Risks from the Nuclear Fuel Cycle

While most of the focus on nuclear safety relates to the possibility of a meltdown in the reactor, there are other concerns relating to rest of the nuclear fuel cycle as well. At the front end, there have been a number of accidents at military and civilian uranium conversion or enrichment plants.[672] However, the risks from these kinds of accidents are typically no more severe or unique than the risks posed by large oil refineries, the shipment of petroleum in supertankers, the off-loading and regasification of liquefied natural gas, or a number of other elements of the fossil fuel energy system. The risks from accidents during the storage or reprocessing of spent fuel, however, are unique in that they are potentially as severe as those associated with some types of reactor accidents and can lead to the same kind of long-term contamination of vast areas.

When discharged from the reactor, the decay heat from the spent fuel is so high that it must be cooled under water for at least five years before it can be moved to other types of storage. At a typical U.S. reactor, the spent fuel pool are between 30 and 60 feet long, between 20 and 40 feet wide, and approximately 40 feet deep. The pools have steel-lined concrete walls that are four to six feet thick, while the fuel is stored under at least 20 feet of water to ensure adequate cooling as well as to provide shielding from the intense radiation. At Pressurized Water Reactors the cooling pools are located in independent structures on or partially below the ground. At most Boiling Water Reactors, on the other hand, the cooling pools are located near the reactors themselves, and are elevated off the ground making them significantly more vulnerable to terrorist attack.[673]

The cooling pools were originally designed to serve only as temporary storage, and it was believed that the waste would soon be removed for either reprocessing or disposal in a geologic repository. The high cost, coupled with the opposition of both the Ford and Carter administrations on proliferation grounds led to the end of commercial reprocessing in the United States. In addition, the repository program is far behind schedule and the possible date of its completion remains highly uncertain due to a number of technical and regulatory concerns (see Section 5.2). As a result, the spent fuel has continued to build up in these onsite reactor pools. The utilities have chosen to repack the waste more densely in order to increase the pool's capacity. As of the end of 2002, nearly 90 percent of the spent fuel that had been discharged from the reactors in the U.S. was still being stored in cooling pools. The remaining waste was mostly stored in air-cooled dry casks at reactors whose cooling pools have completely filled up.[674]

As a result of the utilities' decision to leave the waste in the pools as long as possible, the cooling pools at many older reactors contain several times more cesium-137 and strontium-90 than is contained in the reactor's core. If the cooling water was lost and the fuel was uncovered, the zirconium cladding could ignite and result in a fire that would release large amounts of these fission products. For example, a 1997 study conducted by Brookhaven National Laboratory for the NRC estimated that a severe cooling pool fire that released between 8 and 80 million curies of cesium-137 near a highly populated area such as New York City could cause 54,000 to 143,000 excess cancer deaths, render unusable 2,000 to 7,000 square kilometers of agricultural land, cause the displacement of 1.6 to 7.6 million people, and cause $117 to $566 billion in damage.[675] A later analysis conducted by independent experts estimated that a cooling pool fire that released between 3.5 and 35 million curies of cesium-137 could cause 50,000 to 250,000 deaths, severely contaminate 180 to 6,000 square kilometers of land, and cause $50 to $700 billion in damages.[676] Finally, a third study estimated that a fire in the fuel pool at the Millstone nuclear plant in Connecticut could lead to fallout over an area up to 75,000 square kilometers, an area more than five times the size of Connecticut itself.[677] For comparison, the estimated release of cesium-137 from the April 1986 explosion at the Chernobyl nuclear power plant was on the order of one to nearly three million curies.[678]

Given that the release of radiation from spent fuel pools requires the loss of a significant amount of cooling water, it is believed that terrorist attacks are a more likely initiating event than an accident, however, accidents remain possible. A review of cooling pool safety published in 2003 identified a number of accidents or acts of sabotage that could potentially result in the loss of water in a spent fuel pool.[679] Following the publication of this review, the National Research Council of the U.S. National Academy of Sciences was asked by Congress to review the safety of the spent fuel cooling pools. The committee concluded that, "under some conditions," an attack on a spent fuel pool would be capable of starting a propagating fire with a resulting release of "large quantities of radioactive material to the environment."[680] They concluded, however, that the magnitude of this release could not be quantified conclusively since the available models have not been validated for simulating the conditions in a pool fire. The committee also concluded that, while the release of radiation could likely be prevented if adequate intervention occurred in time, the "damage to the pool and high radiation fields could make it difficult to take some of these mitigative measures."[681] In summary, the National Research Council committee concluded "that it is not prudent to dismiss nuclear plants, including their spent fuel storage facilities, as undesirable targets for attacks by terrorists."[682]

The risks from the storage of spent fuel could be significantly reduced by moving more of the waste into dry cask storage. A typical vertical cask is 20 feet tall, 9 feet in diameter, and can weight up to 125 tons when loaded with spent fuel.[683] While it is possible that an attack against such a cask could still release radiation, the National Research Council committee concluded that the release would be "relatively small" and that "[t]hese releases are not easily dispersed in the environment."[684] These risks could be even further reduced by spacing out the casks and moving them into underground silos. This type of configuration, known as Hardened On-Site Storage, would make it even more difficult to successfully attack and rupture a cask and would serve to help contain the radionuclides in the event of such an attack (see Section 5.5).

Estimates for the cost to transfer 35,000 tons of older fuel to dry casks while leaving at least 5 years of recently discharged fuel in the cooling pools would amount to only 0.03 to 0.06 cents per kWh which is less than one percent of the expected cost of electricity generation from new nuclear plants.[685] Some amount of freshly discharged fuel would have to be left in the cooling pools, but moving the majority of the older waste to dry casks and spacing out the remaining fuel in the pools would greatly reduce both the risk of a fire occurring as well as reducing the impacts should such a fire occur.

In addition to the risk of fires in spent fuel cooling pools, the other serious safety risk associated with the nuclear fuel cycle is the high-level waste generated by reprocessing. Reprocessing plants in Britain (Sellafield) and France (La Hague) routinely discharged liquid high-level waste into the North Sea. This contamination has been detected in marine life throughout the region and has led to conflicts between Britain, France, Norway, and Ireland over the continuation of this dumping.[686] In addition to these routine discharges, the majority of the liquid high-level waste must be stored before it can be vitrified. This waste must be cooled continuously in order to prevent a serious explosion from occurring. Finally, at both the Hanford and Savannah River Site complexes a number of the carbon steel high-level waste tanks are corroding and several have already begun to leak. These leaks now pose a long-term threat to the local ground and surface water resources.[687]

By far the most severe accident to have occurred at either a civilian or military reprocessing plant was the explosion of a high-level waste tank at the Chelyabinsk-65 nuclear weapons complex near the town of Kyshtym in the Southern Ural mountains. Due to the failure of its cooling and exhaust systems, a waste tank holding 70 to 80 tons of high level waste from reprocessing exploded on September 29, 1957, with the force of between 5 and 100 tons of TNT. While the CIA knew of the explosion as early as 1959, the accident was not publicly acknowledged by any government until June 1989. An estimated 20 million curies was released in the explosion with approximately 2 million curies being deposited offsite. As many as 270,000 people are believed to have been exposed to the fallout which covered more than 15,000 square kilometers (an area larger than the state of Connecticut). Of these affected people, 10,180 eventually had to be evacuated as a result of the accident.[688] Similar explosions in the high-level waste tanks at either Hanford or the Savannah River Site are also possible if they were to lose cooling.[689]

Even less severe accidents at reprocessing plants can result in serious levels of contamination. For example, during plutonium extraction at the Siberian Chemical Combined Works, a military reprocessing plant at the Tomsk-7 facility, a processing cell exploded on April 6, 1993, releasing approximately 32 grams of plutonium and more than 877 kilograms of uranium. The explosion blew off the concrete lid of the cell damaging the facility's roof and part of one wall. The release of radiation to the environment contaminated an area more than 120 square kilometers in total size.[690]

Finally, the use of plutonium as a reactor fuel can increase the consequences of a severe reactor accidents beyond those discussed in Sections 4.2.1 and 4.2.2. This increase in risk is due to the fact that irradiated MOX fuel has a larger inventory of highly toxic long-lived transuranic elements like americium-241 compared to spent uranium fuel. For example, an analysis of MOX usage in Japan estimated that the area impacted by fallout would increase by roughly three to four times if MOX fuel was used compared to fresh uranium oxide fuel.[691]

Many of these risks associated with reprocessing were explicitly cited as concerns by the authors of the MIT study. Specifically they noted that

> We are concerned about the safety of reprocessing plants, because of large radioactive material inventories, and because the record of accidents, such as the waste tank explosion at Chelyabinsk in the FSU [Former Soviet Union], the Hanford waste tank leakages in the United States and the discharges to the environment at the Sellafield plant in the United Kingdom. Releases due to explosion or fire can be sudden and widespread. Although releases due to leakage may take place slowly, they can have serious long-term public health consequences, if they are not promptly brought under control.[692]

In addition to economic and proliferation concerns, these safety considerations supported the decision to oppose the future use of reprocessing. The safety concerns over the reprocessing of spent fuel can most easily be avoided by choosing not to reprocess the fuel. IEER shares the scientific consensus that the least worst solution for the spent fuel already in existence would be its direct disposal in a mined geologic repository.[693]

Section 4.2.4 – Safety and Public Opinion

The partial core meltdown at Three Mile Island in 1979 and the steam explosion at Chernobyl in 1986 demonstrate that an accident in any country can affect the acceptance of nuclear plants around the world. These two accidents served to heighten the public awareness of the risks inherent in nuclear power and forced the NRC to, at least temporarily, tighten its regulations and oversight. The impact of these accidents on public opinion was summarized by the authors of the MIT study as follows

> Since the accident at the Three Mile Island power plant in 1979, 60 percent of the American public has opposed and 35 percent have supported construction of new nuclear power plants, although the intensity of public opposition has lessened in recent years. Large majorities strongly oppose the location of a nuclear power plant within 25 miles of their home. In many European countries, large majorities now oppose the use of nuclear power. Recent Eurobarometer surveys show that 40 percent of Europeans feel that their country should abandon

nuclear power because it poses unacceptable risks, compared with 16 percent who feel it is "worthwhile to develop nuclear power."[694]

Austria, Belgium, Demark, Germany, Italy, the Netherlands, and Sweden have all made official commitments to prohibit or phase out nuclear power. In addition, the MIT study notes that the opposition to nuclear power is growing in other industrialized countries like Japan and Taiwan as well. Adding to this list, we note that New Zealand has declared itself to be free of both nuclear weapons and nuclear power.[695]

Despite the growing concerns over global warming, the opposition to nuclear power continues. In October 2005, the International Atomic Energy Agency, a body explicitly charged with promoting the spread of civilian nuclear technologies, released a report on public opinion in 18 countries. In their survey, the IAEA found that, overall, nearly three out of every five people interviewed opposed the construction of new nuclear plants. In only one country, South Korea, was a majority in favor of building new reactors.[696] It is expected that such opposition is likely to spread if attempts were made to expand nuclear power into the Global South.[697]

Despite the importance of accidents in capturing the public's imagination, opposition to nuclear power had already become widespread within the United States long before the Chernobyl disaster, and even before the Three Mile Island meltdown. During the late 1970s and early 80s there were numerous protests and acts of direct action that took place at nuclear power plant construction sites. In 1977 there were more than 1,400 people arrested in demonstrations against the Seabrook nuclear plant under construction in New Hampshire.[698] On one weekend in 1979 alone there were 45 protests against nuclear power in states such as Oklahoma, Arkansas, Massachusetts, Connecticut, Georgia, Illinois, Michigan, Virginia, and Colorado. These demonstrations resulted in more than 370 arrests.[699] In 1981, a total of nearly 1,600 people were arrested at protests opposing the Diablo Canyon power plant which was being built next to an active offshore fault that had been discovered during construction.[700] In addition, concerns over the ability to successfully execute an emergency evacuation led to the permanent closure of the Shoreham nuclear plant in New York before it had generated a single kWh of electricity for commercial sale. This closure occurred despite the Long Island Lighting Company having already spent approximately $5.5 billion to complete the construction of the plant.[701]

These protests are important to consider because, even if new plants can be built without significant disruption, it is unlikely that public opposition could be avoided in the wake of a serious accident or successful large scale terrorist attack on a nuclear power facility. If nuclear power is in the process of being expanded and accounts for a significant amount of capacity in terms of either absolute generation or a percentage of overall electricity usage, then public pressure to shutdown existing plants would leave open far fewer energy options (particularly in terms of greenhouse emissions). The options that would be available to achieve a rapid phase-out of nuclear power would likely come at a very high price considering both the sunken capital in the completed nuclear plants as well as the cost of ad hoc measures that would be needed to rapidly replace the off-lined baseload nuclear capacity. On the other hand, however, if long-term plans to phase out nuclear power were already being carried out when the accident or attack occurred, there would likely be far more options available and those options could be accelerated with far less serious disruptions to the overall economy.

Predicting the public's reaction to a major nuclear accident is in no way certain. In fact, the opposition to nuclear power has been changing recently. However, taking the history of public opinion as a guide, the risks of serious disruption to an energy system which is heavily reliant on nuclear power following a major accident or successful terrorist attack should not be ignored. As summarized by Dr. Russell Peterson, one of the commissioners appointed by President Carter to investigate the accident at Three Mile Island, prior to the Chernobyl disaster

> As a final comment, I wish to emphasize my conviction, strongly reinforced by this investigation, that the complexity of a nuclear power plant -- coupled with the normal shortcomings of human beings so well illustrated in the TMI accident -- will lead to a much more serious accident somewhere, sometime. The unprecedented worldwide fear and concern caused by the TMI-2 "near-miss" foretell the probable reaction to an accident where a major release of radioactivity occurs over a wide area. It appears essential to provide humanity with alternate choices of energy supply.[702]

Section 4.3 – Probabilistic Risk Assessments

That the impacts from a major accident involving a commercial reactor, a spent fuel pool, or a high-level waste tank could be extremely severe if they occurred near a major population center is no longer in debate. While the details vary, and there remain some important areas of disagreement, the results of analyses from Sandia National Laboratory and many previous and subsequent works have consistently demonstrated the high cost that could accompany the most serious accidents at these kinds of nuclear facilities. Unlike the question of consequences, however, the likelihood that such accidents might occur remains a highly contentious issue. This is a particularly important area of debate since what is most relevant is not just the consequences of a nuclear accident, but what the risks of nuclear power are. The concept of risk takes into account both the consequences of an accident as well as the probability that it will occur. This section will focus primarily on the technical issues surrounding the efforts that have been made to quantify the probability of a major accident, and by extension the risk.

Once the estimate of the probability is made, however, an even more contentious question arises; namely what level of risk should be considered acceptable. A particularly complicated element of this question is how to properly compare the risks between different types of activities. For example, it is difficult to adequately take into account the unique features of low-probability, high consequence events that affect many people all at once (such as an airplane crash) and compare them to the impacts of more routine events with similar, but highly diffuse effects, (such as automobile accidents). For instance, focusing only on the risk as given by the probability of an event multiplied by its consequences would lead to the conclusion that a voluntary activity that killed one person each week somewhere in the U.S. was the same as an event that suddenly killed 5,200 people in one area, but occurred only once every one hundred years. The risk associated with these two activities, however, would likely be viewed quite differently by society. As an example, the number of people killed during the September 11, 2001 terrorist attacks (3,028) accounted for just 15 percent of the U.S. homicide victims in that year while the number of non-terrorism related homicides in the U.S. actually grew from 16,765 in 2000 to 17,638 in 2002.[703] However, the public's view on the relative risk posed by terrorism and by general violent crime is not always consistent with these raw numbers.

Related to the question of what level of risk is acceptable is the question of what level of uncertainty is acceptable. This is a particularly important issue for low-probability, high consequence events because the risk is extremely sensitive to the underlying assumptions that are made in the analysis. This sensitivity can lead to a significant level of uncertainty in the final results. Finally, it is clear that the resolution of some of these issues is only partly a technical question, and one that instead deals more directly with issues of societal values as well as individual psychology. It is only through an open, honest, and active debate in which all relevant information is available that the question of "acceptable risk" can be justly answered.

In trying to determine the likelihood of an accident occurring, the simplest way is to look at the operating history of the industry. While the true failure rate of high-reliability systems like nuclear plants will not be revealed until a very large number of operating hours have been logged and representative plants have passed through both the shakedown and wear-out phases, past experience can serve to place some useful bounds on the actual accident probabilities. While Table 4.1 lists seven accidents that have so far led to the release of radioactivity from the core, only one of these accidents (the meltdown at Three Mile Island) occurred at a commercial light-water reactor in the United States.

Through 2003, the 104 licensed nuclear power reactors (including the currently mothballed Browns Ferry 1 reactor) had accumulated a total of 2,360 years of operating experience. An additional 385 years of operating experience had been accumulated by reactors that are now permanently shutdown and undergoing decommissioning. While all of these reactors are similar in fundamental respects, it is important to note that they are not all identical and that there are a number of unique safety concerns among them. Among the plants that have been built in the United States, there have been two main types of reactor concepts (PWR and BWR), four different reactor vendors (General Electric, Westinghouse, Babcock & Wilcox, and Combustion Engineering), and 80 different detailed reactor designs.[704] In the 2,745 total reactor-years of experience accumulated at these commercial plants there has been one loss of coolant accident that resulted in a partial core meltdown as well as a number of near misses and close calls. Using past experience to estimate the underlying average accident rate is an inherently probabilistic problem. Assuming for simplicity that the failure rate remains constant over time (see Section 4.1.1), we can estimate that the probability of a TMI level accident occurring at currently operating reactors is between 1 in 8,440 to 1 in 630 per year.[705]

Aside from historical experience, the other widely used method is to determine the likelihood of an accident's occurring is an analytic technique known as Probabilistic Risk Assessment (PRA). In conducting a PRA analysis the engineers seek to identify all possible ways in which a serious accident could occur and each pathway is mapped out from initiating event to ultimate consequence. Once these so-called "fault trees" are created, probabilities are assigned to each failure along each path under the assumption that each individual failure is random and independent of other failures. The entire tree is then combined with the resulting number meant to represent the overall likelihood that an accident would occur. PRAs conducted for the currently operating reactors typically predict accident probabilities on the order of 1 in 10,000 per year which is below the lower range of what would be expected from historical experience.[706] This type of quantitative risk analysis was developed for use in high-reliability systems such as the space program and was first applied to accidents at nuclear power plants in the 1970s when the Atomic Energy Commission began the preparation of the Rasmussen Report, officially known as WASH-1400 or the *Reactor Safety Study*. Since then the technique has been greatly improved and some of its more important restrictions have been somewhat relaxed. However, the original results of the 1975 Rasmussen Report, can still be found reported in many places today, including, for example, a nuclear engineering textbook used at the Massachusetts Institute of Technology.[707]

Given its importance we will begin this section with a discussion of the Rasmussen Report and the controversies that surrounded it. We will then examine the major limitations of modern Probabilistic Risk Assessments and the impact of these limitations on the resulting uncertainty of the estimated accident frequencies.

Section 4.3.1 – The Rasmussen Report and the History of the PRA Methodology

The earliest official study of the potential impacts from a nuclear power plant accident was conducted at Brookhaven National Laboratory for the Atomic Energy Commission. The report, known as WASH-740, was completed in March 1957, the same year that the Shippingport reactor came online. The WASH-740 report considered the impacts of an accident at a small (100 to 200 MW) nuclear reactor located 30 miles upwind of a major U.S. city. The study concluded that up to 3,400 people could die and as many as 43,000 could be injured in the event of a catastrophic accident at such a plant. The report also concluded that between 18 and 150,000 square miles could be affected by the radioactive fallout, and that the resulting property damage could be as high as $7 billion ($43 billion in 2000 dollars). These results were used by the AEC to argue for the passage of the Price-Anderson Act later that year. As initially passed, the Act limited the private liability from a nuclear plant accident to just $560 million, or less than one-twelfth the worst case property damage calculated by WASH-740.[708]

Significantly, while the consequences of a catastrophic accident could be roughly estimated, there was little that could be said quantitatively about the likelihood of such an accident given the sate of knowledge at the time. As the size of the nuclear plants under construction expanded rapidly, the Atomic Energy Commission sought to update the results of the WASH-740 report to keep pace with the changing technology. It was believed by the nuclear industry and the AEC at the time that, despite the larger size of the newer reactor designs, that the improvements that had been made in the plant's safety features and in the general knowledge of accident conditions would combine to reduce the consequences of any potential accidents making the lack of a quantitative probability estimate less significant.

The analysis for the update of WASH-740 was carried out between 1964 and 1965 at Brookhaven National Laboratory. This work found that, despite the safety improvements that had been made, the results of the 1957 analysis had to be considered to be a realistic estimate of the possible outcome for a serious accident, and not as a conservative worst case estimate as then claimed by the AEC. Illustrative calculations found that an accident at one of the new, larger plants might cause up to 45,000 fatalities with accompanying property damage many times higher than the WASH-740 estimate. Internally, the staff of the AEC expressed concern that, should these results become public, they might make it difficult for the Commission to justify the continued granting of licenses for new nuclear plants. Partly due to these concerns, the results of the WASH-740 update were never put into a final form, and the study was withheld from the public until 1973 when a Freedom of Information Act request filed by the Union of Concerned Scientists forced its release.[709]

With the continued estimation of such severe potential consequences from a major nuclear accident, the AEC turned its attention more fully to determining the probability that such an accident would occur. This effort was initiated in 1972 and became the study known as WASH-1400 or the *Reactor Safety Study*. The work was funded by the AEC and was conducted at the AEC's offices in Germantown, Maryland. The final version of the report, also known as the Rasmussen Report, after Professor Norman Rasmussen who led the study, was issued by the Nuclear Regulatory Commission in October 1975.[710] The methodology chosen was Probabilistic Risk Assessment, which requires a great deal of knowledge about both the possible failures that can lead to a particular accident as well as the likelihood of each of these individual failures. Given the immense complexity of this task, the authors of the Rasmussen Report choose to focus their efforts on just one pressurized water reactor and one boiling water reactor. These results were then extrapolated to all 98 operating reactors without any effort to examine the uncertainties introduced by this approach.[711] With this extrapolation made, the executive summary presented the now infamous figures purporting to show that the risks from living near a nuclear plant were less than being struck by a meteorite.

The flawed logic behind the kind of extrapolation used in the Rasmussen Report was highlighted less than five years after the study was published when the reactor at Three Mile Island suffered a partial core meltdown in March 1979. The particular accident sequence that occurred at TMI had been identified by the Rasmussen Report, however, it was predicted to have a probability of occurring of just once every 100,000 years. This estimate was based upon their analysis of the Surry nuclear plant which was a Westinghouse pressurized water reactor. Following the accident, however, it was realized that if the PRA methodology had been applied to a Babcock and Wilcox reactor like the one at Three Mile Island, than this type of accident would have appeared far more probable than at a reactor like Surry.[712]

Beyond the limitation of looking at only two reactor designs, the Rasmussen Report was widely criticized in both its draft and final forms on a wide range of technical issues. The most important of these independent reviews concluded that the Rasmussen Report was likely underestimating the probability of serious accidents as well as underestimating the consequences of those accidents.[713] In response to the intense criticism of this report, the NRC appointed an expert panel in July 1977 to review the findings of the Rasmussen Report. This committee, known as the Lewis Commission, concluded that, while it was not possible for them to determine if the results of the Rasmussen Report were over or understatements of the risk given the limited scope of the their review, the panel was positive that the uncertainty in the results were understated for a variety of reasons.[714] For example, the Lewis Commission concluded that

> The statistical analysis of in WASH-1400 leaves much to be desired. It suffers from a spectrum of problems, ranging from

> lack of data on which to base input distributions to the invention and use of wrong statistical methods. Even when the analysis is done correctly, it is often presented in so murky a way as to be very hard to decipher.
>
> For a report of this magnitude, confidence in the correctness of the results can only come from a systematic and deep peer review process. The peer review process of WASH-1400 was defective in many ways and the review was inadequate.[715]

In addition, the Lewis Commission concluded "that the Executive Summary is a poor description of the contents of the [Rasmussen] report, and should not be portrayed as such" and that the way in which the summary was written "has lent itself to misuse in the discussion of reactor risks."[716]

In response to the findings of the Lewis Commission, the Nuclear Regulatory Commission officially withdrew "any explicit or implicit endorsement of the Executive Summary" of the Rasmussen Report.[717] Further, the NRC concluded that

> The Commission accepts the Review Group Report's conclusion that absolute values of the risks presented by WASH-1400 should not be used uncritically either in the regulatory process or for public policy purposes and has taken and will continue to take steps to assure that any such use in the past will be corrected as appropriate. *In particular, in light of the Review Group conclusions on accident probabilities, the Commission does not regard as reliable the Reactor Safety Study's numerical estimate of the overall risk of reactor accidents.*[718]

Since this time, the techniques of Probabilistic Risk Assessments have been greatly improved and a number of plant specific analyses have been carried out. However, as the coming sections will show, there remain many important uncertainties with these newer estimates, and that many of the concerns revealed in the reviews of the Rasmussen Report remain important limitations of the methodology today.

Section 4.3.2 – Issues of General Completeness

The most fundamental uncertainty in the results of Probabilistic Risk Assessments is whether or not all important accident scenarios have been identified and properly included in the fault-trees. The question of determining completeness is a particularly difficult one because it requires one to know what one does not know; namely, are there important pathways for an accident to occur that have not been foreseen and included in the engineering model being used. The Lewis Commission recognized this as an "inherent limitation" of the methodology that will always give rise to some level of uncertainty in the results.[719]

Interestingly, the report of the Lewis Commission inadvertently provided an excellent example of the kinds of uncertainties that can be introduced by a lack of completeness. In their discussion of the Rasmussen Report, the panel noted that for "sufficiently simple systems," no experimental data is required to accept a theoretical probability calculation, and that these results can be considered to be "entirely reliable." As an example of such a system, they noted that one can conclude that a symmetrical coin will have a 50 percent chance of landing heads and a 50 percent chance of landing tails without ever having to flip such a coin.[720] At first glance this seems to be a reasonable conclusion, however, even in this simplest and most classic example of probability theory, the authors failed to consider the complete set of possible outcomes. Specifically, they ignored the fact that real coins can land on their edge as well as on either face. For some types of coins, this probability can be non-trivial. For example, one experiment found that the probability of a one pound British coin landing on its edge and remaining upright was as much as 3 chances in 500 while the probability of a U.S. nickel landing on edge was predicted to be approximately 1 chance in 6,000.[721] If the low probability event of the coin landing on edge was accompanied by a set of very serious consequences we would find that even for such a seemingly simple system, probabilistic assessments based on theoretical assumptions or partial experimental knowledge could miss a potentially important contribution to the overall risk while seeming for all intents and purposes to be both fully complete and wholly accurate.

The extrapolation of data and analyses from one plant to another and the resulting uncertainties over completeness continue to affect the probabilistic risk assessments performed by the nuclear industry to date. As noted by the NRC in 1998

> Data often are not available on important initiating event frequencies and component reliability, and their specific applicability and usefulness may vary somewhat plant to plant. Thus, while a comprehensive plant-specific data analysis is within the current capabilities, it sometimes is not performed because of the lack of basic failure data for a plant, as well as the costs and resource allocations required.[722]

This lack of plant of specific information may lead to important accident scenarios being missed or probabilities being underestimated.

An additional limitation is that the PRA methodology assumes that the plant is always operating as designed. However, the GAO noted that as of the late-1990s "some utilities do not have current and accurate design information for their nuclear power plants."[723] While the number of violations has been falling since the late 1980s, there were still more than 1,400 instances in 1997 where plants failed to meet their technical specifications or regulatory requirements. Some of these violations continued for long periods and have been found to have the potential to adversely affect important safety systems. For example, in August 1998 the operators of the Big Rock Point nuclear plant informed the NRC that one of its safety systems, the Standby Liquid Control System, had been completely out of order for somewhere between 13 and 18 years.[724] These violations can lead to new types of accidents that would not necessarily be foreseen by those who assumed the plant was operating as designed. In addition, they can also render completely inoperable important safety equipment that the risk assessments are assuming will fail randomly.

A particularly striking example of a plant failing to meet its technical specifications was the extensive corrosion of the reactor vessel head discovered at the Davis-Besse plant in early 2002 (see Section 4.1.2). Boric acid leaking from within the reactor had corroded the top of the carbon steel vessel over time leaving only a 0.125 inch thick stainless steel liner to hold in the high-pressure cooling water. In 1993, the nuclear industry and the NRC had concluded that the likelihood of serious vessel head corrosion going unnoticed was low because nozzle leaks would be detected long before any degradation of the steel could occur. Subsequent experiments, however, found that this conclusion was likely to be incorrect on a number of counts. As summarized by the NRC in the wake of the discoveries at Davis-Besse

> Specifically, predictions regarding boric acid-induced corrosion rates, for in-plant boric acid leaks, have not been reliable in all cases. Operating experience reveals instances in which corrosion rates were significantly underestimated for identified boric acid leaks because of erroneous assumptions regarding the nature of the leakage, environmental conditions, the relationship between the actual leakage and experimental data, or other factors. As a consequence, in some instances, carbon steel components have been corroded to a much greater extent than anticipated. A number of these events occurred even though the underlying leakage had been previously identified

by licensees, as they deferred material wastage assessments and repairs on the basis of the assumption that the corrosion rates would be inconsequential.[725]

The Probabilistic Risk Assessment used by the NRC to allow Davis-Besse to continue operating beyond the December 31, 2001 deadline set by the NRC for conducting inspections was seriously incomplete because it did not consider the possibility of vessel head corrosion. More than two years after its discovery, the NRC had yet to produce a final analysis for the probability of a loss-of-coolant accident associated with this failure mode.[726] An additional area in which the initial Davis-Besse PRA was found to be incomplete was that the NRC considered only the case of a control rod ejection at high-power. In the case of an ejection at low-power, the control rods would be deeper inside the reactor, and the rapid removal of one rod could result in a much larger power excursion that could bring the reactor to well above its rated power before enough coolant had boiled out of the resulting hole to shutdown the chain reaction.[727]

A further example where an important aging related accident pathway may be overlooked is the fact that the so-called "maximum credible accident" stops short of considering the rupture of the reactor vessel itself. This type of failure would be particularly catastrophic given that no emergency actions of any kind would be capable of restoring cooling water to the fuel. It is recognized that there is the possibility that the intense neutron radiation experienced by the steel vessel might lead to its becoming increasingly brittle over time. If this embrittlement occurs, the vessel would be more likely to rupture in the event that cold, high pressure water was injected into the core during a loss-of-coolant accident. Despite the potential importance of this possibility, the rupture of a reactor vessel is not included in the PRAs of commercial reactors. The concern over the exclusion of this failure mode dates back to criticisms of the Rasmussen Report in the late 1970s, and was heightened by the closure of the Yankee Rowe plant in Massachusetts in 1992 due to radiation induced embrittlement of its reactor vessel.[728]

In addition to random accidents, the possibility of terrorism or intentional acts of sabotage to affect the safety of a plant adds to the uncertainty of probabilistic risk assessments. Nuclear power plants and research reactors have been the target of attacks both during construction and after commissioning. These attacks have been conducted by governments, such as Israel's bombing of the Osirak research reactor outside Baghdad in 1981, as well as by sub-national groups, such as in South Africa where the African National Congress claimed credit for four blasts that damaged the Koeberg nuclear power plant in 1982.[729] Following the September 11, 2001 terrorist attacks and the U.S. led invasions of Afghanistan and Iraq, concerns over the security of nuclear facilities have been greatly heightened.[730] For instance, in a 2001 report from the DOE's Energy Information Administration entitled *Impact of U.S. Nuclear Generation on Greenhouse Gas Emissions*, the authors pointed out that

> Any discussion of nuclear power today must acknowledge the impact of the now active war on international terrorism. Nuclear plants have specifically been enumerated among the potential targets of terrorists.[731]

While much of the current focus is on international terrorism, it is important to recall the risk from domestic actors that was demonstrated by the Oklahoma City bombing in 1995. The ability of a group of highly trained and motivated individuals to seriously damage a nuclear reactor or spent fuel storage pool with or without inside help is an important vulnerability about which we do not yet have sufficient information to allow its inclusion in a probabilistic analysis. The issue of completeness in this case is particularly acute given the inherent difficulties in trying to anticipate the types of attacks that might be possible.[732]

Finally, the omission of design and construction defects in probabilistic risk assessments adds to the concerns over their completeness. In a PRA, the accident scenarios are assumed to flow from one failure to the next. In other words, it is assumed that the system as designed and built functions properly and that it is only when equipment breaks or operators make a mistake that an accident can occur. However, in a real system, equipment may function as designed, but simply not be appropriate to the task intended, such as a pump that activated as planned, but was of insufficient power to force water to where it is needed. A recent example of such a problem came to light in October 2005 when it was discovered that the Emergency Core Cooling Systems at Palo Verde Units 2 and 3 have a design defect that may prevent their proper functioning during certain types of loss of coolant accidents. These reactors were shutdown, and were to remain offline until this issue was resolved. Palo Verde Units 2 and 3 were licensed in 1986 and 1987 respectively, and thus were each operated for nearly two decades with a design defect that could have potentially rendered inoperable their most important single emergency system.[733]

Adding further to these complications is the possibility that the equipment, as designed, might be appropriate to the task, but simply installed improperly or maintained poorly so that when activated it would not perform as expected. In the language of the PRA, these components would have a failure probability of 100 percent under the specified conditions. However, once it is known that components are inappropriately designed, fabricated, or installed, the pieces could be replaced or a workaround could be developed. Thus, as with the issue of general completeness, design problems are a question of trying to know what one does not know.[734]

The risk assessments used by the nuclear industry typically deal with these problems by assuming that there are no design or construction problems sufficient to affect safety. This omission is made despite the long and well documented history of their occurrence in nuclear power plants and other high reliability systems. As summarized by the Lewis Commission,

> The history of failures of other complex and presumably safe systems is heavy with instances of design defect and quality assurance failures. (A spectacular case in point is the collapse of the Tacoma Narrows Bridge.)[735]

These are not simply problems from long ago. In a review of 200 incidents that occurred at 10 different plants between November 1996 and January 1998, it was found that 22 percent were attributable to design errors. All of these plants were more than 10 years old when the design defects were finally discovered.[736] Between 1995 and 1997, an average of 38 safety related incidents were reported to the NRC every month as having been "caused by design, construction, installation, [or] fabrication errors."[737] These flaws went undetected during the extensive process of engineering review carried out on nuclear plant designs, in part, because reviewers all tend to follow the same type of logic in approaching a problem, and can thus be more likely to overlook the same failing in the design.[738]

As summarized by Edward Hagen, a development specialist at Oak Ridge National Laboratory and editor of the Control and Instrumentation section of the journal *Nuclear Safety*,

> Mistakes made in the past are not likely to be repeated, but in each new design other mistakes will creep in. The need for vigilance is eternal.
> ...
> No reactor system has ever failed because of a deficiency that could be seen on a designer's flow sheet or an analyst's

model. Such deficiencies have been revealed only via operating experiences.[739]

The issue of completeness is a particular concern for the claims made regarding the safety of new reactor designs that so far exist only on paper or which have only a limited amount of operating experience with full scale systems. Many important unforeseen accident scenarios and design flaws have been discovered during the nearly 3,000 reactor years of operating experience with current designs. Placing too much faith in theoretical estimates for the safety of new designs without a suitable consideration of the uncertainties could be a potentially serious mistake.

Section 4.3.3 – "Human Factors"

A second area that adds to the uncertainty of probabilistic risk assessments is the influence of so-called human error. Unlike the issues of completeness and design defects, the issue in this case is primarily focused on how to accurately model the impact of human mistakes on known failures modes. These mistakes may be initiating events that cause a failure to occur or they may exacerbate an existing failure by responding to it incorrectly and thus making the accident more severe. Such mistakes can take on very wide variety of forms from operator errors to improper maintenance or calibration of equipment to improper design of procedures to the failure to enforce appropriate safety standards. One of the main uncertainties that is introduced by these mistakes is that, as noted by researchers for the NRC, "[t]he percentage of hardware unavailability due to human error as opposed to random hardware failures is not known."[740] As summarized by Edward Hagen from Oak Ridge National Laboratory:

> When dependencies and human factors are considered, there are no analytical techniques presently available for treating them.... There is not now and never will be a "typical" or "average" human being whose performance and reactions to any operating condition, let alone an abnormal operating condition, can be cataloged, qualitatively defined, or quantitatively determined. There are no human robots.[741]

The uncertainties introduced by the ways in which operator errors are treated in the PRA methodology was recognized by the Lewis Commission to be "one of the major contributors to the general problem faced by the RSS [Reactor Safety Study] in making quantitative risk estimates."[742]

This difficulty in integrating human error into PRAs is particularly important given the contribution of these mistakes to the overall failure rate of many systems. In particular, this importance has been found in several reviews of operating events that have occurred at commercial nuclear plants in the U.S. For example, in a review of 200 safety related incidents at 10 plants that took place between November 1996 and January 1998, it was found that 35 percent of these were attributable to worker mistakes.[743] In a separate review of 35 events that occurred between 1992 and 1997, the NRC found that more than 68 percent had "human performance" as one of the significant contributory factors.[744] These findings were supported by a third NRC review. This analysis concluded that human errors contributed to 77 percent of the 48 safety related events examined. Interestingly, this last review found that in incidents where human factors were found to be a contributing factor, there was an average of four separate mistakes made, with many incidents involving between six and eight mistakes.[745]

There are a number of factors that can unpredictably affect human performance which add to the uncertainty in PRAs. The stress of accident situations when the consequences are potentially so high is certainly one such factor. However, even simple fatigue can have a dramatic impact on human reliability. The National Transportation Safety Board has found that 30 percent of all important errors in transportation accidents are attributable to fatigue.[746] There have already been incidents where fatigue has affected both commercial and research reactors as well. For 25 minutes the 4.9 megawatt MIT research reactor located in the middle of Cambridge, Massachusetts, was effectively un-staffed when one of the operators was locked out while the other had fallen asleep. In addition, the first (and only) time that the NRC has ordered an operating commercial power reactor to be shutdown due to safety concerns was on March 31, 1987, when it ordered the Peach Bottom nuclear power plant in Pennsylvania to be shutdown following the discovery of operators sleeping in the control room.[747] Finally, there is also the possibility for drug or alcohol use to affect the performance of operators and maintenance crews. This is known to be a potential problem in other high reliability systems such as air travel. For example, between the mid-1970s and 1990 approximately 1,200 U.S. airline pilots were treated for alcoholism under a special government program. Between 1984 and 1990, 61 pilots lost their licenses to fly due to their having operated a plane under the influence.[748]

Overall, human errors can have a significant impact on the quantitative results of risk assessments. For example, a February 2000 report from the Idaho National Engineering and Environmental Laboratory concluded that

> Most of the significant contributing human performance factors found in this analysis of operating events are missing from the current generation of probabilistic risk assessments (PRAs), including the individual plant examinations (IPEs).
> ...
> In nearly all cases, plant risk more than doubled as a result of the operating event – and in some cases increased by several orders of magnitude over the baseline risk presented in the PRA. This increase was due, in large part, to human performance.[749]

As summarized by the Presidential Commission appointed to investigate the Three Mile Island accident

> We are convinced that if the only problems were equipment problems, this Presidential Commission would never have been created. The equipment was sufficiently good that, except for human failures, the major accident at Three Mile Island would have been a minor incident. But, wherever we looked, we found problems with the human beings who operate the plant, with the management that runs the key organization, and with the agency that is charged with assuring the safety of nuclear power plants.[750]

Nuclear power is unlike any other energy source in that it demands an extremely high level of competence at all times from all levels of the organization -- from the regulators and managers all the way through to the technical and maintenance crews. If the human element of the system falters, then there is the possibility for a severe accident to occur. This element of uncertainty in the risk assessments will remain as important for the new plants as it is for the existing plants, since they all rely on humans as an integral element of operation.

Section 4.3.4 – Computers and Digital Control Systems

The problems of completeness and of how to incorporate the impact of human errors into the PRA methodology both date back to the time of the Rasmussen Report. Since the late 1970s a new issue has emerged that relates to both of these limitations; namely the question of how to incorporate the unique features of programmable devices into the risk assessment methodology. While this concern would be more pronounced in new plants that make more extensive use of other digital systems for performing control, protection, and monitoring functions, it also arises within the existing fleet of reactors due to the ongoing replacement of older mechanical or analog systems with digital devices.

Already a number of problems with digital systems have been reported by reactor operators. In one review of 79 license events reported to the NRC between 1990 and 1993, 38 percent were found to have been caused by software errors while an additional 32 percent were attributable to mistakes resulting from a failure of the operator-computer interface compared to just 11 percent attributable to random component failure. Similar results have been found in other countries as well. For example, a study from Canada's Atomic Energy Control Board of 459 events from 22 reactors over 13 years, found that software and operator-interface failures contributed to 54 percent of the reported problems.[751]

During normal operation, the use of digital systems has many safety advantages due to their wider range of functionality, their lower rates of component failure, and their ability to provide more detailed information about plant conditions. However, experience has also shown that the use of programmable devices can also introduce new types of failure modes and can increase the likelihood of some previously existing failures occurring. In particular, the use of programmable devices can make operator errors more likely under certain conditions and can reduce the level of redundancy provided by backup systems. For example, on March 3, 2006, an operator at the Civaux nuclear plant in France placed a notebook on his keyboard accidentally causing a group of control rods to move out of the reactor. As a result, the reactor exceeded its maximum rated power for one minute and twenty seconds before the operator recognized the problem and reinserted the control rods.[752]

The increased information made available to operators by digital systems can both improve and degrade safety under different conditions. Unlike analog control systems in which all available information was presented to the operator at all times, computerized interfaces generally filter the incoming data and are configured by the operator to display only the portion of the information desired at that time. Under normal operating conditions, this is usually an advantage in that it allows the operator access to more detailed information and can arrange that information into a more comprehensible and less cluttered format than having to try and keep track of a vast array of gauges, dials, readouts, and alarms.

However, during an accident this same feature of digital systems can become a drawback. The greater complexity of the computer interface has the ability to confuse operators and to make it more difficult to see vital information or to complete the needed tasks in the event of an accident. When faced with having to act quickly during the early stages of an accident, an operator must either choose to go ahead with whatever information is available on their screens at the time even though it may not be the best information available for diagnosing the problem or they can choose to spend time reconfiguring their workstation to get access to better information at the expense of delaying their ability to take corrective actions.[753] Further, some computer systems are designed to limit access to very detailed information about plant conditions and instead provide the operator with only high-level information that may not be sufficient during all types of accidents. While the risk of human error during normal operation may be somewhat reduced by the introduction of these computer systems, how the individual operators will respond to these trade-offs between information and action during an accident adds to the uncertainties of quantitative risk assessments. In summary, the National Research Council concluded that

> At this time, there does not seem to be an agreed-upon, effective methodology for designers, owner-operators, maintainers, and regulators to assess the overall impact of computer-based, human-machine interfaces on human performance in nuclear power plants.[754]

As with other types of design errors, the specific limitations of a software interface are difficult to foresee during development, and it is only after the system is implemented and the mistakes are made in the real world that the limitations are revealed. In one review it was found that as many as 92 percent of the accidental, computer-related deaths that have occurred in such safety critical application like air travel and medical treatment were caused by failures of the operator-computer interface.[755] This is a particular concern for nuclear power because there is little room for trial and error given that an important operator mistake can turn a minor incident into a serious accident in a very short time.

In addition to the issues of human performance, the replacement of mechanical or analog systems with programmable devices has the potential to reduce the functional diversity of control and safety systems making it more likely that a single failure will affect all of the related systems. At its simplest level this effect can be seen in systems that rely on multiple computers running the same code. Such systems assume that the software will always function correctly and that there only needs to be protection against the potential failure of the hardware in one of the computers. A particularly dramatic example of an accident caused by such a system was the explosion of the European Space Agency's Ariane 5 rocket on its maiden flight on June 4, 1996. Both flight computers onboard the rocket were running the same flight code which turned out to contain an important programming error. Both systems failed within 0.05 seconds of each other due to this software bug, sending the rocket dangerously off course and resulting in its destruction less than 40 seconds after liftoff.[756]

However, simply using different pieces of software that perform the same task will not necessarily ensure independence, although it will likely result in a significant improvement over the use of the same code. This problem with software is unlike those encountered in mechanical systems, because the failures of software are caused solely by design failures and not the random failure of equipment. This fact makes different pieces of software that perform the same function more likely to share a common failure mode. In its review of the use of digital systems in nuclear power plants, the National Research Council concluded that "there is no way to verify or evaluate the diversity of two software versions or to determine whether they will fail independently."[757] In fact, studies of software that was developed independently by different programmers but performed the same tasks have found that the number of correlated failures were too high to have occurred simply by chance. As a result of this type of work the National Research Council concluded that

> All evidence points to the fact that independently developed software that uses different programmers, programming languages, and algorithms but computes the same function (satisfies the same functional requirements) cannot be assumed to fail in an independent fashion.[758]

In addition to the question of determining the functional independence of redundant software driven systems, Nancy Leveson, a professor in the Aeronautics and Astronautics Department at the Massachusetts Institute of Technology and leading authority on the use of software in high-reliability systems, has pointed out that, like in mechanical systems, the addition of redundancy in digital systems can increase complexity leading to new types of accidents. In a review of five major aerospace accidents attributed to software problems she notes that

> Throughout the accident reports, there is an emphasis on failures as the cause of accidents and redundancy as the solution. Accidents involving software, however, are usually "system accidents" that result from the dysfunctional interactions among components, not from individual component failure. All these accidents (as well as almost all the software-related accidents known to the author) resulted from the software doing something wrong rather than the computer hardware or software failing to operate at all of [sic] In fact, in most cases the software or hardware components operated according to their specifications, that is, they did not fail, but the combined behavior of the components led to disastrous system behavior.[759]

Similar problems emerge from the addition of self-checks to the software in an attempt to improve its reliability. Experiments have shown that self-checks were able to prevent very few errors from occurring and, instead, were found to introduce more failures than they caught.[760]

Determining how to include software failures into PRAs poses a number of problems that are similar to those encountered in dealing with design inadequacies in general. The new feature is that, with the introduction of software, important elements of control and safety systems are now vulnerable to purely design failures. Despite extensive testing and validation programs, problems in the design of software will likely continue to slip through during development. As noted by Professor Leveson,

> Most software-related-accidents [in high reliability systems] have involved situations that were not considered during development or were assumed to be impossible and not handled by the software.[761]

The National Research Council noted that there remains an ongoing "controversy within the software engineering community as to whether an accurate failure probability can be assessed for software or even whether software fails randomly."[762]

The increased reliance on software can have an important impact on the completeness of PRAs as well. As summarized by the National Research Council, "[a]nalog systems are believed to fail in more predictable and obvious ways than do the more hidden and insidious failure mechanisms in software."[763] The possibility for such hidden failure modes is increased in some plants by the much greater complexity of the software systems that have been developed. For example, Ontario Hydro was the first utility in Canada to license a fully computerized plant shutdown system. This system consisted of roughly 6,000 lines of code and contained only software directly related to the shutdown of the plant. On the other hand, Britain's first reactor shutdown system to make use of computer controls consisted of a program with approximately 100,000 lines of code that ran on several hundred microprocessors and was designed to handle functions beyond those related to plant shutdown.[764]

In fact, to date, the inclusion of software failure modes in the PRAs for nuclear plants have been very inconsistent. For example, General Electric's new Advanced Boiling Water Reactor design did not include any possibility of software failures in its risk assessment. In addition, the guidelines for performing PRAs contained in the Electric Power Research Institute's Utility Requirements Document did not include any discussion of how to incorporate software failures.[765] Similar cases of software being omitted entirely or "treated superficially at best" can be found in the risk assessments carried out for other types of high reliability systems as well.[766] On the other hand, Westinghouse did choose to include subjective estimates for software unavailability in its analysis of the AP600 pressurized water reactor's protection and monitoring system.[767] While there remains important questions as to how to properly include digital systems in the PRA methodology, the National Research Council did recognize that "explicitly including software failures in a PRA for a nuclear power plant is preferable to the alternative of ignoring software failures."[768]

Section 4.3.5 – Expert Judgment and Uncertainties of Methodology

The preceding sections have discussed in detail some of the most uncertainties surrounding the quantitative results of probabilistic risk assessments. A further source of uncertainty concerns the use of expert judgment to determine failure modes and failure rates in cases where insufficient experimental or historical information exists to make a determination. Examples of the failure of expert judgment abound in high reliability systems. For example, during the debate over the construction of the Fermi Fast Breeder reactor near Detroit, Hans Bethe, the Nobel Prize winning physicist, predicted that it was not possible for this type of reactor to suffer a melt down. Another expert was somewhat more cautious, but still predicted that at most one fuel subassembly could melt under the worst case accident conditions. In 1966, however, the Fermi reactor suffered a core melt accident that damaged a total of four fuel subassemblies. Two of the subassemblies were found to have been so heavily damaged that were actually partially melted together.[769]

Other examples of the failure of expert judgment can be found in the Space Shuttle program. A DOE analysis of the safety of launching satellites containing plutonium-238 officially estimated the probability of a shuttle failing during the launch phase to be just 1 in 100,000. This would imply that NASA could launch a Space Shuttle every day for nearly 274 years and expect only one disaster on average.[770] Following the destruction of the Challenger in 1986 shortly after lift off, NASA issued its own risk estimates putting the probability of failure during the launch phase at 1 in 248, more than 400 times larger than the DOE's estimate.[771] Similar disagreements over the safety of Shuttle's main engines were also apparent. Engineers at Rocketdyne, the engine's manufacturer, estimated the failure probability as 1 in 10,000, while engineers at the Marshall Space Center estimated a probability of 1 in 300, and an independent engineering consultant hired by NASA thought 1 in 50 to 1 in 100 would be more reasonable.[772]

In addition to the use of expert judgment in estimating failure rates, there are also a number of different ways in which the PRA methodology can be set up and applied, and the choice between these can have an important impact on the results as well. In light of this, the Nuclear Regulatory Commission noted in the early 1980s

> Therefore, the NRC has had programs under way since 1975 to improve the PRA methodology. Progress is being made and the program has provided useful insights on nuclear reactor safety. However, there remain significant uncertainties associated with the overall results of PRAs, and there exists a wide spectrum of expert views on the ability of the PRA methodology to provide reliable estimates of the risk associated with the operation of nuclear power plants. Furthermore, the studies done thus far have not been performed using consistent methodology and assumptions.[773]

In testimony before Congress two years later, NRC Commissioner James Asselstine acknowledged that

> Moreover, there are enormous uncertainties in the estimated core meltdown risks to the public. Scientifically accepted data and methodology are not in hand today to substantially reduce those uncertainties. Thus, I believe it is mandatory to consider

> forthrightly those uncertainties in reaching any conclusion on the acceptability of the core meltdown risks.[774]

Despite ongoing efforts by the NRC, academics, and the nuclear industry, these issues of methodology have yet to be adequately resolved.

The plant specific safety assessments published by the NRC in 1996 estimated core damage frequencies for boiling water reactors that varied by over roughly three orders of magnitude while the estimates for the operating pressurized water reactors varied by over two orders of magnitude. This wide spread in estimated accident frequencies at the various facilities was attributed to "plant design differences", "variability in modeling assumptions", and "differences in data values (including human error probabilities) used in quantifying the models."[775] Of these three reasons, only the first represents a real difference in safety among the plants, while the second two are simply artifacts of the modeling technique.

These artifacts of the methodology can be seen in an even clearer way by comparing the PRAs for very closely related plants. For example, Wolf Creek in Kansas and Callaway in Missouri were built from identical blueprints using the same materials and were brought online just eight months apart. Despite these similarities, however, the PRA for Wolf Creek estimated some accident probabilities that were 10 to 20 times less than the estimates for those same accidents occurring at Callaway. A similar example can be found in a comparison of the risk assessments for the Sequoyah and Watts Bar plants in Tennessee. Both of these plants share the same general design, and both are operated by the Tennessee Valley Authority. Despite these similarities, however, the overall core damage frequency predicted for Sequoyah was 1 in 26,525 per year while the frequency for Watts Bar was just 1 in 3,030 per year (almost 9 times greater). After comparing these results, the TVA recalculated the expected accident frequency for Watts Bar and reduced it to 1 in 12,500 per year. Despite this change, the estimated probability for an accident at Watts Bar was still more than twice that for the Sequoyah plant.[776]

In light of the influence that may be exerted on the results of quantitative risk assessments by the choice of methodology and the use of expert judgment, William Ruckelshaus, the head of the U.S. Environmental Protection Agency under both Presidents Nixon and Reagan summarized the situation as follows

> We should remember that risk assessment data can be like the captured spy: if you torture it long enough, it will tell you anything you want to know.[777]

Section 4.4 – Safety of an Expansion of Nuclear Power

In the preceding sections we examined both the potential consequences of a major accident at a nuclear plant as well as some of the uncertainties surrounding the estimates for the probability of such accidents occurring. With this information we can finally turn to the central question of reactor safety; namely what would be the risk posed by the expansion of nuclear power under the global or steady-state growth scenarios and whether that risk is likely to be acceptable.

Historical experience with nuclear plants in the United States imply that the probability of a core melt accident is likely between 1.2×10^{-4} and 1.6×10^{-3} per year (1 in 8,440 per year to 1 in 633 per year).[778] Typical probabilistic risk assessments for the U.S. nuclear fleet predict a value of 1 in 10,000 per year, just below the lowest end of the historic range.[779] Even assuming that the smaller accident rate predicted by the PRA methodology is correct, the risks that would be posed by 1,000 to 2,500 light-water reactors operating around the world would be unacceptably large. As noted by the authors of the MIT study

> **With regard to implementation of the global growth scenario during the period 2005-2055, both the historical and the PRA [Probabilistic Risk Assessment] data show an unacceptable accident frequency.** The expected number of core damage accidents during the scenario with current technology would be 4. We believe that the number of accidents expected during this period should be 1 or less, which would be comparable with the safety of the current world LWR fleet. A larger number poses potential significant public health risks and, as already noted, would destroy public confidence.[780]

A particular concern with the expansion of nuclear power would accompany the envisioned increase in the number of plants located near large population centers both in the U.S. as well as in densely populated countries like India and China.

The solution put forward by the authors of the MIT to this unacceptable level of risk from the global growth scenario rests entirely upon their belief in the industry's claims that the new generation of nuclear plants can be made significantly safer than current plants. In particular, the authors of the MIT study concluded that the reduction of the core damage probability from 1 in 10,000 per year to 1 in 100,000 per year "is a desirable goal and is also possible, based on claims of advanced LWR designers, that we believe plausible."[781] With this reduction in the estimated accident probability, the risks from the global growth scenario could be reduced to roughly the same level as the risks posed by the current U.S. fleet. In other words, there would still be more than one chance in three of a meltdown occurring somewhere in the world during the lifetime of the reactor fleet. In the case of the steady-state growth scenario, with 2,500 reactors online in 2050, the probability of a meltdown would rise to above 60 percent, even assuming that the claimed safety improvements were actually achieved.

Surprisingly, in discussing the basis for their conclusion regarding the improved safety of new reactor designs, the authors of the MIT report admit that

> Our study has not been able to address each aspect of concern as thoroughly as deserved. One example is safety of nuclear operations. **Accordingly, we report here views of our group that we believe to be sound but that are not supported by adequate analysis.**[782]

This statement is rather shocking from a scientific point of view given the importance of reactor safety to the future of nuclear power. The consequences of a major accident could be catastrophic and long lasting if the reactor's containment was breached. Even in the event of a meltdown that did not release significant quantities of radiation to the environment, the financial loss of the reactor and the impact on public confidence in nuclear power could have dramatic and far reaching impacts on the viability of the industry. Given the numerous uncertainties outlined in the last section concerning the estimates for the accident frequencies, and the fact that the lesson to be learned from past experience is that *a priori* engineering estimates based on paper drawings or prototype plants often miss important failure modes and accident scenarios, it seems difficult to justify the recommendation of such a large expansion of nuclear power based upon an inadequately supported belief of what safety improvements are "plausible."

As discussed in Sections 4.2.1 and 4.2.2, the potential consequences of a worst case nuclear accident are so severe that determining the risk posed by nuclear power is extremely sensitive to the assumptions used in estimating the probability of such accidents occurring. This is sometimes referred to as the "zero-infinity problem," in which the product of low probabilities and high consequences become poorly defined in discussions of risk in the same way that the product of zero and infinity is poorly defined in a mathematical sense. In these situations it is most prudent to pursue a precautionary approach and to carefully consider the range of uncertainties inherent in the overall risk estimates. As concluded by the GAO in 2004,

> PRA estimates for nuclear power plants are subject to significant uncertainties associated with human errors and other common causes of system component failures, and it is important that proper uncertainty analyses be performed for any PRA study.[783]

Thus, rather than focusing upon a single number and basing conclusions on that estimate alone, it is necessary to consider a range of reasonable accident probabilities associated with a range of reasonable consequences.

The need to focus more attention on uncertainties is not a new idea. For example, in his testimony before Congress in the mid-1980s, NRC Commissioner James Asselstine commented that

> The Commission has yet to formulate a coherent approach to addressing the uncertainties. Until this is done, I believe there is no justifiable technical basis for concluding that the core meltdown risks are acceptable for the long term.[784]

Commissioner Asselstine further testified that he felt it was reasonable to assume there was at least a factor of ten in the uncertainty of core damage estimates generated by PRAs.[785] This level of uncertainty is supported by the results of the plant specific risk assessments carried out in the mid-1990s which estimated core damage frequencies for the U.S. fleet of reactors that varied by two to three orders of magnitude (see Section 4.3.5). This uncertainty is also consistent with the range of accident frequencies that can be estimated from historical experience as noted above. The uncertainty in risk assessments conducted for reactor designs that have yet to be built or have only limited operating experience would be expected to be at least as large as the uncertainties for existing plants, and would likely be larger. This is particularly true for designs that make greater use of programmable logic devices and other digital systems than today's reactors.

Given the uncertainties inherent in the PRA methodology, and the lack of significant real-world experience with most of the advanced or evolutionary reactor designs that would be likely candidates for the global or steady-state growth scenarios, we will instead use the experience gained from the existing fleet of light-water reactors to examine the expected accident rates. This is consistent with the caution noted by the authors of the MIT study as well:

> Although safety technology has improved greatly with experience, remaining uncertainties in PRA methods and data bases make it prudent to keep actual historical risk experience in mind when making judgments about safety.[786]

Thus, retaining the assumption from the MIT report that new reactor designs will be about 10 times less likely to suffer a meltdown than the existing fleet, we estimate that new nuclear plants would have an accident probability of between 1.2×10^{-5} and 1.6×10^{-4} per year. From this range of probabilities, we can then estimate the likelihood of at least one meltdown occurring somewhere in the world over then next few decades if the global or steady-state growth scenarios were pursued (see Table 4.6).

Table 4.6: Cumulative probability of at least one accident occurring somewhere in the world by the given date under the global or steady state growth scenarios. The range of accident probabilities for the reactors is taken to be 1.2×10^{-5} to 1.6×10^{-4} per year which is ten times lower than the estimates derived from historical experience with the existing fleet of U.S. reactors.

Year	Global Growth Scenario (1,000 GW in 2050)	Steady-State Growth Scenario (2,500 GW in 2050)
2020	6.6 to 60%	7.2 to 63%
2030	12 to 82%	15 to 88%
2040	19 to 94%	26 to 98%
2050	27 to 98%	42 to 99.9%

From Table 4.6 we find that, by 2030 there would be better than one chance in ten at best of at least one TMI level accident occurring under the global growth scenario. At the upper end of this range there would be more than an 80 percent chance of at least one such accident having occurred. Importantly, these estimates have already taken into account a significant, and as yet unproven, increase in the safety of new reactor designs compared to those in operation today. To illustrate these accident rates another way, Figures 4.1 and 4.2 show the cumulative probability for one or more meltdowns occurring somewhere in the world prior to 2050. These figures use the median accident rate projected from historical experience and include the same ten-fold increase in safety assumed by the authors of the MIT report.[787] The overall likelihood of such an accident occurring would, of course, continue to rise beyond 2050 while the reactors continued to operate until each had reached the end of its 40 year operational lifetime.

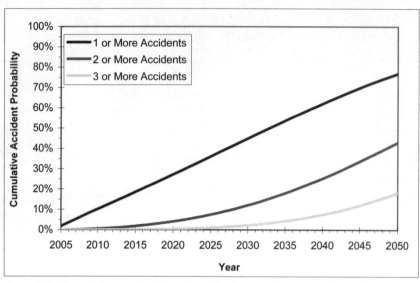

Figure 4.1: Cumulative probability of accidents occurring somewhere in the world between 2005 and 2050 assuming that 1,000 reactors are online at mid-century. The accident probability is taken to be 5.6×10^{-5} per year (i.e. 1 in 17,880 per year) which is ten times lower than the median estimate for the accident rate derived from historical experience with the existing fleet of U.S. reactors.

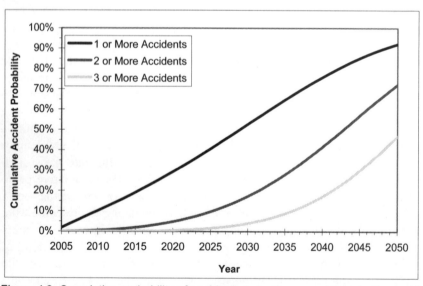

Figure 4.2: Cumulative probability of accidents occurring somewhere in the world between 2005 and 2050 assuming that 2,500 reactors are online at mid-century. The accident probability is taken to be 5.6×10^{-5} per year (i.e. 1 in 17,880 per year) which is ten times lower than the median estimate for the accident rate derived from historical experience with the existing fleet of U.S. reactors.

Luckily, the probability that a accident would occur in which the reactor's containment dome was also breach, would be less than the core damage frequencies shown in Figures 4.1 and 4.2. Despite this fact, however, it is very unlikely that the level of risk would be low enough under the global or steady-state growth scenarios to be considered acceptable by the public. The inclusion of potential terrorist attacks on the scale of those on September 11, 2001 adds even further to the already significant uncertainties inherent in predicting the safety of a large scale expansion of nuclear power. The history of public opposition to nuclear power reveals quite clearly the importance of reactor safety to the acceptability of this technology. In light of the high probabilities of at least one serious accident occurring by mid-century, the possibility that public option could turn sharply against the widespread use of nuclear power following a meltdown is a significant vulnerability in plans that envision a heavy reliance on this energy source.

Section 4.5 - Conclusions

Like the potential for nuclear weapons proliferation, the potential for catastrophic reactor accidents or well coordinated terrorist attacks to release large amounts of radiation make nuclear power a uniquely dangerous source of electricity. Such releases would have severe consequences for human health and the environment, would require expensive cleanup and decontamination efforts, and would leave buildings and land contaminated for generations. The last systematic analysis of accident consequences released by the government was completed nearly a quarter of a century ago. The study, entitled Calculation of Reactor Accident Consequences for U.S. Nuclear Power Plants (CRAC-2), found that a worst case accident at many power plants could result in tens of thousands of deaths from prompt radiation effects and long-term fatal cancers and cause hundred of billions of dollars in damage.

Since CRAC-2 was completed, the population in the U.S. has both grown in size and has grown increasingly concentrated. In addition, the scientific understanding of radiation risks has advanced significantly since the early 1980s. In light of these changes, the Nuclear Regulatory Commission should update the CRAC-2 study to provide the public with a clearer understanding of the risks of nuclear power. The updated study should include the potential impacts from large-scale, well coordinated terrorist attacks which may have consequences that differ from normal accidents. In addition, given the potentially large impacts of such releases, the update should also consider the risks from fires that may occur in the reactor's cooling pools. While the precise details of the accident and attack scenarios are likely to remain classified on security grounds, the full distribution of impacts should be released to the public to facilitate a more informed and democratic debate on the future of nuclear power.

In addition, even if the reactor's secondary containment is not breached, and there are not large offsite releases of radiation, a serious accident at a nuclear plant that damages the core would still cost the utility a very significant amount of money due both to the loss of the reactor and the need to buy replacement power. As summarized by Peter Bradford, a former commissioner of the NRC,

> The abiding lesson that Three Mile Island taught Wall Street was that a group of N.R.C.-licensed reactor operators, as good as any others, could turn a $2 billion asset into a $1 billion cleanup job in about 90 minutes.[788]

To date, there have been at least seven reactor accidents that have resulted in the release of radiation. These accidents have occurred at both military and civilian research and power reactors that employed a variety of different design features. The worst of these accidents was the April 26, 1986 steam explosion at the graphite moderated, water-cooled Chernobyl nuclear power plant. An estimated 220,000 people were eventually forced to relocate following the accident and large areas of agricultural land had to be abandoned. Several thousand people across Europe and the former Soviet Union are expected to ultimately die as a result of this disaster. So far, the one accident to have occurred at a commercial light-water reactor (i.e. the March 1979 partial meltdown at Three Mile Island) is not officially believed to have resulted in the release of large quantities of non-noble gas radionuclides to the environment. However, as Richard Feynman famously noted in relation to the O-ring failures that led to the destruction of the Space Shuttle Challenger, "[w]hen playing Russian roulette, the fact that the first shot got off safely is of little comfort for the next."[789]

Estimates for the likelihood of such accidents occurring have significant uncertainties that greatly complicate projections about the safety of an expanded use of nuclear power. In fact, the authors of the MIT report admit that

> Our study has not been able to address each aspect of concern as thoroughly as deserved. One example is safety of nuclear operations. **Accordingly, we report here views of our group**

that we believe to be sound but that are not supported by adequate analysis.[790]

This statement is rather shocking given the importance of reactor safety to the viability of any nuclear revival.

The Probabilistic Risk Assessments which try to quantify the likelihood of accidents in high reliability systems have numerous methodological weaknesses that limit their reliability. The issues of completeness and how to handle design defects are particularly difficult to handle within the PRA methodology in that they essentially require the analyst to know what they don't know about what could go wrong. If important accident scenarios can be foreseen they would already be included in the analysis, and if design defects were identified they could be fixed. As summarized by Edward Hagen, a development specialist at Oak Ridge National Laboratory and editor of the Control and Instrumentation section of the journal *Nuclear Safety*,

> No reactor system has ever failed because of a deficiency that could be seen on a designer's flow sheet or an analyst's model. Such deficiencies have been revealed only via operating experiences.[791]

This is a particular concern for the safety of new reactor designs that so far exist only on paper or which have only a limited amount of operating experience. Many important unforeseen accident scenarios and design flaws have been discovered during the nearly 3,000 reactor years of operating experience with current reactor designs. Placing too much faith in theoretical estimates for the safety of new designs without a suitable consideration of the many inherent uncertainties involved could be a potentially serious mistake.

Additional concerns arise due to the fact that nuclear power demands an extremely high level of competence at all times from all levels of the organization -- from the regulators and managers all the way through to the technical and maintenance crews. If the human element of the system falters, then there is the possibility for a severe accident to occur. This element of uncertainty in the risk assessments will remain as important for new plants as it is for existing plants, since they all rely on humans as an integral element of operation. Finally, the increased use of computers and other digital systems for performing control, protection, and monitoring functions create important safety tradeoffs with improvements during normal operation, but the potential for unexpected problems to arise during accidents.

In light of the uncertainties inherent in quantitative risk assessments and the influence that may be exerted by the persons conducting the analysis, William Ruckelshaus, the head of the U.S. Environmental Protection Agency under both Presidents Nixon and Reagan cautioned that

> We should remember that risk assessment data can be like the captured spy: if you torture it long enough, it will tell you anything you want to know.[792]

Using historical experience with light-water reactors in the United States as a more reliable starting point for considering the risks of nuclear power, we find an unacceptably high risk of accidents under either the global or steady-state growth scenarios. Using the median accident rate from U.S. experience, and retaining the assumption from the MIT report that future plants will be ten times safer than those in operation today, we find that the probability of at least one TMI level accident occurring somewhere in the world by 2030 would be roughly 45 percent under the global growth scenario and more than 50 percent under the steady-state growth scenario. By 2050, the probability of at least one accident having occurred would be greater than 75 percent under the MIT scenario and over 90 percent under the steady-state growth scenario. In fact, with the construction of 2,500 reactors, there would be nearly a 50-50 chance that three or more accidents will have occurred by mid-century.

In addition to the extremely large economic, human, and environmental impacts that could accompany a serious reactor accident or successful large-scale terrorist attack, the history of public opposition to nuclear power clearly demonstrates the importance of reactor safety to the acceptability of this technology. In light of the high probability that at least one accident like Three Mile Island would occur somewhere in the world between now and 2050, the possibility that public opinion could turn sharply against the widespread use of nuclear power following a meltdown is a significant vulnerability with plans that envision a heavy reliance on this energy source. If a rapid expansion of nuclear power was the only way to avoid the looming threat of climate change, then these risks from reactor accidents would have to be looked at differently given the potentially catastrophic impacts of global climate disruption. However, nuclear power is not the only available option for economically reducing emissions as discussed in Chapter Two. Therefore, to trade one uncertain, but potentially serious health and environmental risk for another is not a sound basis for an energy policy.

Chapter Five: The Legacy of Nuclear Waste

> A worldwide deployment of one thousand 1000 megawatt LWRs [light water reactors] operating on the once-through fuel-cycle with today's fuel management characteristics would generate roughly three times as much spent fuel annually as does today's nuclear power plant fleet. If this fuel was disposed of directly, new repository storage capacity equal to the currently planned capacity of the Yucca Mountain facility would have to be created somewhere in the world roughly every three or four years.[793]
> - *The Future of Nuclear Power* (2003)

> As I reflect on my own involvement in the waste problem, I have these regrets. Most importantly, during my years at ORNL [Oak Ridge National Laboratory] I paid too little attention to the waste problem. Designing and building reactors, not nuclear waste, was what turned me on.... Indeed, as I think about what I would do differently had I to do it over again, it would be to elevate waste disposal to the very top of ORNL's agenda.[794]
> - Alvin Weinberg, Director of Oak Ridge National Laboratory from 1955 to 1973 (1994)

Along with the potential for facilitating nuclear weapons proliferation and concerns over reactor safety, the challenge of managing the radioactive wastes generated by the nuclear fuel cycle is a long standing vulnerability accompanying the use of nuclear power. In addition to its high radiotoxicity, the existence of large quantities of weapons usable plutonium in the spent fuel from commercial power plants complicates the waste management problem by raising concerns over nuclear weapons proliferation as discussed in Chapter Three.[795] This link between nuclear waste and nuclear weapons makes reprocessing technologies highly undesirable, even if those waste management technologies could somehow be made economical and could overcome their other significant environmental problems. Finally, it is important to recognize that the impacts of nuclear waste have so far fallen disproportionately on Indigenous Peoples in the United States and around the world raising concerns about environmental justice. This disproportionate impact is true both for wastes from uranium mining and milling as well as for the management of spent nuclear fuel and high-level waste.[796]

It has been more than 60 years since the first spent fuel and high-level reprocessing wastes were created as part of the Manhattan Project's development of the atom bomb. Since then, more than 440 commercial reactors have been brought online around the world generating well over a hundred thousand metric tons of spent fuel as well as large quantities of liquid high-level waste. The current scientific consensus, which is shared by IEER, is that placement of this waste in a deep geologic repository is the least worst option available for the long-term management of the spent fuel that is already in existence.[797] However, to date no country has yet successfully disposed of a single ton of this waste and all repository programs have encountered unexpected problems.

Through 2050, the proposed expansion of nuclear power under the global growth scenario would lead to roughly a doubling of the average rate at which spent fuel is currently being generated. The characterization and siting of repositories rapidly enough to handle this volume of waste would be a very serious challenge, and one unlikely to be overcome by the development of new technologies or disposal options. The site of the Yucca Mountain repository has been studied for more than two decades and it has been the sole focus of the U.S. Department of Energy since 1987. Despite this effort, as yet no license application has been filed and a key element of the regulations governing the site has been struck down by the courts and re-issued in draft form. It is one of the central failures of nuclear power's expansion to date that those responsible simply had faith that the waste problem was tractable and that as technology advanced it would eventually prove a relatively straightforward task to manage spent fuel safely. The densely packed spent fuel pools spread around the U.S. today and a highly contentious and long-delayed repository program are the result of that faith.

In this chapter we will briefly address "low-level" waste disposal, and then consider in detail the problems surrounding geologic disposal of spent fuel in a mined repository. While different issues will arise among the various international repository programs, we will focus on the proposed Yucca Mountain repository given that it is one of the largest and also one of the farthest along in development. We will then discuss the safety of spent fuel transportation which relates both to permanent disposal in a repository as well as to proposals for reprocessing or for consolidation of the waste at a centralized long-term retrievable storage facility. Finally, we will consider some of the alternative strategies for the management of spent fuel that have been proposed and discuss why it is unlikely that any of them would be able to ease the waste disposal problem associated with a large expansion of nuclear power.

Section 5.1 – Disposal of "Low-Level" Nuclear Waste

In the United States, radioactive waste is not classified according to the level of hazard it represents, but is instead classified by its origin. Spent nuclear fuel and liquid wastes generated by reprocessing are classified as high-level waste. Materials contaminated above a certain threshold with long-lived alpha emitting transuranic isotopes such as plutonium, neptunium, or americium are classified as transuranic (TRU) waste.[798] Both high-level waste and TRU waste in the U.S. are required to be disposed of in deep geologic repositories.[799] The mill tailings and other wastes from uranium mining are classified as 11.e.(2) byproduct waste. All other radioactive wastes fall under the general category of "low-level" waste (LLW) even if those materials actually have a higher activity or are more radiotoxic than some transuranic wastes. The depleted uranium created during the enrichment of nuclear fuel is an example of a "low-level" waste that poses long-term disposal problems that are comparable to those of some TRU wastes in terms of radiation risk.[800]

While the vast majority of the radioactivity in nuclear waste is in spent fuel and reprocessing wastes, the vast majority of the volume is in uranium mill tailings and low-level wastes. For example, at the end of 1994, the estimated volume of mill tailings was more than 300 times the volume of high-level waste and spent fuel combined while the volume of low-level waste was nearly 12 times larger. On the other hand, the radioactivity in the mill tailings and low-level waste together was more than 1,100 times less than the radioactivity in high-level waste and spent fuel.[801] Given its radiological properties, the disposal of most kinds of LLW (with the exception of wastes such as depleted uranium) is typically a far less complicated problem than the management of transuranic or high-level waste. However, despite this fact, the disposal of LLW has also encountered difficulties that should not be overlooked in the context of proposals for the expansion of nuclear power.

The 1980 Low-level Radioactive Waste Policy Act made the disposal of LLW the responsibility of individual states, in contrast to the management of high-level waste which is a federal responsibility. In order to deal with these wastes, the states formed compacts that agreed to cooperate in developing regional disposal facilities.[802] An amendment to the law in 1985 allowed the three then existing commercial disposal sites to stop accepting waste from outside their compacts after the end of 1992. Since then, the commercial facility at Beatty, Nevada, has closed, the facility at Hanford has stopped accepting commercial waste from outside the Northwest and Rocky Mountain compacts, and the disposal site at Barnwell, South Carolina, has announced that it will stop accepting waste from outside the Atlantic compact by the end of 2008. The only other commercial facility in operation today is the Envirocare site at Clive, Utah, which accepts waste from all across the United States. This facility, however, is only licensed to dispose of the lowest class of low-level waste. In February 2005, the new owners of Envirocare announced that they would no longer seek a license amendment to allow the acceptance of higher activity wastes.[803] This announcement was made just days before the Utah House and Senate passed legislation permanently banning the import of such higher activity wastes to the state.

The development of new low-level waste disposal sites has run into a number of difficulties. A disposal facility approved in 1993 for construction in California was stopped when the Department of the Interior refused to transfer ownership of the federal land to the state as expected. Disposal sites in Ohio and Nebraska have been abandoned by the Midwest and Central compacts respectively. In November 2004, the voters in Washington State approved Initiative 297 by more than a 2 to 1 margin. This initiative called for an end to the disposal of waste in unlined trenches at the Hanford reservation as well as for the DOE to not leave behind high-level waste from plutonium production in the tank farms, and for the DOE to ensure that the land and groundwater at Hanford are cleaned up before any additional waste is brought to the site. Currently, the only new disposal facility that is undergoing an advanced state of active development is the Waste Control Specialists (WCS) site in Andrews County, Texas. WCS filed a license application in August 2004 to build two low-level waste disposal units, one for waste from the Texas Compact and a second facility for the disposal of federal low-level waste. The application is currently pending and the facility is not expected to open until 2007 at the earliest.[804] A previous attempt to license a disposal facility in Texas at the Sierra Blanca site failed due to concerns raised by the State, environmental groups, and members of the local communities.

The difficulties encountered in licensing new LLW disposal facilities have the potential to impact, among other things, the cost of decommissioning nuclear reactors and fuel cycle facilities. In addition, the enrichment of uranium to supply the fuel for 1,000 GW of capacity would generate more than 184,300 metric tons of depleted uranium per year plus a proportionately large amount of mill tailings.[805] For comparison this would mean that, under the global growth scenario, as much depleted uranium would be created every two years as was created at all three DOE facilities for all civilian and military programs in the nearly half a century from 1944 through mid-1993.[806] In the case of the steady-state growth scenario, the amount of DU generated would be increased by two and a half times over the global growth scenario. No country has yet developed a long-term disposal facility for the more than one million two-hundred thousand metric tons of depleted uranium already in existence around the world.[807] A recent analysis by IEER estimates that the ultimate cost of managing the depleted uranium tails in the United States would likely range from $20 to $30 per kilogram of DU.[808] Assuming this cost was passed on by the enrichment corporation as part of the annual fuel price, this would add an estimated $3.7 to $5.5 million per year to the fuel costs of each nuclear plant.

The management and disposal of low-level waste will continue to pose challenges in the future (particularly those issues regarding depleted uranium), however, by far the largest concerns regarding waste management relate to the handling of spent fuel given that it contains both the vast majority of the radioactivity and weapons usable plutonium that can be chemically separated by reprocessing.

Section 5.2 – Geologic Disposal of Spent Nuclear Fuel and High-Level Waste

The amount of spent fuel discharged annually from a commercial reactor depends upon both the reactor's capacity factor and the total burnup of the fuel. The higher the capacity factor, the more energy the reactor generates in a year. Thus, higher capacity factors mean more fuel is undergoing fission, which increases the amount of spent fuel that is generated over time. On the other hand, by increasing the total burnup of the fuel, the operator can extract more energy from a given amount of uranium, thus reducing the mass of spent fuel that is generated. In the future, a typical 1,000 MW reactor is expected to generate approximately 20 tons of spent fuel per year.[809] This value is generally consistent with the recent experience at operating reactors in the United States. Between the end of 1998 and the end of 2002, U.S. reactors discharged an average of 22.1 metric tons per GW per year.[810]

The EIA reports that, as of the end of 2002, U.S. reactors had discharged a cumulative total of 47,023 metric tons of spent fuel. This commercial waste is currently stored at 76 sites in 33 states according to the DOE. (A number of these sites have multiple reactors and multiple fuel storage facilities located within the same general complex.) Nearly 90 percent of the spent fuel was stored in cooling pools, while the remainder was stored in independent spent fuel storage installations (ISFSIs), mostly in dry casks located at reactor sites. Assuming a continuation of the recent rate of waste generation, by the end of 2005 there was an estimated 53,100 metric tons of spent fuel stored around the United States. By 2012, which is the earliest date that the Yucca Mountain repository could possibly be opened, the amount of spent fuel discharged from the currently operating reactors would amount to approximately 67,500 MT.[811] Thus, even without any new nuclear construction, by 2012 the inventory of spent fuel in the U.S. would already exceed the statutory limit of the planned Yucca Mountain repository (i.e. 63,000 MT of commercial spent fuel capacity and 7,000 MT of capacity for military high-level waste from plutonium production).

Developing ways to manage spent fuel and reprocessing wastes in such a way that will protect both current and future generations is greatly complicated by the time-scales over which some of the important radionuclides will remain dangerous. For example, the half-life of plutonium-239 is 24,000 years, and thus even after 160,000 years there would still be more than four and a half metric tons of plutonium-239 remaining in the waste.[812] For comparison, this amount of plutonium would be enough to make nearly 570 nuclear bombs if it was recovered. Other radionuclides of concern have even longer half-lives. For example, technetium-99 has a half-life of 212,000 years, cesium-135 has a half-life of 2.3 million years, and iodine-129 has a half-life of 15.7 million years. Due to their long half-lives and other radiological and chemical properties, these radionuclides have been found to be significant contributors to the long-term risks posed by repository disposal. Because harmful radionuclides like these will remain dangerous for such long periods of time, the peak dose from the high-level waste disposed of at Yucca Mountain is not expected to occur for more than a hundred thousand years, assuming that the engineered barriers function as intended.

To put some of these timescales into perspective, we note that they are comparable to the entire amount of time that modern humans are believed to have existed on Earth.[813] Another way to appreciate the magnitude of these timescales is to note that the first evidence of domesticated plants and animals dates back only roughly 10,000 to 12,000 years while the oldest known evidence of human writing goes back at most 5,300 to 5,500 years.[814] In light of such considerations, the National Research Council concluded that there was no scientific or technical basis to support claims for the long-term reliability of such things as markers, monuments, or other written records at preventing future humans from inadvertently intruding upon a repository.[815] In other words, the management of nuclear waste concerns a time scale that is more appropriate to human evolution than human civilization.

Regardless of whether or not new nuclear plants are built in the coming decades, the tens of thousands of metric tons of spent fuel already in existence as well as that which will continue to be generated before the current reactors could be shutdown, will require a significant effort to ensure safe management. IEER agrees that the disposal of this waste in a mined geologic repository is the least worst option available. However, the implementation of this generic concept is extremely complicated. The suitable characterization of a site and the development of engineered barriers specially tailored to the particular features of the site geology will necessarily take a significant amount of time and money.[816] In addition to the technical issues surrounding site selection and repository design, there are a number of societal issues that must be addressed, including the need to receive the informed consent of all affected parties as well as the need to seek an equitable distribution of the benefits and liabilities associated with the past use of nuclear power. The importance of this mixture of social and technical issues in repository development is widely recognized. In its review of spent-fuel management, the National Research Council concluded

> At the present time, probably a relatively small segment of the public and of political decision makers agrees on the definitions of acceptable risk and uncertainty developed by the waste community in its professional and scientific practice. In the opinion of the committee, these notions will require extensive discussion and debate among the various stakeholders over the coming years before any broad consensus can be obtained.[817]

The authors of the MIT study made a similar observation and noted that

> No country has yet established an operating repository for high-level waste, and all have encountered difficulties with their programs. In many countries public and political opposition to proposed nuclear waste facilities and to the transportation of nuclear waste by road or rail has been intense, and public opinion polls reveal deep skepticism around the world about the technical feasibility of safely storing nuclear waste over the long periods for which it will remain hazardous.[818]

Given the difficulty of both the technical and political sides of repository development, it seems reasonable to conclude that a successful program will require several decades to complete.[819]

With the exception of the United States, no country currently plans to have a repository in operation before 2020 at the earliest (see Table 5.1). In the U.S., the opening date for the Yucca Mountain repository has been pushed back by the DOE at least twice from the original deadline of 1998 to at least 2012. It is now certain that the license application will be further delayed, which will again delay the program (see Section 5.2.2). Even if no further delays had occurred, the U.S. repository would still have opened 14 years behind its originally scheduled completion date and more than 25 years after Yucca Mountain was selected by Congress as the sole site for characterization in the United States.

Table 5.1: Estimates for the earliest anticipated dates at which geologic repositories for the disposal of spent fuel and high-level waste may open as summarized by the authors of the MIT report.[820]

Country	Earliest Anticipated Date of Repository Opening
United States	2012[a]
Finland	2020
Sweden	2020
Switzerland	2020 or later
France	2020 or later
Canada	2025 or later
Japan	2030
United Kingdom	later than 2040
Germany	date not yet determined

(a) In February 2006, Secretary of Energy Samuel Bodman stated that the Department of Energy can no longer make an official estimate for when the Yucca Mountain repository might open due to ongoing difficulties faced by the project.[821]

An additional point to note is that only the United States and Finland have officially selected sites for characterization with the expressed intent of actually building their repositories at those sites. In the U.S., the Yucca Mountain site was selected as the sole site for characterization in 1987, and in 2002 the President and U.S. Congress formally approved it as the proposed location for the first repository in the United States. In 2001, the Finnish parliament approved the construction of a test facility at Olkiluoto, in southwestern Finland, which, assuming the site proves suitable, will become the location for that country's permanent repository. In addition to the U.S. and Finnish programs, the French parliament authorized the construction of an underground laboratory at a site in Bure, Meuse in the province of Lorraine in 1999. An alternative site was also supposed to be evaluated, but no laboratory has yet been built in France. The French government is expected to begin debate on the future of their repository program in 2006. On the other hand, Sweden, Switzerland, and Japan are still evaluating potential sites, while Germany and the U.K. have both decided to postpone any additional work on their repository programs for the time being.[822] Given the history in the U.S., and the public concerns over radioactive waste in Japan and Western Europe, it is reasonable to conclude that at least some, and likely many, of these other country's programs will also be pushed back in time.

The rate of construction required to meet the global growth scenario would already be very challenging. Assuming that construction began next year, the global growth scenario would require one plant coming online somewhere in the world every 15 days between 2010 and 2050. To meet the more aggressive steady-state growth scenario, this rate would have to fall to one reactor coming online every six days. Thus, it is likely that the decision to begin the revival of nuclear power, as well as the actual construction of reactors, would need to occur years before a single repository would be opened anywhere in the world. This is not a reasonable situation given the history of nuclear waste management programs to date, and one that is likely to seriously exacerbate a problem that has plagued nuclear power for a half a century. The difficulties encountered in the development of a U.S. repository should serve to caution against the creation of any additional waste from new reactors.

Section 5.2.1 – General Uncertainties Regarding Geologic Disposal

The evaluation of repository performance is highly dependent upon the local geology and other site specific considerations. Thus, the particular uncertainties that are most important will also be highly site dependent. There are, however, three types of uncertainties that will be general to all repository programs. The first relates to the completeness of the models used to simulate the behavior of the engineered barriers and radionuclide transport through the geologic medium. The second relates to the long-term climate variability at the repository site due both to natural and anthropogenic influences. And the third relates to the changes in human behavior and consumption patterns as well as the continued evolution of our species over the very long times to be considered in evaluating repository performance. We will discuss aspects of each of these areas below and then highlight specific examples as they relate to the Yucca Mountain repository in later sections.

Similar to the issue of completeness discussed in the context of probabilistic risk assessments (see Section 4.3.2), the completeness of the conceptual models used to simulate the behavior of the repository presents a particular challenge in that it requires the designer to know what they don't know about the geologic and chemical processes that will be important at a given site. Adding to this difficulty is the fact that due to the very long times being considered, a conceptual model can never be proven correct, but can only be disproved by experimental work. As the National Research Council concluded in its 2001 evaluation of the U.S. repository program

> Radionuclide transport out of the waste form and repository (the near-field environment) through the more distant geological host medium (often called the far-field or the geosphere) to that part of the environment accessible by humans (the accessible environment) is probably the most uncertain area of modeling.[823]

Some of these uncertainties, particularly those associated with the engineered barriers, can be reduced by using only naturally occurring and thermodynamically stable materials for the waste packages and by choosing a site with a chemistry likely to retard radionuclide transport. The U.S. has done the opposite, however, and chosen to use novel, human-made alloys whose long-term behavior is not as well understood experimentally, and a site with a chemical environment favorable to accelerated transport of important radionuclides.

Uncertainties in the geologic behavior of the site and in the mechanisms of radionuclide transport are, in many ways, more complicated to address than those of the engineered barriers, and are largely not under the control of repository designers beyond selection of the site. For example, experiments in permeable media have shown that the results of radionuclide migration experiments are scale dependent and, therefore, that parameters measured on the laboratory scale may not always be appropriate for use at the repository scale.[824] In addition, if important mechanisms for radionuclide transport to the biosphere have been overlooked then the model results may not adequately reflect the true behavior of the repository. As summarized by the National Research Council,

> Simply stated, a transport model is only as good as the conceptualizations of the properties and processes that govern radionuclide transport on which it is based. If the model does not properly account for the physical, hydrogeochemical, and when appropriate, biological processes and system properties that actually control radionuclide migration in both the near- and far-fields of the repository system, then model-derived estimates of radionuclide transport are very likely to have very large -- even orders of magnitude -- systematic errors.[825]

Unfortunately, the history of DOE's past use of models to support decision making reveal a number of examples that demonstrate the dangers inherent in relying on incomplete physical models for decision making.

When many of the sites within the U.S. nuclear weapons complex were originally founded, it was believed that their arid climate and thick unsaturated zones would protect the groundwater for hundreds to thousands of years from any contaminants buried in shallow unlined trenches. Measurements over time, however, have revealed these early assumptions to be in substantial error. For example, the travel time estimated by the DOE for radionuclides to reach the Snake River aquifer under the Idaho National Engineering and Environmental Laboratory (INEEL) in Idaho has fallen from tens of thousands of years in their models from the mid-1960s to just a few tens of years today. This thousand fold increase in the estimated migration rate was prompted by the discovery that plutonium had already reached the groundwater 200 meters beneath the Radioactive Waste Management Complex at INEEL.[826]

A second example is the fact that tritium, a radioactive form of hydrogen, has been found 48 meters below the waste disposal facility at Beatty, Nevada, despite the fact that it was originally predicted that no tritium would migrate from the site at all over this timescale. A third example of this kind of failure was the DOE prediction that the low rain fall and 90 meter thick unsaturated zone below the waste disposal sites at the Hanford reservation in Washington State would prevent any contamination from reaching the groundwater. Unfortunately, however, fission products and other radionuclides from leaking high-level waste tanks have already been found to have reached the water table below Hanford, in some areas, after just 60 years.[827]

Finally, a fourth example of the failure of conceptual models can be found in the DOE's analysis of plutonium migration from the underground nuclear weapons tests conducted at the Nevada Test Site (NTS). Originally, the DOE claimed that the heat from the nuclear explosions as well as the chemical interactions of plutonium with the soil would result in its being fixed in the rock and, therefore, that it would not be transported any significant distance through the environment. However, this analysis failed to recognize the importance of colloids (small particles suspended in water that are typically less than 0.001 millimeters in diameter) in mediating the movement of plutonium and other actinides. Measurements from NTS have found that plutonium from at least one test has already migrated as much as 1.3 km in just 30 years.[828] The researchers who made this initial discovery, concluded that "[m]odels that either predict limited transport or do not allow for colloid-facilitated transport may thus significantly underestimate the extent of radionuclide migration."[829] In all four of these examples, the conceptual models relied upon for decision making failed to accurately predict contaminant transport, and it was only after observations of contamination spreading into the environment were made that they were revised.

The second general uncertainty in of repository performance relates to changes in the climate that will occur over the very long times involved. The climate conditions that will prevail at a site such as temperature and annual rain fall are closely related to the issue of radionuclide transport. Over the next few hundred thousand years, the Earth's climate will go through a number of natural variations that can be roughly inferred from past climate events. The further into the future one looks, however, the larger the associated uncertainties will generally become regarding the timing and intensity and duration of these natural changes. The impact of these uncertainties on repository performance are likely to be more significant if a "dry" site such as Yucca Mountain is selected where keeping water away from the waste is of the highest priority. Other types of climate changes will potentially be important for repositories located in parts of Canada or northern Europe. Most important among these climate effects would be the impact of potential future glaciation.

Additional uncertainties are introduced by the impact of anthropogenic climate change caused by the build up of greenhouse gases. Global warming may influence both the timing and intensity of natural climate changes as well as cause new types of climate states to arise in certain areas. As such, global warming may alter the local meteorological and hydrological conditions in the region of the repository and to affect future human usage of the surrounding area. Despite their potential importance, the U.S. does not currently consider the impacts of anthropogenic climate change in its evaluation of Yucca Mountain. Like the previous failures of some of the DOE's other conceptual models discussed above, this omission may prove to be an important one as more is learned about the extent of the anthropogenic changes that may occur.

Finally, over the very long times involved with nuclear waste, significant changes will occur in human consumption patterns and lifestyles. To understand the importance of this uncertainty, one has only to recall that less than 600 years ago no Europeans were living in permanent colonies on the American continents, that as little as 12,000 years ago no society practiced wide spread agriculture or kept herds of domesticated animals for food, and that 30,000 years ago Neanderthals could still be found living in isolated pockets across Europe and western Asia. In order to try and address these unknowns, it is necessary to set strict standards for repository performance and to focus on conservative scenarios for future human activities. In addition, it is helpful to place a particularly strong focus on robust exposure pathways such as drinking water which are unlikely to change significantly even over very long times.

Demonstrating that these general areas of uncertainties have adequately been taken into account is a very complex task and one that will require a significant amount of time. As noted above, there are a number of choices that can be made in site selection and in the selection of engineered barriers that can help to reduce both the number of unknowns as well as the importance of those unknowns that will remain. While these factors are obviously not the only considerations, they should be an important part of any decision made on the design and siting of a repository. We will see in the next section, however, that politics, more than sound scientific considerations, has influenced the most important choices in the U.S. repository program to date.

Section 5.2.2 – The History of Geologic Disposal in the United States

The U.S. National Academy of Sciences published its first examination of the nuclear waste problem in 1957, the same year that the first commercial reactor in the U.S. came online at Shippingport.[830] At that time, the NAS identified a number of possible disposal strategies, but concluded that geologic disposal of high level waste in deep salt beds was likely to be the most promising option. Little was done initially to pursue these findings, however, as the AEC chose instead to focus on nuclear weapons production and the civilian nuclear complex focused on expanding the number of reactors in operation. Meaningful experiments on the suitability of salt formations for the disposal of spent fuel did not begin until 1965, nearly a decade after the first NAS report was published. The site chosen for characterization was an abandoned salt mine in Lyons, Kansas. This site, however, was soon rejected due both to local and state opposition as well as to important questions that were raised regarding the movement of water through the surrounding geology. Another decade later, in 1975, the Energy Research and Development Administration (the successor to the Atomic Energy Commission and predecessor of the Department of Energy) began an investigation of various geologic formations located in 36 different states for their potential suitability as disposal sites. The decision by the Ford and Carter administrations in 1975-76 to move away from commercial reprocessing in the U.S. due to proliferation concerns (see Section 3.2) gave new importance to the search for a permanent disposal site for the spent fuel that was building up at reactors sites around the U.S. By 1980, the DOE had chosen to limit the search to nine sites in six states: Louisiana, Mississippi, Nevada, Texas, Utah, and Washington State.

The 1982 passage of the Nuclear Waste Policy Act (NWPA) was the watershed event in the government's attempt to deal with problem of spent nuclear fuel. The NWPA specified that commercial spent fuel and high-level nuclear waste from military plutonium production were to be disposed of in mined repositories to be built and operated by the Department of Energy. There would be two such repositories built with one in the western United States (limited to a capacity of 70,000 metric tons which would be made up of 63,000 MT of spent fuel and 7,000 MT of military high-level waste) and a second repository in the east which would accept the remainder of the waste discharged by existing reactors. This division of the waste was imposed in order to ensure an equitable geographic distribution of disposal sites given that the eastern states generate far more nuclear electricity than western states.[831] The construction of the repository was to be paid for, in part, by a fee of 0.1 cents per kWh of electricity generated by nuclear plants. Additional funding for the repositories would be provided by the taxpayers through spending by the DOE. The law also mandated that the Department of Energy would have to meet health and environmental standards set by the Environmental Protection Agency and that the Nuclear Regulatory Commission would be responsible for licensing the repositories and ensuring that the EPA standards were properly enforced. Finally, the 1982 NWPA set a firm, legally binding, timetable for repository development which required the DOE to enter into contracts with utilities requiring the agency to begin accepting spent nuclear fuel for disposal by no later than January 31, 1998.

In response to the passage of the Nuclear Waste Policy Act, the DOE narrowed its list of possible sites to just three: the Hanford reservation in Washington State, the Nevada Test Site, and a location in Deaf Smith County, Texas. The final choice of which site to focus on was not based upon a careful scientific comparison of the relative advantages and disadvantages of the three sites, much less a determination that any of the three were actually suitable for disposal of the waste. Instead the decision was made by politicians in Congress when they amended the NWPA in 1987 to limit the DOE to characterizing only a single site: Yucca Mountain, a volcanic ridge on the Nevada Test Site approximately 160 km (100 miles) to the north and west of Las Vegas. The choice to focus on Yucca Mountain alone was made despite the fact that a report by the National Research Council in 1983 had concluded that the peak doses to the maximally exposed individual from a repository at Yucca Mountain would likely be very high. The Congressional decision makes more sense when it is noted that, at the time the amendment to the NWPA was enacted, Nevada was a politically weak state with a far smaller population than Washington State or Texas, and that it had only two relatively new Senators in Congress. In addition to imposing a decision on the location for the western repository, the 1987 amendment also indefinitely put off any work on a second repository in the east until Yucca Mountain was licensed.

By selecting a single site for the DOE to characterize and simultaneously requiring the department to open a repository on a fixed deadline, Congress, in effect, turned the scientific process on its head in the U.S. repository program. Instead of being able to easily approach the site with an open mind and try to determine *if* it is suitable for disposing of spent fuel, the 1987 amendment essentially sent the DOE out with the mandate to try and prove that Yucca Mountain *was* a suitable site. The pressure on DOE to open a repository was heightened by the fact that laws in California and elsewhere had effectively imposed a moratorium on new nuclear construction in those states until a strategy for the long-term management of the waste had been formally approved by the federal government. This linking of the nuclear waste problem to the development of new reactors put political pressure on the DOE from the nuclear power industry, and created internal pressure as well given DOE's explicit and active role as a promoter of nuclear power. Adding to this, on November 14, 1997, the U.S. Court of Appeals for the District of Columbia ruled that the DOE would be financially liable for damages if it missed the mandated deadline to begin accepting spent fuel. This ruling from the District Appeals Court came shortly after the March 1997 decision by the DOE to delay applying for a repository license application until at least March 2002. As of March 2006, the application had yet to be filed, and the date when it might finally be filed remains uncertain.[832]

In a review of the repository program, the General Accounting Office concluded that

> On the basis of the information we reviewed, DOE is unlikely to achieve its goal of opening a repository at Yucca Mountain by 2010 and currently does not have a reliable estimate of when, and at what cost, such a repository can be opened.[833]

The GAO found that the NRC had requested 293 additional pieces of technical work from the DOE that needed to be completed before the DOE's could support its license application for the repository. By November 2001, however, the DOE had only completed work on about one-fifth of these 293 areas and the NRC had been provided with the information for review on just three-fourths of those that had been completed. As a result of this large amount of outstanding work, Bechtel, the site contractor responsible for constructing the repository, concluded that the earliest the DOE would likely be ready to submit a license application would be January 2006. At that time the DOE rejected Bechtel's conclusion and continued to plan for submitting a license in March 2002. However, the date for filing the application has now been pushed back by the DOE past the end of fiscal year 2007. Based on its analysis, the GAO concluded that the repository was not likely to be opened before 2015, taking into account existing delays and an extended period of review by the NRC. The GAO went on to concluded, that while it was "within his discretion, it may be premature for the Secretary of Energy to make a site recommendation in the near future."[834]

In their January 24, 2002 report, the Nuclear Waste Technical Review Board, a scientific advisory body created as part of the 1987 amendment to the NWPA, whose members are nominated by the National Academy of Sciences and appointed by the President, concluded that

> In evaluating the DOE's technical and scientific work related to individual natural and engineered components of the proposed repository system, the Board finds varying degrees of strength and weakness. Such variability is not surprising, given that the Yucca Mountain project is in many respects a first-of-a-kind, complex undertaking. *When the DOE's technical and scientific work is taken as a whole, the Board's view is that the technical basis for the DOE's repository performance estimates is weak to moderate at this time.*[835]

In their review of models used by the DOE to evaluate the long-term performance of the repository (called a Total System Performance Assessment), the NWTRB went on to conclude that

> Performance assessment is a useful tool because it assesses how well the repository system as a whole, not just the site or the engineered components, might perform. However, gaps in data and basic understanding cause important uncertainties in the concepts and assumptions on which the DOE's performance estimates are now based. *Because of these uncertainties, the Board has limited confidence in current performance estimates generated by the DOE's performance assessment model.*[836]

Similar concerns with the DOE performance assessment have been raised by other scientists as well (see Section 5.2.5).

Despite the concerns over the suitability of the site and the completeness and strength of the DOE's technical work that had been raised by that time, President Bush recommended Yucca Mountain to Congress on February 15, 2002, following the advice of Energy Secretary Spencer Abraham. Note that this was less than three months after the GAO report cited above was published and less than one month after the NWTRB report was submitted. The Governor of Nevada subsequently exercised his right of "veto" under the Nuclear Waste Policy Act which sent the matter back to Congress. The House voted to override the Governor's veto on May 8, 2002, and the Senate followed suit on July 9, 2002. President Bush signed these decisions into law on July 23, 2002, formally selecting the Yucca Mountain site as the location for the United States' first repository.[837]

By law, the DOE had 90 days from the date of President Bush's signature to submit a license application for Yucca Mountain to the NRC. The

NRC would then have three years to review the application with the possibility of extending their review by at most one year if they went before Congress and the Secretary of Energy to explain their reasons for the delay.[838] However, the DOE had not submitted a license application as of May 2006, and the application is likely to be been even further delayed. It has been more than three and a half years since the formal selection of the site or nearly 15 times longer than the time limit allowed by law. It seems clear that the GAO's conclusion was substantially correct, and that the recommendation of the site in 2002 by the DOE and President Bush was, in fact, quite premature.

As a result of its failure to meet the January 31, 1998 deadline for beginning to accept the spent fuel and to the continually slipping date for the opening of the repository, the DOE has been sued by a number of utilities to recover costs associated with the continued storage of waste on-site. The courts have so far found the DOE specifically liable for fuel storage costs at three permanently closed reactors. The total amount of this liability, however, has yet to be agreed upon. This delay in setting the total amount of the financial penalty is due, in part, to the fact that the date when the fuel might finally be accepted by the DOE remains speculative. The utilities involved in the current lawsuits are already claiming as much as $2.4 billion in damages. In a separate lawsuit, Exelon reached an agreement with the government on August 10, 2004 regarding its own claims for reimbursement of ongoing fuel storage costs. In the settlement, Exelon would be paid $300 million from the Federal Judgment Fund if the DOE begins accepting the waste by 2010 and up to $600 million if the DOE does not begin accepting waste until 2015. Claims by approximately 20 other nuclear utilities are currently pending.[839] These lawsuits have led to renewed proposals from some within the government for the construction of centralized interim storage facilities at existing DOE sites and to the resumption of reprocessing commercial fuel (see Section 3.2).

Section 5.2.3 – Ready, Fire, Aim... The DOE Strategy at Yucca Mountain

The DOE strategy at Yucca Mountain has been referred to as the D.A.D. method of decision making: that is Decide, Announce, Defend. This mindset is due, in large part, to the peculiar history of the high-level waste disposal program in the U.S., outlined in the previous section, in which Yucca Mountain was selected by politicians as the only site for characterization despite previous scientific analysis showing that peak doses from site were likely to be quite high. In addition, this mindset grows out of the history of the Department of Energy (and before it the Atomic Energy Commission and the Energy Research and Development Agency) as a self regulating body with a strong culture of secrecy, resulting from its primary focus on the manufacture and maintenance of nuclear weapons. The widespread environmental contamination that occurred across the nuclear weapons complex is, in part, a result of this institutional culture.

At Yucca Mountain, the decision to seek a license has already been made and announced by the DOE. The Department is now seeking to defend its selection of the site to the NRC and more generally to the public. While a complete review of the issues and uncertainties relevant to the Yucca Mountain site is beyond the scope of this report and could fill volumes, we will briefly touch upon some of the more important areas to give a sense of the problems that have been encountered to date. The extent of these uncertainties and their impact on the potential performance of the repository is obviously important to consider. However, the financial liability and political pressures faced by the DOE against further delaying the repository are significant. The negative impact of this type of pressure was noted by the National Research Council when they concluded that

> There is a danger that a legalistic, prescriptive regulatory environment or a project forced to meet deadlines can induce scientists charged with developing a performance assessment to assume that they have no uncertainty in their conceptual models.[840]

Before proceeding, we must first review the characteristics of the proposed disposal site. The repository is to be located in the unsaturated zone of Yucca Mountain which is composed of porous volcanic tuff. The mountain's rock was deposited as hot ash and cracked as it cooled resulting in a highly fractured geology. Over time, the rock was built up in beds which range from a few meters to a few hundred meters in thickness. The footprint of the planned repository (assuming a capacity of 70,000 MT of waste) will be approximately 1,170 acres or an area equivalent to a square roughly 2.2 km on a side. Under current climate conditions, the area over the repository gets approximately 17 centimeters (6.7 inches) of rain per year, while higher elevations nearby get nearly two-thirds more precipitation.[841]

The spent fuel and high-level waste would be placed into five meter long casks with an outer shell made from a nickel based alloy. Each cask will hold 20 to 40 fuel assemblies and titanium drip shields will be erected over the canisters after placement. These casks would be located in tunnels, called drifts, that are 600 meters long and 5.5 meters in diameter. Neighboring tunnels would be separated by 81 meters between their centers.[842] The amount of empty space that would surround each waste cask is determined by the maximum temperature that the designers ultimately choose to allow in the repository. Because the waste will continue to give off substantial amounts of heat for some time after placement due to radioactive decay of the fission products and transuranic elements, the more densely packed the repository is, the higher the peak temperature will be. For higher burnup fuel, the casks would need to be spaced further apart in the tunnels to maintain a given peak temperature, thus limiting the savings achieved in the required repository volume.

Between FY85 and FY05, the expenditures on developing Yucca Mountain by the Office of Civilian Radioactive Waste Management totaled $8.77 billion. Of the nearly nine billion dollars that have been spent so far, approximately 30 percent has come from funds allocated for the disposal of high-level waste generated during the manufacture of plutonium for nuclear weapons. This division of funding is surprising considering the fact that the military high-level waste is currently allocated just 10 percent of the 70,000 MT statutory capacity of Yucca Mountain, and that the military waste is expected to make up just 5 percent of the radioactivity to be disposed of in the repository. The FY06 budget requests an additional $651.4 million for Yucca Mountain, with 54 percent coming from appropriations for military waste.[843]

The latest official estimate for the total life-cycle cost of the repository is $57.5 billion, but that estimate was made before the most recent delays in the program occurred and when only a small number of reactors had received 20 year license extensions.[844] As of the end of November 2005, however, 37 reactors had received license extensions, 12 applications were under review by the NRC, and a number of other operators had notified the NRC of their intent to seek additional license renewals over the coming years.[845] The DOE now estimates that as much as 105,000 metric tons of spent fuel and high-level waste will require disposal by 2035, which is roughly 25 percent more than the spent fuel inventory used in deriving the official estimate for the total life-cycle cost of Yucca Mountain in 2001, and 50 percent more than the current statutory limit of the repository.[846]

With respect to the suitability of the Yucca Mountain site, the first concern is not technical, but political. The land upon which the Nevada Test Site and Yucca Mountain are located is claimed by the Western Shoshone Nation which opposes the placement of the repository. In 1863, the Treaty of Ruby Valley between the Western Shoshone and United States government legally affirmed the boundaries of the Western Shoshone land and gave approval for access to that land for only a limited number of specified purposes. Although the treaty was never formally superseded, the U.S. government has argued that the Western Shoshone lost their rights as far back as 1872 as a result of the encroachment of settlers. However, in 2001, the United Nations Committee on the Elimination of Racial Discrimination, part of the U.N. Commission on Human Rights, expressed their "concern with regard to information on plans for expanding mining and nuclear waste storage on Western Shoshone ancestral land."[847] As previously noted, the impacts of uranium mining and milling have fallen disproportionately upon Native Americans and other Indigenous Peoples around the world. The lack of informed consent from those with a deep cultural and historical connection to the land should alone be sufficient to prevent any further consideration of the Yucca Mountain site. As the rest of this section will demonstrate, however, there are a number of important technical reasons to oppose the choice of Yucca Mountain as well.

In 2003, the International Atomic Energy Agency (IAEA) published its *Scientific and Technical Basis for the Geologic Disposal of Radioactive Wastes* which included its recommendations for the geologic characteristics most suitable for repository development. One of these properties was that the site should have "[l]ong term (millions of years) geological stability, in terms of major earth movements and deformation, faulting, seismicity and heat flow."[848] Yucca Mountain, however, is located in a tectonically active area both seismically and with respect to volcanism. A recent earthquake in the area occurred on June 29, 1992, with an epicenter just 20 km (12 miles) from Yucca Mountain and measured 5.6 on the Richter scale. In the past million years, eight volcanic eruptions are known to have occurred within 50 km (31 miles) of the repository site. In addition, the volcanism in the Yucca Mountain area is not yet fully understood making it more difficult to evaluate the accuracy and completeness of the models used to predict the impact of future eruptions on repository performance over the very long times involved.[849]

A second geologic property recommended by the IAEA in choosing a repository site was that it should have a "[s]table geochemical or hydrochemical conditions at depth, mainly described by a reducing environment and a composition controlled by equilibrium between water and rock forming minerals."[850] The importance of maintaining a reducing chemical environment in the repository was summarized by the authors of the MIT report as follows:

> In siting a repository, it is important to select a geochemical and hydrological environment that will ensure the lowest possible solubility and mobility of the waste radionuclides. The geochemical conditions in the repository host rock and surrounding environment strongly affect radionuclide transport behavior. For example, several long-lived radionuclides that are potentially important contributors to long-term dose, including technetium-99 and neptunium-237, are orders of magnitude less soluble in groundwater in reducing environments than under oxidizing conditions.[851]

In addition, the behavior of the uranium oxide fuel itself is strongly influenced by repository chemistry. Despite their dominance of the radioactivity at the time of emplacement, the fission products only make up about three to four percent of the mass of the spent fuel while the transuranics make up about one percent. The remaining mass of the spent fuel is unfissioned uranium dioxide (UO_2).[852] The uranium dioxide fuel is thermodynamically unstable in an oxidizing environment where moisture is present and will corrode over time. On the other hand, UO_2 is orders of magnitude more stable in a reducing environment and would thus corrode much less quickly.[853]

These reinforcing considerations (increased UO_2 stability and decreased radionuclide mobility) have led every country except the United States to pursue a repository located in a saturated, reducing environment. Yucca Mountain is the only proposed repository site that will be located in unsaturated rock above the water table in which the presence of both air and moisture will maintain an oxidizing environment around the waste.[854] This choice greatly increases both the number and importance of the uncertainties remaining with the proposed U.S. repository design.

Most importantly, the choice to locate the repository in an oxidizing environment creates the need to keep the waste as dry as possible to limit corrosion and subsequent migration of the radionuclides. This requirement places a great burden on the accuracy of the models used to predict the flow of water through the host rock as well as on the models for the performance of the engineered barriers. The earliest estimates for the rate of water infiltration at the Yucca Mountain site were based on expert opinion rather than detailed site characterizations. For example, estimates from 1984 assumed that there would be little flow of water through the fractures in the rocks until they were nearly saturated. Under this assumption, the geology of the Yucca Mountain site would limit water infiltration at the surface to 0.0045 meters per year and to just 0.0002 meters per year through the rocks found at the repository depth (approximately 250 to 300 meters below ground). For these early estimates of the infiltration rate, it was believed that it would take between 500 and 20,000 years for water to migrate from the surface into the mountain and reach the repository depth. On the basis of these assumptions the DOE's 1986 site assessment concluded that there would be no release of radionuclides from Yucca Mountain to the biosphere within the first 10,000 years even without the use of engineered barriers.[855] In other words, it was assumed that the arid environment and natural geology alone would combine to ensure effective waste isolation.

As with the previous examples of DOE reliance on an arid climate and thick unsaturated zone to protect the ground water discussed in Section 5.2.1, experiments at Yucca Mountain conducted in the late 1990s began to show indications that the conceptual models was incomplete. Chlorine-36 is a naturally occurring and anthropogenic radionuclide that was distributed around the globe in large amounts as a result of above ground nuclear tests conducted in the Pacific Ocean. The vast majority of this radionuclide found in the environment today is known to have been generated in 1950s and 1960s. In direct contradiction to the expected results based on the estimated rate of water infiltration, Cl-36 from fallout was found in some of the water samples collected at the repository depth from experimental tunnels at Yucca Mountain. This finding implied that there may be "fast" water pathways which had allowed water to migrate from the surface down to the level of the repository in as little as 40 to 50 years rather than the hundreds to thousands of years previously assumed. Subsequent investigations at Yucca Mountain and related sites have provided strong additional evidence for the existence of fast water pathways through the rock.[856] Thus, what was supposed to be a "dry" repository, was now found to be potentially far wetter than expected.

These fast pathways are likely to be related to faults and connected fissures which provide highly conductive channels that allow water to move quickly through the rock. In addition, faults that come near the surface are more likely to erode and create local depressions in the rock that can collect surface water and increase the rate of infiltration. These considerations connect to uncertainties relating to the seismicity of the site since earthquakes can lead to the formation of new fissures in the rock as well as enlarging and connecting existing cracks. To date there is no single conceptual model that has been proposed which is capable of completely explaining all of the data that has been collected on the nature of fluid flow through the unsaturated zone at Yucca Mountain.[857] In November 2003, the Nuclear Waste Technical Review Board concluded that

> The active fracture model may be a reasonable approach to this very challenging problem [of simulating flow through the unsaturated zone], but it has never been tested adequately, and the key controlling active fracture geometric parameter is not measurable using any presently known technique.[858]

The DOE's current reliance on engineered barriers rather than the site geology for containment is due, in large part, to the questions that have been raised by these findings of fast water pathways.

In addition to its shift to a focus on engineered barriers, the DOE has also claimed that two natural compensatory features will work to keep the water that is migrating through the mountain from coming into contact with the waste. These are the so-called vaporization and capillary barriers. The current design of the repository calls for the waste to be placed such that the decay heat will keep the temperature in the repository above the boiling point of water for approximately 1,000 years, reaching a high temperature of 160 to 180 °C (320 to 356 °F) several decades after closure. This initial period is referred to as the thermal pulse.[859] It is believed by the DOE that during the thermal pulse, most of the water approaching the drifts would be vaporized and driven back into the rock preventing its penetration into the tunnels around the waste. However, the Nuclear Waste Technical Review Board has highlighted a number of uncertainties remaining in the DOE's analysis of the vaporization barrier.

Recent field data and laboratory analyses all point to a lower average value for the thermal conductivity of the rocks surrounding the repository than is currently used by the DOE. A lower value for the conductivity means that less heat is conducted away by the rocks resulting in a higher temperature in the tunnels. In addition, the drifts are likely to degrade over time, resulting in rocks landing on some of the waste packages. These rocks would act to insulate the waste packages, leading to higher local temperatures and other disturbances in the distribution of heat in the tunnel. Finally, the influence of natural ventilation and air circulation following the closure of the repository is not currently taken into account, which could result in a cooler overall repository temperature. Because these physical processes operate in different directions, no conclusion could be made as to how the combination might ultimately affect the DOE's projected temperatures. The NWTRB did, however, conclude that "the DOE has not demonstrated that the conditions required for a pervasive vaporization barrier to form will occur everywhere" and that "[t]he DOE's view is based on an insufficient analysis."[860]

The other natural barrier to water infiltration claimed by the DOE is the so-called capillary barrier. The DOE claims that water migrating towards the drifts will be diverted around the tunnel by capillary forces in the rock. However, the Nuclear Waste Technical Review Board concluded that the irregular profiles of the tunnel walls combined with drift degradation over time and the natural voids present in the rock, would act to limit the extent of this barrier and could allow more water into the tunnel than DOE currently assumes. As with the vaporization barrier, the NWTRB concluded that "the DOE has not demonstrated that the conditions required for a capillary barrier to form are satisfied throughout the drifts" and that "[t]he DOE's view is based on insufficient data and modeling."[861]

Further uncertainties are introduced by the impacts of climate change. Over the next several hundred thousand years, the climate of Nevada will pass through a number of natural changes in climate in addition to experiencing the impacts of anthropogenic climate change. Currently the impacts of global warming are considered too uncertain by the DOE to be amenable to long-term prediction, and therefore anthropogenic changes are largely left out of the most recent performance assessments. Of the four naturally occurring climate states that have been identified for the Yucca Mountain area, the modern climate has the least effective moisture. The dominant climate states have been both wetter and colder than present, resulting in more water on the ground and lower evaporation rates.[862] Thus, all of the other climate periods likely to be experienced at Yucca Mountain will have an increased amount of water infiltration into the repository. As summarized by Jane Long of the University of Nevada, Reno and Rodney Ewing of the University of Michigan

> If climate change were to produce a larger influx of water, saturation in the mountain could increase. Permeability under any proposed model increases nonlinearly with saturation. Small increases in percolation flux could significantly increase fluid flow through the repository horizon. This nonlinear re-

sponse is one of the greatest challenges in predicting the behavior of hydrologic systems over long periods.[863]

Finally, there is also the potential for larger amounts of water to infiltrate into the mountain during the severe storm events that occur in the area every few thousand years. The frequency and intensity of these storms would be influenced heavily by the prevailing climate state and could be affected by anthropogenic climate changes as well.[864]

Because of their potential impact on water infiltration and other elements of repository performance, predicting the onset and duration of different climate states introduces important uncertainties into the assessment of long-term repository performance. Different methods of interpreting the past climate record result in starting dates that differ by more than 1,500 years even for climate shifts that are predicted to occur within the next 10,000 years. Predictions for the start date and duration of later climate changes vary by even larger amounts the further into the future one goes. The discrepancies between the results from different methodologies can approach ten thousand years over the timescales of interest for repository performance.[865]

Given that the amount of water infiltration is sufficient over the very long-term to corrode the waste packages and release the radionuclides from the fuel, the accuracy of the models governing their transport through the rock to the biosphere is obviously of significant importance. Experiments conducted since the late 1990s have found that the mobility of plutonium and other transuranics under certain conditions could be much higher than earlier models would have predicted, and that these radionuclides could be transported over long distances from the repository via adsorption to colloids.[866] The models used by the DOE in their performance assessment of Yucca Mountain do include some consideration of colloid mediated transport. However, despite the fact that multiple colloidal phases have been found in the groundwater at the Nevada Test Site, including some that are known to have a particularly high capacity for incorporating actinides, the DOE considers only two types of colloids in their analyses.[867]

The importance of these uncertainties is exacerbated by the very long times involved. Initially this problem was dealt with by simply limiting consideration of the impacts of Yucca Mountain to the next 10,000 years. While this timescale is far better than the much shorter times being considered in current discussions regarding the potential overall impacts of anthropogenic climate change, it is not sufficient to capture the true dangers of the very long-lived radionuclides present in high-level nuclear waste. In its 1995 assessment of the Yucca Mountain standards, the National Research Council concluded that

> The current EPA standard contains a time limit of 10,000 years for the purpose of assessing compliance. We find that there is no scientific basis for limiting the time period of an individual-risk standard in this way. We believe that compliance assessment is feasible for most physical and geologic aspects of repository performance on the time scale of the long-term stability of the fundamental geologic regime — a time scale that is on the order of 10^6 years at Yucca Mountain — and that at least some potentially important exposures might not occur until after several hundred thousand years. <u>For these reasons, we recommend that compliance assessment be conducted for the time when the greatest risk occurs, within the limits imposed by long-term stability of the geologic environment.</u>[868]

Both the EPA and NRC initially rejected this conclusion due to their belief the uncertainties would grow too large if the time of peak dose was considered. In 2004, the U.S. Court of Appeals struck down the time limit within the EPA's guidelines, and required the agency to draft new regulations that are consistent with the 1995 recommendations of the National Research Council cited above.[869]

Continuing the NRC and DOE tradition of changing regulations that might prevent further development of the Yucca Mountain, the EPA issued a draft of its new Yucca Mountain standard in August 2005. Surprisingly, this proposed rule would be the least protective by far of any repository regulation anywhere in the world.[870] The draft rule seeks to create a two tiered regulation for the peak dose with different standards in effect at different times. Within the first 10,000 years, the EPA retained its previous dose limit of 15 millirem per year with a 4 millirem per year sub-limit for the drinking water pathway. Between 10,000 and one million years, however, the dose limit jumps to 350 millirem per year without any further restrictions on the dose from drinking water.[871] This later dose limit is three and a half times the current dose limit for the general population from all anthropogenic radiation sources, except medical exposures.[872] For comparison, the French repository program has a limit of just 25 millirem per year at the time of peak dose.[873] To put the proposed Yucca Mountain dose limit into perspective another way, if a person is exposed to 350 mrem per year for 70 years, the excess risk of developing a cancer according to the BEIR VII Committee estimates would be approximately 1 in 45 for men and 1 in 30 for women. The risk of developing a fatal cancer would be about half that.[874] This is a completely unprecedented level of risk for the EPA to put forth as acceptable.

The disparity between the pre and post 10,000 year dose limits proposed for Yucca Mountain (already a factor of more than 23) is made even more significant by the fact that they refer to different statistical quantities. In its license application to the NRC, the DOE will rely on the results of an integrated computer model for repository performance. The model, known as Total System Performance Assessment (TSPA), discussed in more detail in Section 5.2.5, attempts to incorporate uncertainty in repository performance by generating a distribution of possible doses and assigning each dose estimate a probability of occurring. Over the first 10,000 year, the 15 millirem per year dose limit applies to the average of all doses estimated by the TSPA. Between 10,000 and one million years, however, the 350 millirem dose limit is compared against the median dose (i.e. the dose for which half of the estimates fall above it and half below).[875] For the distributions of projected doses generated by DOE, the peak mean dose is always higher than the peak median dose. This is because the average gives a greater weight to very high estimates of the potential dose than the median does which simply counts the number of estimates above or below a certain value irrespective of how far above or below that value they are. It is telling to note that the version of the DOE models published in the Yucca Mountain Environmental Impact Statement predict a median dose after 10,000 years that would be in compliance with the new 350 millirem standard proposed by the EPA, but that the mean dose over that time would not be able to meet the stricter 100 millirem per year dose limit proposed by the nuclear industry.[876]

Using the DOE's estimates, we find the peak mean dose after 10,000 years is about a factor of three to four times higher than the peak median dose.[877] Taking this into account, the actual dose limit proposed by the EPA for times after 10,000 years is between 70 and 90 times larger than the 15 millirem per year dose limit in effect before 10,000 years. The DOE estimates that the 95th percentile peak dose (i.e. the dose for which only 5 percent of the estimates were higher) is approximately a factor of four higher than the mean dose.[878] Thus, a repository design that could be shown to be in compliance with the EPA's proposed standard could, in fact, have an average peak dose of more than one rem per year and a 95th percentile peak dose on the order of four to five rem per year. These upper level doses are comparable to the current worker dose limit, and would be extremely dangerous to the population. For example, if a person was exposed to four rem per year for 70 years, the risk of dying from cancer would be 1 in 6 for men, and nearly than 1 in 4 for women.[879] In such a case, living near the repository would be about like playing Russian Roulette for the most highly exposed individuals.

In addition to the creation of a very unprotective two tiered dose limit, the proposed EPA rule also concluded that, in performing the assessment beyond 10,000 years,

> The DOE should not project changes in society, the biosphere (other than climate), human biology, or increases or decreases of human knowledge or technology.[880]

The dramatic changes in lifestyle and global population distributions over the last 10,000 years make clear the significant uncertainties inherent in such restrictions. In addition, with respect to natural changes in the climate, the EPA concluded that

> The DOE must assess the effects of climate change. The climate change analysis may be limited to the effects of increased water flow through the repository as a result of climate change, and the resulting transport and release of radionuclides to the accessible environment. *The nature and degree of climate change may be represented by constant climate conditions.* The analysis may commence at 10,000 years after disposal and shall extend to the period of geologic stability [approximately one million years].[881]

In other words, the only effect of climate change that the DOE needs to consider is the net influx of water, and even this parameter is to be held constant beyond 10,000 years. Thus, in response to the court's requirement that the EPA follow the National Research Council's recommendations and set limits for the time of peak dose, the agency has chosen to propose a rule that will allow the DOE to leave its models for climate and human behavior fixed and to successfully license a repository that their own models could show as having a peak mean dose in excess of one rem per year. While clearly unacceptable as a scientifically defensible standard, the proposed EPA rule is consistent with the history of Yucca Mountain in that any regulatory obstacle to the repository's development has been recast in such a way that the DOE would be able to move forward with the project.

As a general rule, with increasing complexity in the geology of a repository site comes increasing uncertainty in the predictions of its long-term performance. While far from an exhaustive treatment, the above discussion serves to highlight the fact that Yucca Mountain is a highly complex site which is geologically unique in many important ways from any other sites being considered by any other repository programs anywhere else in the world. While universal acceptance of a particular location is not likely, for a repository program to gain the informed consent of the communities it will affect, the site should be chosen such that scientists and non-scientists can reach a strong consensus on its appropriateness. This is unlikely ever to be the case with Yucca Mountain. Its overall complexity and the associated uncertainties make it likely that scientists will continue to disagree over the suitability of the site. In fact, Yucca Mountain cannot meet even the most basic requirement of a repository; namely that it maintain the peak dose to an acceptably low level. The proposal put forth by the EPA for regulating the repository beyond 10,000 years is totally unacceptable in its treatment of the long-term impacts of natural and anthropogenic climate change as well as the potentially dangerously high radiation dose limit it contains. In light of these considerations, it is IEER's conclusion that Yucca Mountain cannot be regarded as an appropriate site for repository development, and that the search for an alternative location with a more appropriate geology should be begun as soon as possible (see Section 5.5).

Section 5.2.4 – Engineered Barriers at Yucca Mountain, the Changing Focus

As noted above, the discoveries concerning water infiltration and accelerated radionuclide transport have led the DOE to shift its focus from relying on Yucca Mountain's geology as the primary barrier to contaminant release to relying instead on engineered barriers. This change was facilitated by the NRC in 1996 when the Commission proposed changes to its regulations governing spent fuel disposal. The new NRC rule placed its focus entirely on the results of a single integrated model of repository performance in order to determine compliance with the EPA guidelines. In enacting these changes the NRC deleted the previously existing siting criteria that could have potentially eliminated Yucca Mountain from consideration. The NRC had, in fact, already changed these rules once before. The NRC had originally directed the DOE to select a repository located in the saturated zone beneath the water table, but they modified this criteria once the Yucca Mountain site in the unsaturated zone above the water table was identified. Following the NRC's rule changes, the DOE modified its own guidelines in 2001 to also eliminate the previous siting criteria and to allow it to rely solely on the results of computer models to support its license application.[882]

The importance of the engineered barriers to DOE's claims for the performance of a repository at Yucca Mountain is clearly illustrated by the following. Without the inclusion of engineered barriers, the DOE's system model estimates that the peak dose to people living near the site could rise to as much as 500 millirem in just 2000 years. This is more than 30 times the EPA's proposed limit of 15 millirem allowed for the maximally exposed individual during this time. When the engineered barriers were added back into the system model, the estimated peak dose actually went up to as much as 800 millirem, but it was not predicted to occur for approximately 200,000 years. In fact, the DOE found that the performance of the waste packages are the single most important uncertainty in their existing models. Uncertainties in the long-term properties of the engineered barriers have a larger impact on the projected magnitude and timing of the peak dose than either the rate of water infiltration or the rate of contaminant migration through the rock.[883] As summarized by Dr. Allison Macfarlane, co-founder of the Yucca Mountain Project at the Massachusetts Institute of Technology,

> On siting a repository at Yucca Mountain, the DOE has painted itself into a corner that will be difficult to leave. After touting the natural geologic features of the site to retain radioactive waste, the DOE has abandoned the geology for engineering design. It is now making its case for the site based not on the site itself – the natural geologic features of Yucca Mountain – but on the features that the DOE itself will build. The site, as such, no longer matters.[884]

The engineered barriers proposed for the U.S. repository fall into three main categories. The first barrier is the fuel assembly itself. After enrichment, the uranium is converted into uranium dioxide oxide and fabricated into a ceramic that is then encased in a zirconium cladding before it is placed into the reactor. In the repository, this waste form makes it more difficult for the fission products and transuranic elements to migrate away from the drifts since the fuel has to corrode first. The second barrier is the cask which will hold between 20 and 40 fuel assemblies. These casks will have an inner shell of stainless steel that is 5 cm thick and a 2 cm thick outer shell made from a nickel based alloy commonly called Alloy-22 or C-22. This alloy is made up of 56 percent nickel, 22 percent chromium, 13 percent molybdenum, 3 percent iron, and smaller amounts of several other elements. The waste packages will be set on inverts that will act to further restrict the movement of the waste once it breaches the cask. The third engineered barrier is a 1.5 centimeter thick titanium canopy that will cover the waste packages. This canopy will serve as a "drip shield" to inhibit water that enters the drift from falling directly onto the casks.[885]

As discussed in the previous section, uranium dioxide fuel is thermodynamically unstable in the oxidizing conditions that will prevail at the Yucca Mountain site and will corrode in the presence of moisture. Thus, the effectiveness of this first engineered barrier will be greatly reduced at Yucca Mountain compared to sites which would maintain a reducing environment around the waste. This fact places a much greater burden on the performance of the other two barriers and specifically on their ability to keep the spent fuel dry for very long times. While a thorough examination of the uncertainties surrounding the long-term performance of the waste casks and the drip shield is beyond the scope of this analysis, we will highlight a few of the more important issues.

Because of the need for high corrosion resistance, the DOE has proposed the use of specially made alloys for the outer skin of the casks and for the drip shields. Neither pure titanium metal nor Alloy-22 exist in nature, and both are thermodynamically unstable. Pure titanium metal was first produced in 1910 and Alloy-22 has only been investigated for the last 15 to 25 years. While related nickel-chromium-molybdenum alloys similar to Alloy-22 have been in existence for longer, the experience with all types of stainless steel alloys only goes back at most about 100 years. The complexity of these multi-component materials makes their long-term behavior difficult to project with high confidence. Even being generous by considering related but unique alloys to be suitable analogs, the total length of human experience with these materials is quite short compared to experience with naturally occurring materials like copper, much less to the thousands to tens of thousands of years the engineered barriers will need to remain functional at Yucca Mountain.[886]

Adding to these uncertainties is the fact that there is little experimental evidence available regarding the behavior of Alloy-22 at high temperatures under realistic repository conditions. In fact, all multiyear corrosion studies completed through the end of 2003 had been conducted at temperatures at or below the boiling point of water. In addition, the Nuclear Waste Technical Review Board has also raised concerns over the effect that the very high temperatures reached during the welding of waste casks will have on the properties of Alloy-22. These concerns include such important properties as the alloy's phase stability and its susceptibility to localized corrosion. The NWTRB concluded that the measures proposed by the DOE to try and mitigate the welding damage have limited industrial experience or, when experience does exist, that the measures have not been tested on a prototype waste package to determine their effectiveness.[887] In summary, while noting that it is likely to be technically possible to fabricate materials with the necessary lifetimes, the National Research Council concluded in its 2001 review that

> There is little experience, however, in modeling the behavior of modern materials derived from new compositions and fabrication methods. Quantifying the uncertainty of extrapolations with these models from short-term experiments to tens or hundreds or thousands of years is still a major challenge.[888]

An example of how significant some of these uncertainties could be can be found in the concerns raised by the Nuclear Waste Technical Review Board over the DOE's conclusions regarding the corrosion of waste packages. With respect to the general corrosion of Alloy-22, the NWTRB noted that

> Few data exist, however, at the higher temperatures of the thermal pulse period. Moreover, the nature of the aqueous environments in contact with the waste packages (or drip shields) is not very well known under such conditions. Concentration and nonequilibrium processes of various kinds may lead to aggressive chemistries. Thus, the uncertainties surrounding general corrosion during the thermal pulse remain a concern of the Board.[889]

The Board went on to conclude that

> The DOE's analyses of water chemistries and their corrosive potential are extremely complex and suffer from empirical and theoretical weaknesses. Thus, the Board does not have a high degree of confidence in the DOE's conclusion that any seepage water would be dilute or non-corrosive because the methods the DOE used have significant technical uncertainties.[890]

In addition to these issues regarding generalized corrosion, the NWTRB also raised concerns regarding the possibility that severe localized corrosion might occur which would degrade areas of the waste packages much faster than assumed by the DOE.

As noted above, the NWTRB concluded that the DOE estimates for the repository temperature during the thermal pulse are likely be flawed on a number of accounts. The temperature in the drift is important for a variety of reasons, including the fact that it helps to control the relative humidity in the tunnels. In light of its observations regarding DOE's temperature estimates, the NWTRB expressed concerns that the DOE's estimates about the humidity in the drifts may be inaccurate as well.[891] Of particular concern to the Board was the possibility for localized corrosion to occur as a result of the absorption of atmospheric water vapor by the mineral salts that might be present in the dust deposited on the drip shields and waste packages. As a result, the NWTRB noted that

> The Board is aware of data and inferences that waste packages could be penetrated in less than 100 years under certain conditions that could occur at Yucca Mountain during the thermal pulse. If localized corrosion is initiated, penetration of most of the waste packages during and after the thermal pulse becomes quite probable.[892]

They went on to conclude that

> Although a precise statement about whether, or how much, dose might be increased or the safety margin decreased cannot be made given the existing uncertainties, the Board believes that the implications of the Board's conclusions for repository system performance could be substantial. Therefore, it is incumbent on the DOE to demonstrate unambiguously the reliability and safety of any design concept for Yucca Mountain.[893]

Despite the potential importance of this degradation of the engineered barriers, the NWTRB noted that, as of November 2003, it was "not aware of any studies conducted by the DOE to determine the rate or extent of localized corrosion."[894]

In response to the concerns raised by the Nuclear Waste Technical Review Board, the DOE conducted additional research on the conditions expected in the repository. In later reports, the NWTRB was generally satisfied with this new information, but remained concerned about the DOE's technical basis for eliminating this type of localized corrosion from their overall performance assessment. In addition, the Board raised concerns about other mechanisms for localized corrosion that might occur in the period after temperatures fall below the boiling point of water.[895] The fact that such a potentially important effect as rapid localized corrosion had not been adequately explored prior to late 2003, nearly a year and half after the site was officially recommended by the Bush Administration, highlights the risks inherent in relying on incomplete conceptual models. While this particular issue was resolved to the satisfaction of the NWTRB, it should stand as a warning that caution is needed when dealing with projections for the performance of engineered barriers, particularly given their central importance to the DOE's claims regarding the suitability of the Yucca Mountain site.

While much of the attention has rightly been placed on the long term performance of the waste packages themselves, additional concerns remain about the performance of the drip shields as well. In response to the DOE claim that the drip shields would successfully prevent water entering the drifts from contacting the casks for very long times, the NWTRB concluded that

> The Board believes that the DOE's position is based mostly on assumptions that could be unrealistic and overly optimistic. First, no prototype drip shield has ever been built, and the concept of a long-lasting drip shield in an underground application has never been applied elsewhere. Thus, the DOE's projections of how this structure will perform for thousands of years are speculative.[896]

Beyond the degradation of the drip shields, the NWTRB raised additional concerns over their effectiveness. The Board noted that because the drip shields would be cooler than the waste packages that condensation on their inner surface was possible which "could lead to dripping on the waste packages rather than preventing it."[897]

Finally, even if additional research could show that the materials chosen were suitable for use in the repository and to have a theoretical durability adequate for the long times necessary, there would remain important uncertainties due to the differences between theoretical systems and those that can actually be built in the real world. This concern was explicitly noted by the NWTRB in their discussion of the drip shield's performance. Perhaps most importantly, each of the several thousand casks that will be required must be manufactured correctly and properly welded since the joints are particularly sensitive to corrosion. Finally, these waste packages must then be successfully treated to offset the damage caused by the heat of welding.

Assuring the reliability of each critical component in each of the engineered barriers is a daunting task. This is a particular concern in the case of the U.S. repository program given the DOE's history of poor quality control at Yucca Mountain and numerous other projects. The uncertainties surrounding the manufacture and emplacement of the engineered barriers are aggravated by the fact that the DOE has yet to manufacture a full scale prototype canister for testing in realistic repository conditions. Experience in Sweden, where they have encountered ongoing difficulties in trying to successfully weld their copper canisters, highlights the need for such real world tests.[898] The conclusion that additional in situ testing of full scale waste casks and the drip shield is needed has been expressed by both the International Atomic Energy Agency and by the authors of the MIT report.[899]

In light of the serious challenges inherent in relying on novel human-made alloys to ensure the suitability of a repository, other countries have chosen to avoid these uncertainties and make use of materials that exist in nature. This strategy has the advantage of allowing suitable geologic analogs to be found and focuses on materials with which humans have had longer engineering experience. For example, the Finnish government is planning to use copper for their waste packages since it is highly stable over geologic timescales in a saturated, reducing environment. Specifically, the waste packages proposed for the repository at Olkiluoto will be iron with a 5 cm thick outer shell of copper. These casks will then be surrounded by compacted clay which has low permeability to water and has a pore size small enough to limit colloid mediated transport. The clay liner will also help to maintain a reducing chemical environment around the casks and uranium fuel.[900] While important uncertainties remain in this program, their choice of a favorable chemical environment and of engineered barriers tailored to reinforce the strong points of the site geology reduce the unknowns in significant ways.

Section 5.2.5 – The "Technical" versus "Legal" Limit at Yucca Mountain

So far we have retained the assumption that Yucca Mountain will be limited to accepting 63,000 MT of commercial spent fuel and 7,000 MT of high level reprocessing waste. This is consistent with the current statutory limit as set forth in the 1982 Nuclear Waste Policy Act, and is typically referred to as the "legal limit" or "regulatory limit" of Yucca Mountain. This limit was written into the law when there were plans for a second repository to be developed in the eastern United States. However, the 1987 amendment to the NWPA halted further work on the second repository and the DOE's latest cost estimate assumes that a total quantity of 83,800 MT will eventually be disposed of at Yucca Mountain.[901] Proponents of nuclear power, often argue that Yucca Mountain could technically hold far more waste than is currently allowed by law. This larger capacity, typically estimated to be approximately 150,000 MT, is referred to as the "technical limit" or "physical limit" of Yucca Mountain. Given that by the end of 2012, it is estimated that the existing fleet of nuclear plants will have already discharged more than 67,500 MT of spent fuel, the ability of Yucca Mountain to hold more than 63,000 MT of commercial spent fuel is a crucial one.

The claim that the capacity of Yucca Mountain could be expanded to 150,000 MT or more relies on two major changes to the repository design. The first is to increase the size of the facility by adding more drifts while the second is to decrease the spacing between the waste casks within the tunnels. The feasibility of these changes rests on the claimed accuracy of the DOE's integrated repository model called the Total System Performance Assessment. To date there have been two major versions of this model whose results have been presented by the DOE, the TSPA-VA (Viability Assessment) and the TSPA-SR (Site Recommendation). The later version of this program "incorporates over 1,000 sources of data, approximately 60 scientific models, and more than 400 computer software codes to simulate the performance of the repository."[902]

With respect to the size of the repository, the TSPA-VA concluded that, with a substantial effort at further characterization of the site, it was possible that the footprint of the repository might be able to expand to just under 2,000 acres. This would be an increase of 70 percent over the area proposed for the current repository design. In addition to this increase in size, it is claimed that the TSPA-SR could be used to support an increase in the density of waste disposal from the current 60 metric tons per acre to 75 metric tons per acre. Increases to as much as 90 metric tons per acre have also been discussed for the types of spent fuel discharged from currently operating reactors. If the entire 2,000 acres was developed and packed at 75 MT per acre then up to 150,000 MT of waste could be stored. Similarly, if all 2,000 acres was developed but the first 70,000 MT was disposed of at 60 MT per acre as currently planned while the remaining area was filled to 90 MT per acre, than roughly 145,000 MT of waste could be stored in the repository.[903]

If the amount of waste to be disposed of in the Yucca Mountain repository was more than doubled, then the uncertainties discussed in the previous sections would become even more important, and unexpected new problems would likely arise. Already the peak doses from a repository with 70,000 metric tons of waste are well above the typical EPA dose limit of 15 millirem per year and are therefore unacceptably high. Any increase in the amount of waste would only serve to worsen this situation. In addition, there are long standing concerns over the DOE's plan to allow the temperature in the repository to reach above the boiling point of water. Such high temperatures, while useful in some ways, such as creating a temporary vaporization barrier along part of the drifts, also lead to more complicated geologic behavior, by changing the properties of the engineered materials, and by making it more difficult to find suitable natural analogs. In fact, the Nuclear Waste Technical Review Board has recommended that the DOE consider redesigning the repository so as to maintain peak temperatures below the boiling point of water as a way to lesson these uncertainties.[904] However, by increasing the packing density of the waste from 60 MT per acre to 75 or 90 MT per acre, the peak temperature in the repository would also increase.[905]

All of the uncertainties in the performance of the geology and the engineer barriers are all then rolled together in the Total System Performance Assessment. As summarized by the NRC's Advisory Committee on Nuclear Waste in September 2001, just five months before President Bush recommended the site to Congress, the Yucca Mountain TSPA-SR "relies on modeling assumptions that mask a realistic assessment of risk" and that the "computations and analyses are assumption-based, not evidence-supported."[906] In particularly strong language, the NRC's advisory committee concluded that

> However, based on the Committee's vertical slice review [a focused review on the repository modeling system], the principal findings are that the TSPA-SR does not lead to a realistic risk-informed result, and it does not inspire confidence in the TSPA-SR process. In particular, the TSPA-SR reflects the input and results of models and assumptions that are not founded on a realistic assessment of the evidence.[907]

and that

> The Committee believes that the TSPA-SR is driven more by an attempt to demonstrate compliance with the standards than by the need to provide an assessment designed to answer the questions: What is the risk?[908]

In addition to the problems with a lack of completeness and with the accuracy of the models used, additional concerns with the TSPAs have been raised due to persistent problems with the Energy Department's quality assurance program at Yucca Mountain. As far back as 1988, the GAO identified significant lapses in the DOE's QA program based on issues originally raised by the NRC. The issue of QA problems resurfaced in 1998 when the DOE began to run its initial version of the performance assessment model, the TSPA-VA. Specifically, it was found that (1) the DOE could not ensure that the original source of all critical data could be identified, (2) the DOE had no standardized process to govern the development of models to simulate important geologic processes, and (3) the DOE had no formal procedures in place to ensure that the software developed to implement those models would actually work as intended. Quality assurance issues again became a major focus in 2001 when the DOE began to conduct audits in support of its next version of the model, the TSPA-SR.[909]

Despite the fact that they had not reached any conclusion about their success in resolving these QA problems, and despite the fact that audits conducted in April and September of 2003 had revealed both the continuation of known problems and the existence of previously unidentified problems, the DOE officially closed its corrective action reports in March 2004 for QA problems in both data and software.[910] Shortly before this closure, however, the NRC staff concluded that

> The team believes that, if DOE continues to use their existing policies, procedures, methods, and practices at the same level of implementation and rigor, the LA [license application] may not contain information sufficient to support some technical positions in the LA. This could result in a large volume of requests for additional information in some areas which could extend the review process, and could prevent NRC from making a decision regarding issuing a construction authorization to DOE within the time required by law.[911]

Following the publication of the NRC report, the GAO concluded that

> Entering into the licensing phase of the project without resolving the recurring problems could impede the application process, which at a minimum could lead to time-consuming and expensive delays while weaknesses are corrected and could ultimately prevent DOE from receiving authorization to construct a repository. Moreover, recurring problems could create the risk of introducing unknown errors into the design and construction of the repository that could lead to adverse health and safety consequences. Because of its lack of evidence that its actions have been successful, DOE is not yet in a position to demonstrate to NRC that its quality assurance program can

ensure the safe construction and long-term operation of the repository.[912]

New concerns over the quality assurance program at Yucca Mountain have been raised since the GAO and NRC reviews were published. In March 2005, emails between scientists working on the Yucca Mountain project, at least some of whom were employees of the U.S. Geological Survey, came to light indicating that these scientists were fabricating reports to satisfy quality assurance assessments. These alleged actions included withholding output files that could not be explained, fabricating records of software control, and falsifying procedures for conducting calculations.[913] The studies in question relate to the transport of water through the site, which is one of the central issues surrounding the suitability of the site. Nevada lawmakers are calling for an independent investigation into this potential falsification of information. A criminal investigation by the Department of Energy, the Department of the Interior, and the FBI is ongoing. The DOE has reported that it will not consider submitting its license application to the NRC until the investigations into these allegations are complete.[914]

Section 5.2.6 – Additional Concerns Regarding Yucca Mountain

It is our conclusion that, based on the available scientific evidence, Yucca Mountain is not a suitable site for the development of a repository. At its most basic level, the case against Yucca Mountain boils down to the fact that it is being built on land still claimed by the Western Shoshone people and that it will not be likely to keep peak doses to an acceptably low level. In trying to overcome this second fact, the EPA has relaxed the Safe Drinking Water Act standards at the site and has proposed the most lax radiation protection standard ever considered by a governmental body anywhere in the world. These actions have created a dangerous precedent for the DOE nuclear weapons sites which are currently undergoing a multi-billion dollar cleanup program. Contamination at three of these sites already pose a long-term threat to important regional waterways including the Columbia River, the Snake River, and the Savannah River.

The EPA groundwater standard governing the Yucca Mountain site over the first 10,000 years allows the DOE to take credit for a "controlled area" around the repository in which compliance with the Safe Drinking

Water Act is not required.[915] Specifically, the discussion describing the regulation states that

> If fully employed by DOE, and based on current repository design, the controlled area could extend approximately 18 km in the direction of ground water flow (presently believed to be in a southerly direction) and extend no more than 5 km from the repository footprint in any other direction. Allowing for a nominal repository footprint of a few square kilometers, this results in a rectangle with approximate dimensions of 12 km in an east-west direction and 25 km in a north-south direction, or approximately 300 km^2.[916]

Therefore, the EPA regulation creates what is effectively a leach-field out of the repository by only evaluating the impact on ground water after it has traveled 18 kilometers down gradient. While the Nevada Test Site is a controlled area today, this is not a meaningful fact for exposures that will occur tens to hundreds of thousands of years in the future. Typically institutional control of a site is not assumed to be successful at preventing intrusion for more than a few hundred years at most. Therefore, compliance with the EPA's dose limits should be determined at the boundary of the repository, rather than an arbitrary point 18 kilometers away.

The Yucca Mountain site also has the unique problem of having to consider the possibility of additional ground water contamination in the area caused by the hundreds of underground nuclear weapons and sub-critical tests that have been conducted on the Nevada Test Site. In order to protect the health of future site inhabitants, it is necessary to ensure that the annual drinking water dose limit is met for all radionuclides present, and not just those from the waste in the repository. Despite the evidence for rapid colloid-mediate migration of plutonium at the Nevada Test Site, the impact from these additional sources of contamination is not currently considered in the DOE's evaluations of spent fuel disposal at Yucca Mountain.

The focus on the drinking water pathway is particularly important at Yucca Mountain because the arid climate means that there is little surface water, and that humans are therefore reliant on groundwater alone to meet their primary needs. In fact, the aquifer under Yucca Mountain is already being used for irrigation just 32 km (20 miles) from the site. Thus, the pollution of this aquifer would likely result in extensive exposures of any future populations living in the area. In addition, as we have noted previously, drinking water is one of the most probable pathways for future human exposures over the very long times involved, and so it should be the subject of particularly strict guidelines rather than being weakened by the inclusion of a "controlled area" for the first 10,000 years and being effectively eliminated after that.

Section 5.3 – Transportation of Spent Fuel

The government and the nuclear industry have been transporting nuclear materials, including a modest amount of commercial spent nuclear fuel, for decades without a major incident. Despite this history, however, one of the most widely publicized concern regarding the development of a geologic repository or centralized storage or reprocessing facility is the risk associated with transporting the spent fuel from the 76 sites in 33 states at which it is currently located to a single location. In the case of Yucca Mountain, or the proposed interim storage facility near Salt Lake City, Utah (see Section 5.4.1), these transportation risks are heightened by the long distances over which the waste must be shipped from the reactors in the East to the proposed destinations in the West.

The risks associated with spent fuel transportation have been examined extensively, and a number of full-scale tests of transportation packages have been conducted or are currently planned.[917] However, despite these efforts there remain a number of important ways in which the risks from transporting spent fuel could be further reduced by the inclusion of more realistic maximum credible accident conditions into NRC regulations. In addition, it is necessary for the terrorist threat to be more seriously addressed as part of the spent fuel transport program. The requirement of hundreds to thousands of shipments of spent fuel over a number of decades, the attractiveness of radiation as a weapon of terror, and the inherent unpredictability of terrorist acts all add to the uncertainty of spent fuel transport. We will begin with a discussion of the NRC's technical analysis of spent fuel transportation from the year 2000, and then touch upon the need to consider more realistic acts of terrorism as part of an overall risk assessment.[918]

First, the Sandia National Laboratories analysis as presented by the NRC in its *Reexamination of Spent Fuel Shipment Risk Estimates* did not consider specific transportation routes from the existing nuclear plants to the Yucca Mountain site. Instead, the report considered transporting the waste to "3 hypothetical geologic repositories and 6 hypothetical interim storage facilities."[919] Six of the nine sites were in the eastern or central parts of the United States which reduced the average distance the waste casks had to travel in comparison to transport to Nevada or Utah.[920] In addition, the NRC analysis assumed the same distribution of accident velocities as used in the 1987 Modal Study.[921] The data in the Modal Study was based on information gathered between 1958 and 1967. Since that time, many states have increased the speed limits on their highways and the number of cars and trucks on the road have also increased. Even relatively modest changes in the speed limit significantly alter the distribution of accident velocities. For example, a study conducted by the Insurance Institute for Highway Safety found that in New Mexico the percentage of vehicles exceeding 70 miles per hour went from 5 percent to 36 percent after the speed limit was raised from 55 to 65 mph. Since kinetic energy scales as the square of the velocity, increases in speed have a proportionally larger impact on the severity of accidents.[922]

For a side impact of a waste cask, a 1987 study by Lawrence Livermore National Laboratory found that much lower decelerations were necessary to cause the fuel rods to rupture than those considered by the NRC.[923] The average rupture point for all fuel assembly types in the LLNL study was just below the value which the NRC analysis assumed would lead to no fuel damage. In fact, the NRC expected little damage at decelerations four times higher than the average rupture point in the Livermore study.[924] The difference in these estimates is important because the side impact of a truck cask or monolithic rail cask at 30 miles per hour could lead to decelerations above the average rupture point in the Livermore analysis.[925] This could occur, for example, with the collision of a truck and a bridge abutment or concrete building. Another example of an accident that could lead to such rapid decelerations occurred on May 26, 2002. On that date, a barge collided with the support structures of a bridge on Interstate 40 in Oklahoma causing nearly one-third of the bridge to fall 60 feet into the Arkansas River. Parts of the concrete bridge remained atop the barge providing a hard surface on which some of the falling cars and semi-trucks impacted.[926] A final example occurred on April 25, 2005 when a Japanese commuter train derailed and collided with the corner of a nearby concrete apartment building.[927]

If the fuel ruptures it would release radioactivity into the interior of the transportation cask. If the cask was also damaged in the collision, that radioactivity could then be released into the environment. Current NRC regulations require that spent fuel transportation casks are designed to withstand "a free drop test" in which the cask is dropped from a height of nine meters (29.5 feet) onto a flat and unyielding surface as well as a "puncture test" in which the cask is dropped from a height of one meter (3.3 feet) onto a steel rod approximately 15 centimeters in diameter.[928] As noted, however, there have been accidents in the U.S. and abroad that would have resulted in the waste casks being exposed to more intense impacts that those included in the regulations. The height of the drop test should be increased to more accurately reflect the maximum likely height of bridges over which the waste may travel. The height of the puncture test should also be increased to cover the potential of higher speed impacts.

In addition to violent collisions, accidents involving fires are another area that poses a potential risk to spent fuel casks. In their analysis of transit fires, the NRC only considered the impact of a fire on fuel from pressurized water reactors despite the fact that casks for boiling water reactor fuel can carry more waste, and thus have a greater amount of internal decay heat. In addition, the Lawrence Livermore analysis concluded that, under certain accident conditions, the temperature of the central most fuel rods could be more than two and a quarter times higher than that assumed by the NRC. Further, the NRC study only considered undamaged fuel in its analysis of transportation fires. If the spent fuel rods were damaged by a collision and then engulfed in a fire, the release rates would be adversely affected. Finally, the studies upon which the NRC analysis was based did not consider the impact of fires on higher burnup fuels such as that being discharged by some reactors today. The higher burnup fuel has thinner cladding which makes it more likely to rupture in a violent impact or severe fire.[929]

The NRC's regulations require that casks be able to withstand a temperature of 800 °C (1,470 °F) for thirty minutes. The casks are also tested at a temperature of 1,000 °C (1,830 °F) to simulate the effects of a diesel fuel fire. There are, however, many commonly transported materials that burn much hotter such as methyl alcohol, propane, vinyl chloride, and gasoline.[930] The duration and intensity of this regulatory fire is thus less than those that could occur in some types of major accidents. The current NRC testing program should be improved to take into account more realistic conditions that could be reached in credible types of transportation accidents involving fires in combination with collisions.

The most well known transportation fire to have occurred in recent years is the CSX Railroad Tunnel Fire that began in Baltimore, Maryland, on July 18, 2001. The Howard Street Tunnel is the longest underground train route on the East Coast and carries up to 40 freight trains per day.

At 3:07 PM a CSX Transportation train derailed in the tunnel and subsequently sparked a serious fire that sent black smoke out both ends of the tunnel and out manholes along nearby streets. The exact cause of the fire is not known, but the train was carrying flammable liquids such as tripropylene and numerous cars of wood pulp and paper products. Firefighters entering the tunnel lost all vision within about 90 meters, and it was not until nearly seven hours after the fire had begun that they successfully reached the burning cars.[931] A simulation of this fire found that during the first three hours, the peak temperature in the tunnel reached 1000 °C in the area surrounding the flames and 500 °C when averaged over the length of the surrounding three to four rail cars. The peak temperature of the tunnel walls was estimated to have reached 800 °C with an average of 400 °C.[932] The accident, however, could have been much worse.

As it turned out, the fire caused the rupture of a 40 inch water main running through the ground above the tunnel, damaging several city streets and knocking out electricity to more than a thousand customers. The broken water main also flooded the tunnel and was found to have had a positive effect in cooling the fire after the first three hours.[933] The possibility of accidents in more remote areas of the U.S. raise concerns over the availability of properly equipped firefighting and hazmat teams. Approximately 60 million gallons of water was necessary to battle the CSX tunnel fire.[934] In more remote areas the transportation of this much water could pose a potential barrier to fighting such a serious fire.

The study of this accident by the Federal Emergency Management Agency concluded that the tunnel fire created conditions that "surpassed the NRC's design criteria for containers that would hold atomic waste."[935] Instead of immediately recognizing this accident as a warning and a reason to reevaluate their generic requirements for accidents involving shipping casks, however, the NRC conducted a simulation of the conditions in the CSX train fire and concluded that it would not likely have posed a serious risk to the public had a spent fuel cask been involved.[936] The fact that the baseline regulatory requirements were exceeded on a well traveled rail line in a densely populated city in the middle of a weekday afternoon with a major sports stadium nearby should have triggered a through reexamination of the design criteria whether or not this particular accident would have resulted in a breach of containment and the release of radiation.

A number of other accidents and near misses involving serious truck fires have also occurred. A series of accidents on the East Coast in 2004 serve to illustrate the point. On January 13, 2004, a truck carrying 8,900 gallons of fuel crashed through a barrier on Interstate 895 in Maryland and fell 30 feet onto Interstate 95. A tractor trailer, a large truck, a pickup truck, and a car all crashed into the gasoline tanker which subsequently exploded. The concussion from the primary blast and a series of subsequent explosions was felt as far away as half a mile.[937] Just two months later, on March 25, a truck carrying 12,000 gallons of heating oil hit a concrete barrier and burst into flames on a busy stretch of highway near Bridgeport, Connecticut. The truck fire burned for two hours and the bridge span sagged as the 30 inch steel beams softened and began to melt in the 1,100 °C heat. In this accident, both the heat and the duration of the fire exceeded the regulatory requirements for a spent fuel shipping cask. Seventeen hours after the crash in Bridgeport, three tractor trailers and a dump truck collided on a bridge connecting Staten Island, New York to New Jersey. One of the drivers was killed, but a much more serious situation was averted because the liquid oxygen being carried by one of the trucks did not contribute to an explosion.[938]

In their 2006 review of the safety of spent fuel transport, the National Research Council acknowledged this concern and noted that "recently published work suggests that extreme accident scenarios involving very-long-duration, fully engulfing fires might produce thermal loading conditions sufficient to compromise containment effectiveness."[939] As such the Committee recommended that

> The Nuclear Regulatory Commission should build on recent progress in understanding package performance in very-long-duration fires. To this end, the agency should undertake additional analyses of very-long-duration fire scenarios that bound expected real-world accident conditions for a representative set of package designs that are likely to be used in future large-quantity shipping programs.... Strong consideration should also be given to performing well instrumented tests for improving and validating the computer models used out these analyses, perhaps as part of the full-scale test planned by the Nuclear Regulatory Commission for its package performance study.[940]

Finally, the potential exists for a well coordinated terrorist attack to cause far more extensive damage to the shipping casks than all but the most extreme accidents. The Energy Policy Act of 2005 requires the NRC to consider a realistic threat that includes "the potential for attacks on spent fuel shipments by multiple coordinated teams of a large number of individuals."[941] This analysis should include considerations of advanced weapons such as shoulder fired missiles in combination with shaped charges as well as lower tech attacks such as large truck bombs. The bomb that killed Judge Giovanni Falcone near Palermo, Italy in May 1992 by blowing up a section of the road on which he was traveling serves as stark example of what kind of damage can be inflicted upon even rapidly moving targets.[942] The widespread use of camouflaged roadside bombs and other improvised explosive devices in Iraq and the toll they have taken on U.S. and British military forces adds further support to these considerations.[943] While there will always be a level of uncertainty surrounding the potential risk associated with terrorist attacks, the importance of this uncertainty can be reduced by making the design basis threat as robust as possible and by limiting the transport of spent fuel and high-level waste near dense population centers or important shipping routes.

The National Research Council also recognized the threat posed by terrorists and concluded that "[m]alevolent acts against spent fuel and high-level waste shipments are a major technical and societal concern, especially following September 11, 2001 terrorist attacks upon the United States."[944] As such the committee recommended that

> An independent examination of the security of spent fuel and high-level waste transportation should be carried out prior to the commencement of large-quantity shipments to a federal repository or to interim storage. This examination should provide an integrated evaluation of the threat environment, the response of packages to credible malevolent acts, and operational security requirements for protecting spent fuel and high-level waste while in transport. **This examination should be carried out by a technically knowledgeable group that is independent of the government and free from institutional and financial conflicts of interest.** This group should be given full access to the necessary classified and Safeguards Information documents to carry out this task. The findings and recommendations from this examination should be made available to the public to the fullest extent possible.[945]

The need to release as much information to the public as possible, consistent with the requirements of security, is particularly important given the need to receive the informed consent of the many communities through which the spent fuel and high-level waste will have to travel.

Improving the safety of spent fuel transportation is important not only to protect public health and the environment during the large number of shipments that will be needed, but also because an accident or attack that released radiation could have a significant economic impact due to extended cleanup times and lingering public concerns over residual contamination. The bridge over Interstate 40 that collapsed when struck by a barge is one of the main east-west shipping routes carrying an average of 20,000 vehicles per day.[946] The Howard Street Tunnel where the CSX train fire occurred is one of the most important rail tunnels along the East Coast.[947] Interstate 95 where the two truck fires occurred is the main north-south route along the East Coast carrying an average of 120,000 to 200,000 vehicles per day in the areas affected by the accidents.[948] In the western U.S., there are few alternate routes adding to the potential disruption despite the lower traffic density. In light of the potential impacts to both human health and the economy, it is important to find ways to further limit the risks of transportation by improving the realism of the testing programs currently taking place and of the modeling and simulations of shipping cask durability by taking into account the more severe accidents that have occurred to date and by considering appropriately severe terrorist attacks.

Section 5.4 – Alternative Waste Management Strategies

In 2050, the global growth scenario would lead to a tripling of the current nuclear capacity while the steady-state growth scenario would lead to an increase of roughly seven fold. With this expansion would come a proportional increase in the generation of nuclear waste. While going to higher burnup with the fuel would reduce the mass discharged, its higher decay heat would require it to be spaced further apart in a repository, and thus the savings in disposal volume would be less significant. As summarized by the authors of the MIT report,

> A worldwide deployment of one thousand 1000 megawatt LWRs operating on the once-through fuel-cycle with today's fuel management characteristics would generate roughly three times as much spent fuel annually as does today's nuclear power plant fleet. **If this fuel was disposed of directly, new repository storage capacity equal to the currently planned capacity of the Yucca Mountain facility would have to be created somewhere in the world roughly every three or four years.** For the United States, a three-fold increase in nuclear generating capacity would create a requirement for a Yucca Mountain equivalent of storage capacity roughly every

12 years (or every 25 years if the physical rather than the legal capacity limit of Yucca Mountain is assumed.)[949]

We have already discussed the issues surrounding the potential expansion of Yucca Mountain in Section 5.2.5, and will therefore continue to focus on the current statutory limit of 70,000 MT.

Trying to characterize sites and then design, license, and construct repositories this quickly would be a virtual impossibility no matter what level of effort was expended. Adding to this difficulty is the fact that many of the repositories under consideration around the world (see Table 5.1) are smaller than Yucca Mountain due to the smaller volumes of waste that have been generated in other countries. For example, the proposed Swedish repository is designed to hold approximately 7,800 MT of waste which is just one-ninth the statutory capacity of Yucca Mountain.[950] Thus, as long as the disposal of spent fuel is left to individual countries, the actual number of repositories required under the global or steady-state growth scenarios would have to be far larger.

To try to overcome the need for many countries to have to dispose of relatively small amounts of waste, the establishment of international spent fuel repositories have been proposed. These facilities would have the added non-proliferation benefit of placing the plutonium in the waste under international control, thereby reducing the risk of its being recovered through reprocessing.[951] These plans, however, raise serious concerns that weaker, cash strapped countries or those with corrupt governments might end up the dumping grounds for richer more powerful countries in the Global North. It is not a just or sustainable option for one group to receive all the benefits from an activity while another receives the liabilities. Beyond these considerations, however, we have already noted that a large, centralized repository the size of Yucca Mountain would still need to come online every few years to meet the needs of the global growth or steady-state growth scenarios.

Finally, there are new concerns raised by the revelations that U.S. utilities and the NRC have already begun to have trouble keeping track of the spent fuel rods discharged from the existing reactors. These troubles appear to have grown, in part, out of the fact that in 1988 the NRC ended its routine inspection of licensee compliance with spent fuel accounting and control regulations believing that it was unlikely that fuel assemblies could be lost or stolen due to their high radioactivity.[952] In November of 2000, however, during preparations at the Millstone plant in Connecticut for moving their spent fuel to dry cask storage, it was discovered that two damaged fuel rods had been missing for as much as 20 years without being noticed. The spent fuel rods were never found despite an extensive search of the plant and of its records. It is believed that the fuel was most likely illegally shipped to the Barnwell low-level waste disposal facility in South Carolina. The NRC fined the operators of the Millstone plant $288,000 for failing to account for the two rods and for not reporting the loss sooner. As a result of this incident the NRC began a review of spent fuel accounting at other operating facilities. In April 2004, it was discovered that two spent fuel rod pieces were missing at the Vermont Yankee plant. These pieces were located approximately three months later after several million dollars had been spent on the search and on updating the plant's materials control and accounting system. Finally, in July 2004, three spent fuel rod pieces were found to be missing at the Humboldt Bay plant in California. These fuel rod pieces were not definitively accounted for nearly a year later despite a thorough search of the plant and its records. The NRC is also aware of a number of previous instances of spent fuel that was either temporarily unaccounted for or lost at other nuclear installations, and the GAO has concluded that there is still the potential for discoveries of missing fuel at other plants.[953] The difficulties of accounting for spent fuel would obviously increase should a large number of new nuclear plants be built.

In light of the very large number of geologic repositories that would be required to handle the waste from 1,000 to 2,500 nuclear plants, as well as the difficulties that have been encountered with efforts to safely store the waste at individual reactor sites over long times, a number of alternative waste management strategies have been proposed which we will discuss in the following sections.

Section 5.4.1 – Monitored Retrievable Storage (MRS)

While not by itself an option for the long-term management of spent fuel, many alternative proposals envision interim storage of the waste at one or more centralized facilities. On June 25, 1997, a license application for the construction of a privately owned interim storage facility was filed with the NRC. The proposed facility would be developed by a consortium of eight utilities. This company, called Private Fuel Storage (PFS), would build the storage facility on land belonging to the Skull Valley Band of Goshute Indians 110 kilometers (about 70 miles) southwest of Salt Lake City. This facility would be able to store up to 40,000 metric tons of spent fuel in 4,000 casks on 98 acres of land. The lease with the tribe would run for 25 years with a possible renewal for an additional 25 years. On February 24, 2005, the Atomic Safety and Licensing Board recommended that the NRC grant a license for this facility in light of the continued delays in the Yucca Mountain project.[954] On September 9, 2005, the full Commission authorized the NRC Staff to grant PFS a license once the Staff had completed its reviews.[955] On February 21, 2006 the Commission granted a 20 year license to PFS, however, the consortium still required approval from "the Bureau of Land Management, the Bureau of Indian Affairs, and the Surface Transportation Board" before construction could begin.[956]

Despite the NRC rulings which cleared that way for the facility to go ahead, two developments have cast serious doubts on the future of PFS. First, the fiscal year 2006 National Defense Authorization Act, signed into law on January 6, 2006, included a provision to turn an area around the site into a protected wilderness area. While the waste could still be brought in on trucks, the new law cut off the ability of the consortium to build a rail spur that had been their preferred method for transporting the spent fuel to the site.[957] Second, in early December, Southern Company announced that it was dropping out of the PFS consortium and both Xcel Energy and Florida Power and Light announced that they would indefinitely suspend their support for the project. Together, these three companies held more than 56 percent of the PFS shares.[958]

Beyond the proposed Private Fuel Storage consortium, there have been other proposals for the development of monitored retrievable storage facilities. For example, the authors of the MIT report support a combination of at reactor storage, and, where needed, centralized interim storage facilities, as a means of addressing the short term concerns over spent fuel safety and as a means to buy time to explore alternatives to geologic disposal such as deep boreholes (see Section 5.4.3).[959] On the other hand, the U.S. House of Representatives Committee on Appropriations has put forward a proposal for the development of interim storage facilities at Department of Energy sites as part of a broader effort to resume reprocessing and the development of the MOX fuel cycle (see Section 3.2).[960]

While it is true that the continued storage of spent fuel in densely packed cooling pools near large population centers presents a very real safety risk as discussed in Section 4.2.3, no detailed quantitative analysis has ever been presented on the relative implications of leaving the waste onsite until a final disposal site is ready compared to transporting it all to a single interim storage site and then again to a permanent repository.[961] Adding to the complexity of this issue is that fact that, given the recent proposals from the House of Representatives, such a comparative analysis would have to include the much more complicated problem of analyzing the added costs and risks of reprocessing and the use of MOX fuel. Without such an analysis, the well known safety and proliferation risks of reprocessing, combined with the risk that a temporary facility on Native American land could become an ad hoc permanent disposal site, argue strongly against the hasty movement of thousands of metric tons of waste along the country's roads and rail lines to any interim storage site.

Section 5.4.2 – Separation, Transmutation, and MOX Fuel

One of the oldest alternatives to the direct disposal of spent fuel is reprocessing. This technology could be linked either to plans for reusing the plutonium in mixed-oxide (MOX) fuels or transmuting the long-lived transuranics into shorter-lived isotopes in specially designed reactors or linear accelerators. Countries such as France and Japan continue to actively pursue the MOX fuel cycle. There have also been proposals put forward to separate the shorter-lived fission products such as cesium-137 (half-life 30 years) and strontium-90 (half-life 28 years) from the very long-lived radionuclides. The longer-lived radionuclides would then be put into a suitable waste form and disposed of in a geologic repository while the shorter-lived elements would be placed into a disposal facility that would rely more heavily on institutional controls. The advantage of such a scheme would be to reduce the decay heat of the waste destined for geologic disposal, allowing it to be packaged more densely in the repository while the design of the disposal facility that would handle the higher activity fission products would be somewhat simplified by their shorter half-lives.

The first, and most important, argument against all of these proposals is that they require the reprocessing of the spent fuel which can result in the separation of weapons usable plutonium. The potential for nuclear weapons proliferation that would accompany the spread of reprocessing technologies is unlikely to be overcome given that the current nuclear weapons states plan to retain their weapons for the indefinite future and to continue to rely on them as a central part of their military posture. The proliferation dangers associated with reprocessing are discussed at greater length in Sections 3.2 and 3.4.

The second argument against these reprocessing schemes is their cost. We have already discussed the high cost of MOX fuel in the introduction to Chapter Two. According to the MIT report, the cost of MOX fuel is roughly four and half times the cost of low-enriched uranium fuel. As such, the authors of the MIT study concluded that "even the most economical partitioning and transmutation schemes are likely to add significantly to the cost of the once-through fuel cycle."[962] Similar results were found by a study conducted at the Kennedy School of Government at Harvard University as well. The authors of this study stated that the cost of separations and transmutation schemes would "be substantially higher than the cost of traditional reprocessing and recycling."[963] They found that transmutation schemes would not be economical unless the cost of repository disposal rose to more than seven and a half times the current estimated costs.[964] Given the already high cost of generating electricity from nuclear power (see Chapter Two), adding to the cost is a serious disadvantage for any waste management proposal.

Third, there are important safety and environmental concerns associated with reprocessing as well. Some of the concerns over the use of MOX fuel have already been discussed in Sections 3.2 and 4.2.3. As noted, the reprocessing of spent fuel via the PUREX process generates large volumes of liquid high-level waste that must be carefully stored prior to vitrification. Table 5.2 shows the amount of high-level waste that is stored at civilian and military reprocessing sites in the United States. Of the four locations where this waste is stored, only West Valley reprocessed commercial spent fuel.

Table 5.2: Amount of high-level waste in the U.S. from commercial and military reprocessing as of FY1996. These estimates include both the waste that was stored in the tanks and that which had been removed but was still stored on-site.[965]

Location	Total Volume (thousands of cubic meters)	Total Radioactivity[a] (millions of curies)
West Valley Demonstration Project	2.0	23.6
Idaho National Laboratory	10.5	48.4
Hanford	207.3	332.1
Savannah River Site	148.3	498.0
Total	368.1	902.1

(a) The vast majority of the radioactivity in the high-level waste is attributable to cesium-137, strontium-90, and their short-lived daughter products. Therefore the current amount of radioactivity in the high-level waste would be about 20 percent less than these estimates due to radioactive decay.

To give a sense of scale for the volumes listed in Table 5.2, the high-level waste stored at the Savannah River Site would cover a professional football field to a depth of nearly 27 meters (90 feet) while the waste at Hanford would reach nearly 38 meters high (more than 120 feet).[966] In France and Britain, a portion of the liquid high-level waste generated at their reprocessing facilities has been discharged directly into the English Channel and Irish sea respectively. These releases have caused the contamination of sea life in the area, and have led the governments of Ireland and Norway to seek an end to the discharges.[967] The majority of the liquid high-level waste generated by reprocessing is stored in tanks that must be cooled and properly monitored in order to prevent a catastrophic explosion. The 1957 explosion of a waste tank at the Chelyabinsk-65 military reprocessing plant near the town of Kyshtym in the former Soviet Union contaminated an area larger than the state of Connecticut and led to the evacuation of more than 10,000 people.[968] Similar explosions in the high-level waste tanks at Hanford or the Savannah River Site are possible if they were to lose cooling.[969]

Fourth, while reprocessing and the use of MOX fuel would simplify somewhat the type of geologic repository needed, the volume of waste to be disposed of would not be significantly reduced. When the waste packages are included, the volume of vitrified high-level waste from reprocessing is between 50 and 100 percent of the volume of the spent fuel from which it came. While the heat rate of the high-level waste is lower than that of the initial spent fuel, the heat rate of the spent MOX fuel is far greater than that of spent uranium fuel. Even if the MOX fuel is recycled in specially designed reactors the overall heat output of the high-level waste will be larger than if the spent fuel had been disposed of directly.[970] In addition, investigations at Yucca Mountain and at the site of the Finnish repository have found that the long-lived fission products technetium-99 and iodine-129 are important contributors to the long-term risk from spent fuel disposal, and it is not yet clear whether these radionuclides would be effectively reduced under many of the partitioning and transmutation schemes that have been proposed.[971]

Finally, roughly 94 percent of the mass of spent fuel is unfissioned uranium. After reprocessing this uranium is contaminated with plutonium, fission products, and other transuranic elements. Due both to its own radiological characteristics as well as the residual contamination, this recycled uranium would eventually have to be disposed of in a repository similar to the Waste Isolation Pilot Plant (WIPP) in New Mexico which currently accepts transuranic waste.[972] WIPP required several decades to explore and characterize. While the repository construction was completed in 1989, it was nearly a decade later before the first waste was accepted due to difficulties encountered in licensing the repository and overcoming legal obstacles.[973] In Europe, the uranium wastes from reprocessing are already recognized as requiring disposal in a geologic repository (although currently much of the recycled and depleted uranium in Europe is being stored or sent to Russia).[974] Finally, the amount of plutonium bearing wastes generated during the decommissioning of a reprocessing plant are expected to be quite large and "to increase significantly the estimates of volume of wastes generated per ton of fuel processed."[975]

In light of these considerations, the authors of the MIT report concluded that

> The trade-off between reduced risk over very long time scales and increased risk and cost in the short term is an issue on which reasonable people can disagree.... *Nevertheless, taking all these factors into account, we do not believe that a convincing case can be made on the basis of waste management considerations alone that the benefits of advanced fuel cycle schemes featuring waste partitioning and transmutation will outweigh the attendant risks and costs.*[976]

Given the many serious drawbacks of reprocessing schemes,[977] the authors of the MIT report proposed expanding the DOE's research on spent fuel management to consider other types of geologic disposal. The central focus of this proposed research program would be placed on deep borehole disposal discussed in the following section.

Section 5.4.3 – Deep Boreholes

The rate of waste generation that would take place under the global growth scenario would make it very unlikely that disposal sites for mined geologic repositories like Yucca Mountain could be identified and developed rapidly enough. It has been nearly 50 years since the first commercial reactor was brought online, and more than 20 years since the DOE began site investigations at Yucca Mountain, and to date there are no operating repositories anywhere in the world. The possible alternative approach proposed by the authors of the MIT report is to place the spent fuel into deep boreholes.[978] The authors describe this option as follows

An alternative to building geologic repositories a few hundred meters below the earth's surface is to place waste canisters in boreholes drilled into stable crystalline rock several kilometers deep. Canisters containing spent fuel or high-level waste would be lowered into the bottom section of the borehole, and the upper section – several hundred meters or more in height – would be filled with sealant materials such as clay, asphalt, or concrete. At depths of several kilometers, vast areas of crystalline basement rock are known to be extremely stable, having experienced no tectonic, volcanic or seismic activity for billions of years.[979]

An advantage of deep boreholes is that the waste would be in a saturated, reducing environment far below the human accessible environment, and the water present at that depth appears to be highly saline and remain isolated from water systems closer to the surface over long timescales.[980]

A typical borehole as considered by the Swedish nuclear waste agency would be 4 kilometers deep, 80 centimeters in diameter. The bottom half of the hole would be filled with waste while the top half would be sealed. Deeper bore holes such as a 5 kilometer deep borehole with the bottom 3 kilometers filled with waste has been considered in the MIT study. While deeper holes are technically feasible, the Swedish analysis found that holes less than 4 kilometers deep were less likely to "to be troublesome to drill" due to the fracturing and degradation of holes at greater depths brought about by the internal stresses within the rock. The amount of waste that could be placed into these boreholes would be between 220 and 420 metric tons per hole in the case of the Swedish design and 300 metric tons per hole in the case of the MIT design. The range for the proposed Swedish design is due to their consideration of both a hotter borehole where the fuel rods would be densely packed in the canisters, and a cooler design with less densely packed waste.[981] With these characteristics a single borehole would be able to hold 11 to 21 years worth of waste from a typical 1000 MW plant. Therefore, just two to four holes would be enough to store the waste generated over the nominal 40 year lifetime of each plant.

The deep borehole concept remains speculative, however, since it has not been intensely investigated given that the scientific consensus settled on mined geologic repositories as the least worst option. As summarized by the authors of the MIT report themselves

> Implementing the deep borehole scheme would require the development of a new set of standards and regulations, a time-

consuming and costly process. A major consideration would be the difficulty of retrieving waste from boreholes if a problem should develop.... Moreover, at the great depths involved, knowledge of in situ conditions (e.g., geochemistry, stress distributions, fracturing, water flow, and the corrosion behavior of different materials) will never be as comprehensive as in shallower mined repository environments. Recovery from accidents occurring during waste emplacement – for example, stuck canisters, or a collapse of the borehole wall – is also likely to be more difficult than for corresponding events in mined repositories.[982]

The uncertainties in gathering data in very deep boreholes and the potential for fractures in the rock caused by the drilling to affect the migration of the radionuclides were issues raised by the Swedish feasibility assessments as far back as 1989.[983]

One of the major concerns that arise with these proposals maintaining reversibility. In its review of spent fuel management options, the National Research Council of the U.S. National Academy of Sciences concluded that

> Maintaining a capability for reversibility of steps during the long processes leading to a closed and sealed repository is a major factor in enhancing public confidence.[984]

With respect to the specific question of boreholes versus mined repositories they concluded that "[r]etrievability and sealing the boreholes (because they may be numerous) are greater technical challenges than for mined repositories."[985] Swedish studies of this issue have concluded that it may be possible to drill the canister out and retrieve them from a borehole using similar techniques to those used in oil drilling to recover from overshoots. These schemes, however, do not address the issue of what would happen if the drill overshot during recovery and ruptured a waste canister, nor do they address the possibility that the canisters might settle over time and thus might not be oriented properly for the recovery scheme to work.[986]

Finally, the time required to locate, drill, and place the waste in a borehole and the total cost could pose additional obstacles. Estimates for the cost of drilling a 4 kilometer deep borehole based upon experience with smaller diameter holes range from $4.66 million to $6.58 million (in 2004 dollars).[987] Significant additional costs, however, would be encountered for such things as the initial site investigation, test drillings to determine the properties of the rock at depth, obtaining a license, transporting, packaging and placement of waste, and finally sealing the borehole and monitoring the site. In summarizing these costs, the National Research Council concluded that "[f]or large amounts of waste, drilling many deep boreholes from the surface is probably more expensive than a single mined repository."[988]

While it is possible that deep boreholes could eventually prove to be an acceptable alternative to mined geologic repositories in some countries which have a smaller amount of waste to dispose of, this cannot yet be determined given today's level of knowledge. As summarized by the U.K. waste management agency Nirex in June 2004

> It is important to emphasise that, although consideration has been given to this disposal concept over a period of many years, no practical demonstration of the application of this concept has taken place. It is also likely that considerable sums of money would be required before it could be brought up to the same level of understanding that already exists for the several different types of mined geological disposal concept that are currently proposed by waste disposal organisations world-wide.[989]

The proposal from the authors of the MIT study to move to interim storage while investigating deep boreholes and simultaneously committing to a large increase in the rate of waste generation would repeat the central error of past nuclear waste management programs. The concept for mined repositories was laid out as likely to be technically feasible by the U.S. National Academy of Sciences as far back as 1957. With this general idea, and the belief in commercial reprocessing and the plutonium economy, hundreds of nuclear plants proliferated across the U.S. and around the world. However, turning the idea of mined repositories into a reality has proven exceptionally difficult and the solution to the waste problem remains elusive as the discussion in this chapter has shown. It is thus a very unsound policy to make the same mistake again with respect to deep borehole disposal and a proposed revival of nuclear power.

Section 5.5 – Conclusions

The radioactive wastes generated by the nuclear fuel cycle span a wide range of volumes and hazards. While the vast majority of the radioactivity is in spent fuel and reprocessing wastes, the vast majority of the volume is in uranium mill tailings and low-level wastes. Unlike the management of high-level waste, which is a federal responsibility, the disposal of low-level waste is the responsibility of individual states. The development of low-level waste disposal sites have encountered a number of problems and several facilities have either been denied a license or have been otherwise abandoned. The management of low-level waste will likely continue to pose a challenge in the future. Of particular concern is the need to dispose of the large volumes of depleted uranium that is generated by enrichment plants. The disposal of depleted uranium poses similar long-term radiological hazards to the disposal of some types of transuranic wastes, and will likely require the development of a repository comparable to the Waste Isolation Pilot Plant in New Mexico.

By far the largest concerns regarding radioactive waste management relate to the handling of spent fuel. Of particular concern are the very long half-lives of some of the radionuclides present in this waste (for example plutonium-239 – half-life of 24,000 years, technetium-99 – half-life of 212,000 years, cesium-135 – half-life of 2.3 million years, and iodine-129 – half-life of 15.7 million years). In addition, the spent fuel presents security risks due to the fact that it contains weapons useable plutonium that can be chemically separated by reprocessing. By the end of 2005 there was already an estimated 53,100 metric tons of spent fuel stored around the United States. By 2012, which is the very earliest date that the Yucca Mountain repository could possibly be opened, the amount of spent fuel discharged from the currently operating reactors would amount to more than 67,500 MT. Thus, even without any new nuclear construction, by 2012 the inventory of spent fuel in the U.S. would already exceed the 63,000 MT statutory limit for the amount of commercial fuel that can be sent to Yucca Mountain.

Between 2005 and 2050, on average, the proposed expansion of nuclear power under the global growth scenario would lead to nearly a doubling of the rate at which spent fuel is currently being generated with proportionally larger increases under the steady-state growth scenario. Assuming a constant rate of growth through mid-century for the global growth scenario and that Yucca Mountain itself was built and operated, a new repository with the capacity of Yucca Mountain would have to come online every six years in order to handle the amount of waste that would be generated. For the steady state growth scenario a Yucca Mountain sized repository would need to be opened somewhere in the world every three years on average.

The characterization and siting of repositories rapidly enough to handle this volume of waste would be a very serious challenge, and one unlikely to be overcome by the development of new technologies or disposal options. The site of the Yucca Mountain repository has been studied for more than two decades, and it has been the sole focus of the Department of Energy since 1987, however, despite this effort and nearly $9 billion in expenditures, as yet no license application has been filed and a key element of the regulations governing the site has been struck down by the courts and re-issued in draft form. As acknowledged in January 2006 by Ernest Moniz and John Deutch, the two co-chairs of the MIT study, "it is unclear whether Yucca Mountain will ever receive a license from the Nuclear Regulatory Commission."[990] In fact, in February 2006, Secretary of Energy Samuel Bodman admitted that the Department of Energy can no longer make an official estimate for when the Yucca Mountain repository might open due to ongoing difficulties faced by the project. No other country currently plans to have a repository in operation before 2020 at the very earliest, and all of these programs have encountered some level of difficulty.

Even if the U.S. repository program had not been plagued by delays and poor quality control and quality assurance practices, serious concerns would remain over the U.S. approach to the management of spent fuel. First, the land upon which the Nevada Test Site and Yucca Mountain are located is claimed by the Western Shoshone Nation which opposes the placement of the repository. The lack of informed consent from those with a deep cultural and historical connection to the land should alone be sufficient to prevent any further consideration of the Yucca Mountain site. Second, as a general rule, the more complex the geology of a repository site, the greater the uncertainty there is in predictions for its long-term performance. Yucca Mountain is a highly complex site which is geologically unique in many important ways from any other sites that are being considered around the world. Finally, and most importantly, Yucca Mountain cannot meet even the most basic requirement of a repository; namely that it maintain the peak dose to an acceptably low level. It is therefore IEER's conclusion that Yucca Mountain cannot be regarded as an appropriate site for repository development, and that the search for an alternative location with a more appropriate geology should begin as soon as possible.

Alternatives to geologic disposal in a mined repository are unlikely to overcome the many challenges posed by the amount of waste that would be generated under the global or steady-state growth scenarios. Proposals to reprocess the spent fuel and to use the resulting plutonium in MOX would not only not solve the waste problem, but would greatly increase the vulnerabilities of a nuclear revival. Using existing technology, reprocessing schemes are expensive and create significant environmental risks due to the generation of liquid high-level waste. Routine discharges and accidents at commercial and military reprocessing facilities have contaminated large areas in Russia, the English Channel, and the Irish sea. In addition, reprocessing generates a large amount of waste that would still require geologic disposal. Vitrified high-level waste and spent MOX fuel, which is not generally reprocessed, would require a repository similar to that required for unreprocessed spent fuel. As with the depleted uranium, the unfissioned uranium separated by reprocessing would eventually have to be disposed of in a repository similar to the Waste Isolation Pilot Plant in New Mexico. The amounts of plutonium bearing wastes generated during the decommissioning of the reprocessing plants is also expected to be quite large, and to add significantly to the volume of waste destined for repository disposal. Most importantly, reprocessing results in the separation of plutonium that can be used to make nuclear weapons. While future reprocessing technologies like UREX+ or pyroprocessing, if successfully developed and commercialized, could have some nonproliferation benefits compared to current reprocessing technologies, they would still pose a significant proliferation risk if deployed on a large scale (see Section 3.2).

The authors of the MIT study acknowledge the high cost as well as the health, environmental, and security risks of reprocessing and, as such, advocate against its use. Instead they advocate for interim storage where needed and for expanded research on deep borehole disposal as a potential alternative to mined repositories. While it is possible that deep boreholes might possibly prove to be an acceptable alternative in some countries which have a smaller amount of waste to manage, this cannot yet be determined given today's level of knowledge. Committing to a large increase in the rate of waste generation based only on the plausibility of a potential waste management option would be to repeat the central error of past nuclear power programs. The concept for mined geologic repositories dates back to at least 1957, however, turning this idea into a reality has proven quite difficult and a solution to the waste problem remains elusive to this date.

Irrespective of future nuclear power development, there will need to be a long-term effort to manage the waste that is already stored around the world, and that which will continue to be generated by the existing fleet of reactors. A solution to this problem cannot simply be to transfer of the liability of spent fuel from the private utilities to the federal government as a means of allowing new plants to be built. This does nothing to resolve the risks to society posed by this very hazardous and long-lived material. To manage this waste IEER proposes that the existing spent fuel be removed from the cooling pools as soon as possible and placed into Hardened On-Site Storage (HOSS).[991] This would consist of first placing the spent fuel into dry casks which include an outer shell of Alloy-22, the corrosion resistant alloy proposed for use in the U.S. repository. This strategy would allow greater experience to be gained with this relatively new alloy, and to allow longer-term measurements of its properties under real world conditions to be made. All spent fuel older than five year should be removed from the cooling pools and stored in these types of dry casks. The dry casks should then be placed into hardened or underground structures onsite that would reduce the chance of a terrorist attack successfully damaging the casks and mitigate the impacts of any release of radioactivity that could accompany such an attack. This strategy would substantially reduce the risks posed by the storage of spent fuel in densely packed cooling pools, and allow sufficient time for a more sound repository program to be developed.

Given that the timescales over which spent fuel must be managed are more akin to human evolution than human civilization, even the best geologic repositories will always have meaningful uncertainty in their performance. In light of the many difficulties that arise in disposing of high-level waste, the public must have a high level of confidence in the agency that is overseeing the repository's development. It is our conclusion that the Department of Energy has demonstrated by its performance on Yucca Mountain and other projects that it is not the right agency for the job. A new, highly transparent agency with strict public oversight and no institutional conflict of interest concerning the promotion of new reactors should be created to manage the repository program in the United States.

While the risks of nuclear waste must be viewed in relation to the potentially catastrophic impacts of global climate change, greatly expanding the production of highly radioactive, long-lived waste which also contains weapons useable plutonium at a time when not one spent fuel rod has yet been permanently disposed of anywhere in the world, is not the basis for a sound energy policy when a clear set of robust and economically viable alternatives are available. The future production of spent fuel should thus be minimized to the maximum extent practicable, and the existing waste should be managed as we have recommended above.

Chapter Six: Looking Back, Moving Forward

> By the end of the century, climate change and its impacts may be the dominant direct driver of biodiversity loss and changes in ecosystem services globally.... The balance of scientific evidence suggests that there will be a significant net harmful impact on ecosystem services worldwide if global mean surface temperature increases more than 2° Celsius above preindustrial levels or at rates greater than 0.2° Celsius per decade (medium certainty).[992]
> - United Nations Millennium Ecosystem Assessment (2005)

> The potential impact on the public from safety or waste management failure and the link to nuclear explosives technology are unique to nuclear energy among energy supply options. These characteristics and the fact that nuclear is more costly, make it **impossible** today to make a credible case for the immediate expanded use of nuclear power.[993]
> - *The Future of Nuclear Power* (2003)

We began this work with the recognition that climate change is by far the most serious vulnerability associated with the world's current energy system. Outside of full scale thermonuclear war, it is perhaps the largest single environmental threat of any kind confronting humanity today. While there are significant uncertainties surrounding the potential consequences of global warming, the possible outcomes are so varied and potentially so severe in their ecological and human impacts that immediate precautionary action is called for in order to try and mitigate the damage being done to the Earth's climate. Definitive proof will only come following a catastrophe, and by then it will be too late to effectively take action. The potential impacts of global warming, combined with our rapidly evolving understanding of the climate system, provides a strong motivation to prioritize mitigation strategies that will have the largest likelihood of making significant contributions in the near to medium term while not jeopardizing the future implementation of more equitable and sustainable long-term strategies. It is in this light that we have examined the question of what strategies might play a role in combating the threat of global climate disruption.

Compared to the other major energy sources used to generate base load electricity such as coal, oil, and natural gas, nuclear power plants emit far lower levels of greenhouse gases even when mining, enrichment, and fuel fabrication are taken into consideration.[994] As a result of this fact, some have come to believe that nuclear power may be able to play an important role in efforts to reduce emissions from the electricity sector. However, we have found that the large number of reactors required for nuclear power to play any meaningful role in reducing emissions would greatly complicate the efforts required to deal with its unique vulnerabilities. These include the potential for the nuclear fuel cycle to enable nuclear weapons proliferation, the risks from catastrophic reactor accidents, and the difficulties of managing long-lived and highly radiotoxic nuclear waste. The rapid rate of construction required to meet the global or steady-state growth scenarios would also put great pressures on the nuclear industry as well as on regulatory bodies and make it more difficult to achieve or sustain the improvements in cost that have been envisioned by nuclear proponents.

As more is learned about the functioning of the Earth's climate, the more likely it appears that reductions in greenhouse gas emissions on the order of 60 to 80 percent will be required by 2050 in order to avoid the more serious consequences of global warming. As such, it is likely that a number of aggressive policies will be needed in the coming decades to curb and then reverse the growth of CO_2 emissions. Adding to the complexity of this already extremely difficult problem is the fact that these reductions will have to occur at a time of increasing electricity demand throughout the Global South. Of particular note is the projected increase in electricity consumption in the world's two most populous countries, India and China.

Given that both time and resources are limited, a choice must necessarily be made as to which alternatives should be pursued aggressively and which should play only a small role or be put aside all together. Given the immediacy of the problem, it will be necessary to consider both options that are available for immediate use as well as those that can confidently be brought online within then next five to fifteen years. The best mix of alternatives will vary according to local, regional, and country specific resources and needs. As such, the details of the future energy system cannot yet be completely foreseen in all cases.[995] However, in making a choice among the available alternatives, the following considerations should serve to guide the selection: (1) the options pursued must be capable of making a significant contribution to a reduction in greenhouse gas emissions, with a preference given to options that achieve more rapid reductions; (2) the options should be economically competitive to facilitate their rapid entry into the market; (3) the options should, to the extent consistent with the goals of reducing the threat from climate change, minimize other environmental and security impacts; and finally (4) the options should, to the maximum extent possible, be compatible with a longer term vision of creating an equitable and truly sustainable global energy system. It is within this context that the future of nuclear power must be judged. As such we carefully considered not only the cost of electricity from new nuclear plants, but also the environmental, health, and security implications of the global and steady-state growth scenarios in order to determine how the nuclear option compares to other available alternatives.

While concerns over catastrophic accidents and long-term waste management have received more public attention, the largest single vulnerability associated with an expansion of nuclear power is likely to be its potential connection to the proliferation of nuclear weapons. In order to fuel the number of nuclear plants envisioned under the global or steady-state growth scenarios, increases in the world's uranium enrichment capacity of approximately two and half to six times would be required.[996] Just one percent of the enrichment capacity required by the global growth scenario alone would be enough to supply the highly-enriched uranium for nearly 210 nuclear weapons every year.[997] The risks from such an increase in enrichment capacity are such that even the authors of the MIT report concluded that "[n]uclear power should not expand unless the risk of proliferation from operation of the commercial nuclear fuel cycle is made acceptably small."[998]

As discussed in Chapter Three, the proposals that have been put forth to try and reduce the risks of nuclear weapons proliferation are very unlikely to be successful in a world where the five acknowledged nuclear weapons states seek to retain their arsenals indefinitely. The institutionalization of a system in which some states are allowed to possess nuclear weapons while dictating intrusive inspections and restricting what activities other states may pursue is not likely to be sustainable. As summarized by Mohamed ElBaradei

> We must abandon the unworkable notion that it is morally reprehensible for some countries to pursue weapons of mass destruction yet morally acceptable for others to rely on them for security -- indeed to continue to refine their capacities and postulate plans for their use.[999]

Without a concrete, verifiable program to irreversibly eliminate the tens of thousands of existing nuclear weapons, no nonproliferation strategy is likely to be successful no matter how strong it would otherwise be. As such, the link to nuclear weapons is likely to prove to be one of the most difficult obstacles to overcome in any attempt to revive the nuclear power industry.

In addition to its link to nuclear weapons proliferation, the potential for a catastrophic reactor accident or well coordinated terrorist attack to release a large amount of radiation adds to the unique dangers of nuclear power. Such a release could have extremely severe consequences for human health and the environment, would require very expensive cleanup and decontamination efforts, and would leave buildings and land dangerously contaminated well into the future. The CRAC-2 study conducted by Sandia National Laboratories estimated that a worst case accident at some of the existing nuclear plants in the U.S. could result in tens of thousands of prompt and long-term deaths and cause hundreds of billions of dollars in damages.[1000] Even if a reactor's secondary containment was not breached, however, and there were not dangerously large offsite releases of radiation, a serious accident would still cost the utility a great deal due both to the loss of the reactor and the need to buy replacement power. As summarized by Peter Bradford, a former commissioner of the Nuclear Regulatory Commission,

> The abiding lesson that Three Mile Island taught Wall Street was that a group of N.R.C.-licensed reactor operators, as good as any others, could turn a $2 billion asset into a $1 billion cleanup job in about 90 minutes.[1001]

The history of public opposition to nuclear power clearly demonstrates the importance of reactor safety to the acceptability of this technology. In light of the high probabilities that we found in Chapter Four for at least one meltdown occurring somewhere in the world between now and 2050 under with the global or steady-state growth scenario, the possibility that public opinion could turn sharply against the widespread use of nuclear power is a significant vulnerability with plans that envision a heavy reliance on this energy source. If nuclear power is in the process of being expanded, public pressure to shutdown existing plants following an accident would leave open few options (particularly in terms of reducing greenhouse emissions). On the other hand, if long-term plans to phase out nuclear power were already being carried out when an accident or attack occurred, there would be far more options available and those options could be accelerated with less disruption to the overall economy.

Finally, the difficulty of managing the radioactive wastes generated by the nuclear fuel cycle is one of a longest standing challenges accompanying the use of nuclear power. In addition to its high radiotoxicity, the existence of large quantities of weapons usable plutonium in the spent fuel complicates the waste management problem by raising concerns over nuclear weapons proliferation.[1002] While the management of low-level waste will continue to pose a challenge, by far the largest concern regarding radioactive waste management is how to handle the spent nuclear fuel. Greatly complicating this task are the very long half-lives of some of the radionuclides present in this waste (for example plutonium-239 – half-life of 24,000 years, technetium-99 – half-life of 212,000 years, cesium-135 – half-life of 2.3 million years, and iodine-129 – half-life of 15.7 million years).

Through 2050, the expansion of nuclear power under the global growth scenario would lead to nearly a doubling of the average rate at which spent fuel is currently generated with proportionally larger increases under the steady-state growth scenario. Assuming a constant growth rate for nuclear plant construction, and that Yucca Mountain itself was successfully licensed and built, a new repository with the capacity of Yucca Mountain would have to come online somewhere in the world every six years in order to handle the amount of waste that would be generated under the global growth scenario. For the steady state growth scenario a new Yucca Mountain sized repository would need to be opened every three years on average just to keep up with the waste being generated.[1003]

The characterization and siting of repositories rapidly enough to handle the volumes of waste that would be generated by a nuclear revival would be a very serious challenge. The site of the Yucca Mountain repository has been studied for more than two decades, and it has been the sole focus of the Department of Energy since 1987. However, despite this effort, and nearly $9 billion in expenditures, as yet no license application has been filed and a key element of the regulations governing the site has been struck down by the courts and re-issued in draft form. Adding to the uncertainty about the repository's future is the fact that the draft standard proposed by the EPA in August 2005 would be the least protective by far of any repository regulation anywhere in the world, and will therefore likely face future challenges.[1004]

As discussed in Chapter Five, alternatives to disposal in a mined repository are unlikely to overcome the many challenges posed by the amount of waste that would be generated under either the global or steady-state growth scenarios. Proposals to reprocess the spent fuel would not only not solve the waste problem, but would greatly increase the vulnerabilities of a nuclear revival. Reprocessing schemes are expensive and create a number of serious environmental risks. In addition, reprocessing results in the separation of weapons useable plutonium which adds significantly to the risks of proliferation. While future reprocessing technologies like UREX+ or pyroprocessing, if successfully developed and eventually commercialized, could have some nonproliferation benefits, they would still pose a significant risk if deployed on a large scale.

The authors of the MIT study acknowledge the high cost as well as the negative health, environmental, and security impacts of reprocessing and, as such, advocate against its use. Instead they propose interim storage and expanded research on deep borehole disposal. While it is possible that deep boreholes might prove to be an acceptable alternative to mined repositories in countries which have a smaller amount of waste to manage, this cannot yet be determined. Committing to a large increase in the rate of waste generation based only on the potential plausibility of a future waste management option would be to repeat the central error of nuclear power's past. The concept for mined geologic repositories dates back to at least 1957, however, turning this idea into a reality has proven quite difficult, and a solution to the waste problem remains elusive to this date. Significantly expanding the production of highly radioactive, long-lived waste which also contains weapons useable plutonium at a time when not one spent fuel rod has been permanently disposed of anywhere in the world, is not a sound proposal.

Overall, our analysis has shown that nuclear power is a uniquely dangerous source of electricity that would create a number of serious risks if employed on a large scale. In addition, we have also found that it is likely to be an expensive source of electricity with costs in the range of six to seven cents per kWh for new reactors. While a number of potential cost reductions have been considered we showed in Chapter Two that it is unlikely that plants not heavily subsidized by the federal government would be able to achieve any further improvements to the cost of nuclear power.[1005] This is particularly true given that any improvements would have to be maintained under the very demanding timetables set by the global or steady-state growth scenario. In considering other options that may be available to help reduce greenhouse gas emissions from the electricity sector, we found that there are a number of alternatives that are either ready for immediate implementation or that could be commercialized within the next five to fifteen years. Most importantly, we found in Chapter Two that, when projected over this same timeframe, the cost of each of these options tended to fall roughly within or below the range of six to seven cents per kWh.[1006] Thus, the question of cost becomes less important in choosing between the available alternatives, and the deciding factors instead hinge on the rapidity with which the options can be brought online and on their relative environmental impacts.

Of the alternatives available in the near-term, the two most promising options are efforts to increase the efficiency of electricity generation and use and a large-scale expansion of wind power at favorable sites. Improvements in efficiency as well as a reduction in demand through conservation have the potential for significant benefits throughout the Global North and to enable countries in the Global South to leapfrog over older, dirtier technologies. Unlike programs focused on simply increasing supply, demand side options can result in low or negative cost reductions in greenhouse gas emissions while simultaneously providing new jobs and opening new avenues for economic growth. Combined with these efforts, the expanded use of renewable energy, particularly wind power, offers the most economically attractive option for supplying the required near-term incremental growth in generating capacity. At approximately four to six cents per kWh, wind power at favorable sites is already competitive with natural gas or new nuclear power. With the proper priorities on investment in transmission and distribution infrastructure and changes to the ways in which the electricity sector is regulated, wind power could rapidly make a significant contribution. In fact, without any major changes to the existing grid, wind power could expand in the near term to make up 15 to 20 percent of the U.S. electricity supply as compared to less than one-half of one percent today. This expansion could be achieved without having any negative impacts on the overall stability or reliability of the transmission grid. Similar potential for these alternatives exist throughout the Global North. As summarized by the British Department of Trade and Industry

> Energy efficiency is likely to be the cheapest and safest way of addressing all four objectives [i.e. a reduction of greenhouse gas emissions, the maintenance of a reliable energy supply, promotion of competitive markets, and an assurance of ade-

> quate and affordable heat to every home]. Renewable energy will also play an important part in reducing carbon emissions, while also strengthening energy security and improving our industrial competitiveness as we develop cleaner technologies, products, and processes.[1007]

While it will require a significant effort to achieve the implementation of new efficiency programs and to develop the necessary infrastructure to support a large increase in the contribution of wind power, it is important to compare those efforts to the difficulties that would be encountered in restarting a nuclear power industry that last hasn't had a new order placed in the U.S. in more than 25 years and hasn't opened a single new plant built in the last ten years.[1008] Including interest payments on debt, each new nuclear plant is expected to cost nearly $2.6 billion to build under the MIT base case assumptions, and dozens of such plants would have to be started in the next five to fifteen years in order to remain on track to meet the global or steady-state growth scenarios.[1009] In addition, we note that the current fossil fuel based energy system is also very expensive to maintain. For example, the International Energy Agency estimates that the amount of investment in oil and gas between 2001 and 2030 will total nearly $6.1 trillion, with 72 percent of that investment going towards new exploration and development efforts.[1010] Finally, unlike the decision to build new nuclear power plants, it is important to note that there is already strong and sustained public support for expanding energy efficiency efforts and for expanding the use of renewable resources which would help to facilitate their rapid implementation.

While many improvements can be made in the near term, a significant potential will continue to exist for increasing energy efficiency throughout this half century. For example, as the current building stock turns over, older, less efficient buildings can be replaced by buildings that incorporate advanced features such as passive solar systems for lighting and water heating, greatly improved insulation, and high efficiency heating and cooling systems such as earth source heat pumps. In addition to the replacement of buildings, the IPCC has identified what it calls "robust policies" for reducing greenhouse gas emissions over the longer term that include "social efficiency improvements such as public transport introduction, dematerialization promoted by lifestyle changes and the introduction of recycling systems."[1011] As recommended by Dr. Arjun Makhijani in 2001

> Public transportation in urban areas should be regarded as a utility, much like water, electricity or telephones. A diverse

system of transport that includes cars, motorized and rail public transport, bicycle lanes and sidewalks would reduce vulnerabilities to terrorism by diversifying the modes by which people could travel in cities. By making public transportation safe, efficient, economical, frequent, and convenient, energy use as well as time for commuting could be greatly reduced with all the attendant social, economic, and environmental benefits.[1012]

In addition, to continuing improvements in energy efficiency the utilization of wind power, thin-film solar cells, advanced hydropower, and some types of sustainable biomass could allow renewables to make up an increasingly significant proportion of the electricity supply over the medium-term. This expansion could be facilitated through the development of a robust mix of technologies that have different types of intermittency and variability, the development of strengthened regional grids to help stabilize the contribution of wind and solar power through geographic distribution, the use of pumped hydropower systems to store excess electricity during times of low demand, and the tighter integration of large scale wind farms with natural gas fired capacity. Beyond its potential contribution in the Global North, the development of cost effective solar power could also have a profound impact on the development of electricity systems in the Global South where there is currently a lack of robust transmission and distribution infrastructures in many areas.

The continued expansion of both efficiency efforts and renewable energy have few negative environmental or security impacts compared to our present energy system and, in fact, have many important advantages. As a result, these options should be pursued to the maximum extent possible. However, in order to stabilize the climate by mid-century, it appears likely that some transition technologies which have more significant health and environmental tradeoffs will also be needed over the coming decades. In much the same way that a cancer patient may choose to undergo chemotherapy despite its toxic side effects, we will have to make a number of difficult choices now in order to avoid the potentially catastrophic consequences of not dramatically reducing carbon emissions by mid-century.[1013] In this vein, two of the more important transition strategies available are likely to be an increased reliance on the use of liquefied natural gas (LNG) and the use of integrated coal gasification plants with sequestration of carbon in geologic formations.

Compared to pulverized coal plants, combined cycle natural gas plants emits about 55 percent less CO_2 for the same amount of generation.[1014]

If efficiency improvements to the energy system as a whole and an expanded liquidification and regasification infrastructure can stabilize the long-term price of natural gas at the cost of imported LNG, then the use of combined cycle generating plants is likely to remain an economically reasonable choice for replacing some of the highly inefficient coal fired plants in operation today. For example, the levelized gas prices in the moderate to high price scenarios from the MIT study ($4.42 to $6.72 per million BTU) are consistent with the recent average import price of LNG in the United States ($4.37 per MMBtu between 2000 and 2004), the average price for LNG imports to Japan and South Korea over the past decade (~$4 per MMBtu), and the expected price for future LNG imports to India ($4.10 per MMBtu).[1015]

With respect to coal, the use of gasification technologies would greatly reduce the emissions of mercury, particulates, and sulfur and nitrogen oxides. In addition, the higher efficiency of IGCC plants compared to pulverized coal plants would also reduce, somewhat, the carbon emissions from these newer plants. However, for coal to be considered as a potential transition technology, it must be accompanied by carbon sequestration. Experience in the U.S. with carbon dioxide injection has been gained since at least 1972. Overall, about 43 million tons per year of carbon dioxide is currently being injected each year at 65 enhanced oil recovery programs in the United States alone.[1016] A related source of experience has been gained through the sequestration of acid gas from natural gas production in depleted gas fields and nearby saline aquifers. The sequestration of carbon dioxide has also been demonstrated at the Sleipner gas fields in the North Sea and at the In Salah natural gas fields in Algeria.[1017] While the costs of such strategies are more uncertain than those of other mitigation options, our central estimates for the cost of electricity from natural gas or gasified coal plants with carbon sequestration still fall within the range of six to seven cents we found for other options.

Some of the most troubling aspects of these transition technologies, such as mountain top removal mining for coal, would be mitigated by the reduction in demand that would be achieved through an increase in efficiency and the rapid expansion of alternative energy sources. In addition, it appears quite likely that coal gasification and carbon sequestration would be better suited to the Western United States where mine-mouth coal could be used given the greater access to oil and gas fields which have already been explored and which offer the potential for added economic benefits from enhanced oil and gas recovery. On the other hand, the Eastern U.S., mountain top removal mining is currently practiced, would appear better suited for an expanded use of liquefied natural gas as a transition strategy given the existing regasification capacity, the well developed gas distribution system, and the shorter transportation distances from the Caribbean, Venezuela, and Western Africa.

While the continued use of fossil fuels during the transition period will have many serious drawbacks, these must be weighed against the potentially catastrophic damage that could result from global climate change and against the uniquely serious risks that accompany the use of nuclear power, such as the potential for nuclear weapons proliferation and the risks of catastrophic reactor accidents, and the difficulties of safely managing long-lived radioactive waste. Proposals for a revival of nuclear power and its widespread use over the coming decades, would take the already deeply complicated problem of how to reduce global greenhouse gas emissions while expanding access to electricity in the Global South and make it even more difficult to deal with. As we have said throughout this work, trading one uncertain, but potentially catastrophic health, environmental, and security threat for another is not a sensible basis for an energy policy.

Just as the claim that nuclear power would one day be "too cheap to meter" was known to be a myth well before ground was broken on the first civilian reactor in the United States, and the link between the nuclear fuel cycle and the potential to manufacture nuclear weapons was widely acknowledged before President Eisenhower first voiced his vision for the "Atoms-for-Peace" program, a careful examination today reveals that the expense and unique vulnerabilities associated with nuclear power would make it a very risky, unsustainable, and uncertain option for reducing greenhouse gas emissions. As the authors of the MIT report themselves conclude

> The potential impact on the public from safety or waste management failure and the link to nuclear explosives technology are unique to nuclear energy among energy supply options. **These characteristics and the fact that nuclear is more costly, make it impossible today to make a credible case for the immediate expanded use of nuclear power.**[1018]

As we have shown in great detail throughout this work, it is very unlikely that these problems can be successfully overcome given the large number of reactors that would be required if nuclear power were to play a significant role in reducing greenhouse gas emissions. This is particularly true given the urgent need to begin reducing emissions as soon as possible if we are to avoid the more serious consequences of global climate change. It has now been more than 50 years since the birth of the civilian nuclear power industry and more than 25 years since the last reactor order was placed in the United States. It is time for the global community to move on from a belief in the nuclear option and to begin focusing its efforts on developing more rapid, more robust, and more sustainable options for addressing the most pressing environmental concern of our day. The alternatives are available if we have the will to make them a reality. If not, it will be our children and our grandchildren who will have to live with the consequences of our failure.

Appendix A: Uranium Supply and Demand

From the earliest days of the nuclear enterprise, there have been concerns raised over whether the available uranium supply is adequate to meet projected demand. In the first four decades of nuclear power's expansion, there was a widespread belief that uranium was a fairly rare commodity and that the supplies which could be mined at a reasonable cost would be insufficient to provide fuel for the large number of reactors that were then envisioned. The belief that supplies of reasonably priced uranium were limited was a major factor in the decision to aggressively pursue fast-breeder reactors that would burn plutonium in a closed fuel cycle. This belief was heightened when, following the first Arab oil embargo, the projected demand for nuclear power pushed uranium prices to record highs. By 1977, the spot-market price for uranium had reached a peak of $300 per kilogram (in 2003 dollars).[1019] During this same time, however, concerns over the spread of reprocessing technology and the widespread separation of weapons usable plutonium were brought to the foreground by the Indian nuclear test in 1974. The decisions by Presidents Ford and Carter to end commercial reprocessing in the U.S. in 1976-77, and the subsequent formation of the Nuclear Suppliers Group to limit the spread of sensitive fuel cycle technologies were made largely as a result of concerns over the proliferation of nuclear weapons.

Escalating costs of nuclear construction, the accidents at Three Mile Island in 1979 and Chernobyl in 1986, and a slower rate of growth in electricity demand as a result of policies coming in the wake of the first energy crisis, lead to a far smaller number of reactors coming online than had been predicted earlier. This drove down expectations about future fuel requirements and with it the price of uranium fell. The price for uranium on the spot-market declined fairly steadily during the 1980s, and has generally remained between $30 and $40 per kilogram since 1989.[1020] With the fall of the Berlin Wall and the end of the Cold War, substantial amounts of uranium previously held by governments (primarily the U.S. and Russia) were made available to the commercial market. This so-called secondary uranium, has helped to hold down the price of uranium despite the fact that production of primary uranium from operating mines has been insufficient to meet demand for many years.[1021]

In light of this history, it is important to address the question of uranium supplies when projecting such large increases in nuclear capacity as are envisioned under the global-growth or steady-state growth scenarios. If the increased fuel requirements were to lead to renewed concerns over the adequacy of the affordably priced uranium, then the balance of support may shift to a greater focus on reprocessing spent commercial fuel and the use of the MOX or fast-breeder reactor fuel cycles. This shift would have serious repercussions regarding the potential for nuclear weapons proliferation as discussed in Chapter Three. We will begin this review by examining estimates that have been made for the magnitude of remaining uranium resources as well as the difficulties that are likely to be encountered in exploiting these resources within the next few decades. We will then compare these estimates for production to the cumulative demand of the global-growth and steady-state growth scenarios. Finally, we will examine the other links between uranium supply and nuclear weapons proliferation.

Section A.1 – Estimates of Uranium Resources

Estimates of conventionally recoverable uranium resources are divided into broad four categories based on the level of knowledge about the deposits and the confidence in the accuracy of that information. As summarized by the IAEA, these four categories are:

> Reasonably assured resources (RAR) refers to uranium that occurs in known mineral deposits of delineated size, grade and configuration such that the quantities which could be recovered within the given production cost ranges with currently proven mining and processing technology can be specified....
>
> Estimated additional resources category I (EAR-I) refers to uranium in addition to RAR that is inferred to occur, mostly on the basis of direct geological evidence, in extensions of well explored deposits, or in deposits in which geological continuity has been established but where specific data, including measurements of the deposits and knowledge of the deposits' characteristics, are considered to be inadequate to classify the resource as RAR....
>
> Estimated additional resources category II (EAR-II) refers to uranium in addition to EAR-I that is expected to occur in deposits for which the evidence is mainly indirect and which are believed to exist in well defined geological trends or areas of mineralization with known deposits....
>
> Speculative resources (SR) refers to uranium, in addition to EAR-II, that is thought to exist, mostly on the basis of indirect evidence and geological extrapolations, in deposits discoverable with existing exploration techniques. The location of de-

posits envisaged in this category could generally be specified only as being somewhere within a given region or geological trend....[1022]

For convenience, these four categories are sometimes grouped into two super-categories; "Known Conventional Resources," which is the sum of the Reasonably Assured Resources and Estimated Additional Resources Category I, and "Reported Undiscovered Conventional Resources," which is the sum of the Estimated Additional Resources Category II and Speculative Resources.

Like all other extractive resources, the amount of uranium that may be recovered from a particular deposit depends on the price. The higher the price, the larger the amount that can be economically extracted. A typical upper value used for estimating available resources is $130 per kilogram of uranium. This is well above the recent price of about $30 to $40 per kilogram, but it is not so high as to make it unreasonable to exploit if demand was to rise sharply in the near to medium term. Three recent estimates for the amount of uranium recoverable at less than $130 per kilogram made by the OECD Nuclear Energy Agency and the International Atomic Energy Agency are summarized in Table A.1.

Table A.1: Comparison of three recent estimates for the amount of conventionally recoverable uranium (in million of tons). The first two categories make up the so-called known conventional resources while the second two are the reported undiscovered conventional resources.[1023]

	RAR	EAR-I	EAR-II	SR	Total
IAEA Estimate (2001)	4.34	0.88	2.23	3.99	11.44
OECD/IAEA Estimate (2002)	2.85	1.08	2.33	4.44	10.70
OECD/IAEA Estimate (2004)	3.17	1.42	2.25	4.44	11.28

On top of these estimated resources there is 3.10 to 4.68 million tons of additional speculative resources available at undetermined price as well as substantial amounts of very low-grade resources. These very low-grade resources include an estimated 22 million tons of uranium available in phosphate deposits (at concentrations of approximately 6 to 600 parts per million) and up to 4 to 4.5 billion tons of uranium in seawater (at a concentration of approximately 3 parts per billion).[1024] While the very low-grade deposits are not likely to be exploited directly within the coming decades under any scenario, co-recovery of uranium from phosphate as a byproduct of fertilizer production could yield an estimated 3,700 tons of uranium per year. This would be a fairly modest level of annual production equal to just 5.4 percent of the global uranium requirement in 2003.[1025]

While the estimates summarized in Table A.1 are generally considered to be the most authoritative estimates of current resources, there are factors that tend to make the OECD/IAEA estimates closer to a lower bound rather than an upper bound, even when the speculative resources are included. The first such factor is that many countries do not report estimates of their resources in the higher cost, lower confidence categories. For example, Australia does not report any estimate for its speculative resources because of its large known conventional resources.[1026]

Second, Table A.1 only includes the resources that have been estimated as a result of past exploration efforts. The consistently low price of uranium since the early 1990s coupled with the low growth in nuclear capacity and ample secondary supplies of uranium have all weakened the economic incentive for companies to spend money on new exploration. If the price of uranium rises due to increasing demand, it is expected that a significant amount of resources would be found that are not included in the current estimates. As an indication of this effect we note that, despite the modest level of recent investments in exploration, the 2003 OECD/IAEA estimate for conventional resources recoverable at below $130 per kilogram increased by 5.4 percent over the equivalent estimate from 2001. A similar increase in total estimated resources available at any price (5.2 percent) was also seen between the 1999 and 2001 OECD/IAEA estimates.[1027] A more striking example of the increase in resources that can accompany an increase in exploration occurred at the Canadian McArthur River mine. This deposit, initially discovered in 1988, has the highest known ore concentrations in the world at nearly 25 percent U_3O_8. Since its discovery, additional exploration has lead the estimated amount of recoverable uranium at this site to increase by 35 percent in 1999, and then again by a further 50 percent in 2001.[1028]

The past experience with other extractive resources lends further support to the conclusion that uranium resources will likely expand significantly if prices increase. This is both because higher prices will lead to new exploration and because new, more expensive technologies will increase the efficiency with which lower grade ores can be exploited. For example, the U.S. Geological Survey reports that the composite mineral price index for copper, gold, iron, lead, zinc and cement, clay, crushed stone, lime, phosphate rock, salt, and sand/gravel actually decreased between 1900 and 2000 despite a significant increase in the demand for these resources.[1029]

A recent attempt to incorporate an estimate for the expansion of recoverable resources with increasing price into an estimate for future uranium supplies was made by researchers at Harvard University. Starting from the 2002 OECD/IAEA estimates for presently known uranium resources, they developed a simple model for how the amount of recoverable uranium could be expected to increase based on geologic considerations. The authors incorporated estimates for the elasticity of the uranium supply from different sources, including a high estimate from the uranium industry (the Uranium Information Centre in Australia) and a low estimate from advocates of reprocessing (the DOE's Generation IV Fuel Cycle Crosscut Group). At prices up to $80 per kilogram, the Harvard study estimated that the available resources might total between 11 and 21 million tons. At prices up to $130 per kilogram, the authors estimated that uranium reserves might total between 34 and 105 million tons. Other models that have been proposed for the relationship between price and resource discovery would lead to somewhat lower estimates, but the authors of the Harvard study considered their results to be reasonable given the historical experience with other extractive industries.[1030]

Finally, both the MIT and Harvard analyses have advocated that extensive government support of uranium exploration efforts should be given in the event of any decision to expand nuclear power.[1031] Specifically, the authors of the MIT study recommended that a total of $250 million be spent by the U.S. government over the next 5 years to develop "a global uranium resource evaluation program" which would "include geological exploration studies to determine with greater confidence the uranium resource base around the world."[1032] This subsidy would amount to roughly 40 percent of the total spending by all public and private entities in all countries on both domestic and international uranium exploration in the five years between 1998 and 2002.[1033] Such a large increase in the funding of uranium exploration along with the renewed interest such a large government investment would bring would add significantly to the likelihood that new deposits would be found as demand increased. Thus, it is reasonable to conclude that the conventional resources available at less than $130 per kilogram would likely grow to at least several times the current estimates by 2050 under any major expansion of nuclear power.

Section A.2 – Estimates of Uranium Production Capacity

The previous section discussed the amount of uranium in the ground around the world, however, for that uranium to be useful as fuel for reactors the deposits must be found and adequately characterized and the uranium must then be extracted, processed, enriched, and fabricated into fuel rods. Thus, it is not the shear magnitude of available resources that is important, but the magnitude of production that will determine whether the once-through fuel cycle is sustainable through mid-century or not. In 2001, existing and committed productions facilities had a combined capacity of approximately 45,310 tons of uranium per year. This was just 70 percent of the total uranium demand in that year. By 2005, the primary production capacity was expected to increase to between 48,319 and 56,074 tons of uranium per year, which is still well below recent demand.[1034] In fact, even at prices up to $80 per kilogram, the OECD and IAEA estimate that the peak annual production between now and 2020 from all existing, committed, planned, and prospective facilities would not be sufficient to satisfy even today's annual demand for uranium.[1035] This imbalance between primary production and end-use consumption has existed since the early 1990s, while the deficit has been made up by the stockpiles of secondary uranium released at the end of the Cold War.[1036]

In order to satisfy the demand for uranium that would accompany a large increase in light-water reactors operating on the once-through fuel cycle, a significant effort to explore and develop new deposits will be required.[1037] There are two main issues with new exploration and development, however, that will have a direct bearing on the ability of industry to find and exploit the available uranium resources in a timely manner. The first is the long lead-times that are associated with uranium extraction projects. Estimates from the OECD International Energy Agency, the OECD Nuclear Energy Agency, and the International Atomic Energy Agency agree that it will likely require between 10 and 20 years to bring new uranium projects into operation. The reasons for this long lead-time include the requirement for the completion of extensive environmental reviews before construction, the need to design and construct the facility so as to comply with radiation health, safety, and environmental regulations during operation in addition to the general technical difficulties encountered during exploration and development.[1038] As summarized by the IAEA, recent experience has shown that "long lead times will be the rule rather than the exception."[1039]

The health and environmental impacts of uranium production for nuclear weapons has been extensively reviewed in *Nuclear Wastelands: A Global Guide to Nuclear Weapons Productions and Its Health and Environmental Effects* edited by Arjun Makhijani, Howard Hu, and Katherine Yih. In this review, it was found that the impacts from past uranium production have fallen disproportionately upon indigenous populations around the world.[1040] The majority of uranium continues to be produced in underground or open pit mines, but the percentage of open pit mining has decreased since 1998 as the percentage of production from in situ leaching has increased (see Table A.2). It is expected, however, that "the use of conventional [mining] techniques is likely to increase in the future, particularly underground mining."[1041]

Table A.2: Percentage of uranium produced by different mining methods between 1998 and 2003 showing the shift away from open pit mining to in situ leaching and other techniques.[1042]

Technique	1998	1999	2000	2001	2002	2003 (estimated)
Underground Mine	40	36	43	44.2	43.1	39.5
Open Pit Mine	39	35	28	26.1	26.8	27.9
In Situ Leaching	13	17	15	15.5	18.3	20.7
Other	8	12	14	14.2	11.8	11.9

Given the importance of long lead-times at restricting the amount of uranium production, the IAEA identified resources that they believed were at particular risk of delay due to environmental opposition. The found that nearly 415,000 tons of uranium resources in the highest confidence category associated with 31 different projects was "potentially subject to such opposition." This amount was nearly 15 percent of the total amount of uranium in this category that they estimated to be recoverable by 2050. The two countries with the largest number of at risk projects were the United States and Australia. These two countries accounted for more than four-fifths of the total projects facing potential opposition on environmental grounds. Significantly, the IAEA also projected that these same two countries would have to have the highest levels of production at mid-century in order to meet the projected demand of a large increase in nuclear power.[1043]

Another major issue affecting the ability of industry to exploit the available uranium resources in the given timeframe is the financial uncertainties involved with making investments in countries that have political or economic situations which put projects at greater risk of failure. This is considered to be a particular concern for the exploitation of the large resources present in the countries of the former Soviet Union where the arrangement of adequate financing could add to the lead-time of these projects.[1044] These delays could have a significant impact on the rate of production given that, in 2001, roughly 15 percent of known and estimated conventional resources recoverable for less than $130 per kilogram were estimated to be in Kazakhstan while another 7.4 percent was estimated to be in Russia.[1045] The need to aggressively exploit resources in new countries and expand operations in historically small producers is evident in the projections of the IAEA shown in Figure A.1.

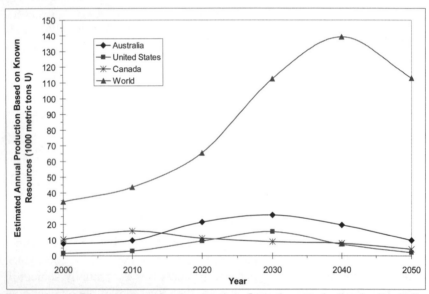

Figure A.1: IAEA projections for the amount of uranium production by the U.S., Australia, and Canada through 2050 under a nuclear revival scenario similar to the global growth scenario. Beyond 2020 there would need to be a very large increase in production outside these three countries.[1046]

The United States, Canada, and Australia accounted for about 55 percent of production in 2000 and together accounted for approximately 40 percent of all historical production through that year. Under a nuclear growth scenario with a similar uranium demand to the MIT global growth scenario (see below), the IAEA projection shown in Figure A.1 has the contribution of these three countries falling to less than 25 percent of global production by 2040 and to less than 15 percent of total production by 2050.[1047]

While the IAEA did not make any significant allowance for an increase in the available resources that would likely accompany a rising uranium price rise as discussed in Section A.1, they did attempt to account for such factors as "cost, technical feasibility, and environmental and political risk" in making estimates for how what could be extracted by 2050.[1048] Table A.3 summarizes their results.

Table A.3: IAEA estimates for the amount of uranium resources that could be extracted and brought to market by 2050 in each of the three highest confidence categories. Decreasing production capacity accompanies decreasing confidence, with a sharp drop between EAR-I resources which are estimated based on direct geological evidence and EAR-II which is estimated from indirect evidence.[1049]

	RAR	EAR-I	EAR-II	Total
Available Resources (IAEA 2001 Estimate)	4.34	0.88	2.23	7.45
Production Estimated Through 2050	3.85	0.74	0.71	5.30
Percentage of Available Resources Exploited	88.7%	84.1%	31.8%	71.1%

As expected the ability to recover resources is highest in the highest confidence category where much of the exploration and characterization has already been done and the percentage decreases sharply as lower confidence ores are considered. The IAEA did not include an explicit estimate for the amount of speculative resources that they believed could be developed by 2050 given the uncertainty inherent in such projections, but they did include an illustrative example in which 15 percent of the total speculative resource would be available for production. This example is consistent with the falling recovery rate for ores that require a greater effort for exploration, characterization, and development.[1050]

Adding 15 percent of the estimated speculative resources to the IAEA's forecast would result in an estimate of 5.9 million tons of total production possible through 2050. This amount of uranium production would be equal to just under 52 percent of the total currently known and unknown conventional resources. We can therefore use this percentage as a rough guide in our calculations for how much of the uranium in the ground is likely to be recoverable within the next four and a half decades. Taking the estimate of 34 to 105 million tons of uranium available at less than $130 per kilogram derived by the researchers at Harvard, we estimate that, if a significant amount of effort was expended on exploration and resource exploitation, perhaps as much as 17.5 to 54 million tons of uranium could be delivered to the market through 2050 at this price. While this estimated range has a significant level of uncertainty, it is likely to be on the correct order of magnitude for comparison to the projected uranium demands of the global or steady-state growth scenario.

Section A.3 – Stretching Uranium Resources

Before attempting to estimate the uranium requirements of the global-growth or steady-state growth scenarios, however, it is necessary to first examine the impact that choices made at the enrichment or reactor operations stage may have on uranium demand. The first such connection is that, in light-water reactors, the level of enrichment of the fuel is linked to the maximum burnup that can be achieved. For example, in pressurized water reactors (PWRs) a burnup of 33 gigawatt-days (GWd) per ton requires an enrichment of about 3 percent U-235 while a burnup of 55 GWd per ton would require uranium enriched to about 5 percent U-235.[1051] Thus, despite the increased amount of energy generated per unit of mass, increasing the burnup of the fuel by this amount will actually slightly increase the requirement for natural uranium per kilowatt-hour of generation given the higher level of enrichment necessary.

The second connection relates to the percentage of U-235 that is left in the depleted uranium waste stream (known as the tails assay). When the tails assay is lowered, less natural uranium feed material is needed since more of the initial U-235 is being recovered, but this requires an increasing amount of enrichment services (see Table A.4).

Table A.4: Amount of natural uranium and enrichment services required to produce one kilogram of low-enriched uranium (4 percent U-235). As the percentage of U-235 left in the depleted uranium tails decreases the amount of feed material and the amount of enrichment required change in opposite directions.

Tails Assay (% U-235)	Natural Uranium Feed (kg U)	Enrichment Services (kg SWU)
0.30	9.00	5.28
0.25	8.13	5.83
0.20	7.44	6.54
0.15	6.86	7.51

From Table A.4 it is clear that, when uranium prices are low compared to enrichment services (as they are today), than it is more economical to have a higher tail assay, while the opposite would be true if uranium prices were higher. Typical enrichment plants today produce depleted uranium with tails around 0.3 percent U-235, however, levels as low as 0.15 percent were considered by the IAEA in their estimate of future uranium needs. To date no country has yet disposed of the large amounts of depleted uranium that has been generated by the existing enrichment facilities. More than one million tons of DU had already accumulated around the world by the end of 1995 and since then the stockpiles have continued to grow.[1052] A recent analysis by the Institute for Energy and Environmental Research concluded that this DU waste will require deep geologic disposal in a repository similar to that required for wastes contaminated with long-lived alpha emitting transuranic elements like plutonium. The cost of safely handling and disposing of this waste is likely to total between $20 and $30 per kilogram of depleted uranium, and would therefore add significantly to the economic incentive to lower the enrichment tails as a means of reducing the amount of waste created.[1053] Table A.5 shows the break even point at which a given reduction in tail assays would become economically justified with and without the estimated cost of depleted uranium disposal included.

Table A.5: Comparison of the price for natural uranium at which it becomes economically advantageous to reduce the amount of U-235 that is left in the depleted uranium stream. The impact of including the IEER estimate for the cost of safely disposing of the depleted uranium on this crossover point is also shown.

Reduction in Tail Assay (percent U-235 in DU stream)	Uranium cost at breakeven point without DU dispositioning costs[a] ($ per kg U)	Uranium cost at breakeven point with IEER estimates for DU dispositioning costs[a] ($ per kg U)
0.30% to 0.25%	63 to 76	33 to 56
0.30% to 0.20%	81 to 97	51 to 77
0.30% to 0.15%	104 to 125	74 to 105

(a) The price per SWU is variable, but the range of $100 to $120 per SWU used in this table is reasonable for U.S. or European suppliers. The monthly spot price between 2002 and 2004 generally varied between $105 and $110 per SWU with a long-term price estimate of $107 per SWU as of October 2004. This long-term cost has remained fairly stable equaling $109 per SWU in April 2005.[1054] Finally, both the MIT study and the report from Harvard University used an enrichment cost of $100 per SWU in their analyses.[1055]

Even without taking the estimated cost for disposing of the depleted uranium into account, uranium prices in the range of those in Table A.5 are projected to be reached well before 2050 by the IAEA under both of its nuclear growth scenarios.[1056] One important consequence of reducing the tails assays, however, will be the associated increase in the number of enrichment plants required. For LEU of 4% U-235, a decrease of the tails from 0.30 percent to 0.25 percent under the global growth scenario would require the construction of an additional four commercial scale enrichment plants to satisfy the demand for reactor fuel.[1057] For a decrease in the tails assays from 0.30 percent to 0.20 percent under the more aggressive steady-state growth scenario, an additional 22 plants the size of the proposed National Enrichment Facility would be required. The need for more enrichment plants around the world would add to the concerns over the spread of advanced centrifuge technology and the associated potential for nuclear weapons proliferation discussed in Chapter Three.

Section A.4 – Estimates for Cumulative Uranium Demand

A number of estimates have been made for the uranium requirements of a future expansion of nuclear power. The two most important complications in these estimates are the fact that different reactor designs have different uranium requirements (i.e. light-water reactors, heavy-water reactors, graphite moderated reactors, etc.) and the fact that countries like France and Japan continue to envision the use of the MOX fuel cycle. The base case of the IAEA projections of uranium demand through 2050 assumes that 80 percent of the future nuclear reactors will be light-water reactors and that the usage of MOX fuel will continue at the maximum rate supportable by the existing reprocessing facilities in Europe.[1058] However, in 2003, light-water reactors accounted for 87 percent of the global nuclear capacity and are projected to be the main choice for new generation (see Chapter Two).[1059] In addition, the large British commercial reprocessing plant at Sellafield (Throp) is scheduled to cease reprocessing operations in 2010 and begin cleanup activities due to a lack of commercial interest in MOX fuel.[1060] In light of these facts, and the serious proliferation concerns that are raised by the continued separation of weapons usable plutonium through reprocessing, we have chosen to estimate the uranium demands of the global-growth and steady-state growth scenarios assuming that all of the capacity would be light-water reactors using the once-through fuel cycle (see Table A.6).

Table A.6: Comparison of the cumulative uranium demands through 2050 and through the lifetime of the reactors under the global and steady-state growth scenarios as well as the IAEA middle and high growth scenarios. The two scenarios we have chosen to focus on have similar uranium demands to those of the IAEA and, in fact, bound the agency's projections.[1061]

Scenario	Base-Case (0.3% Tails)		Reduced Tails	
	Through 2050 (million tons U)	Over Lifetime of the Plants (million tons U)	Through 2050 (million tons U)	Over Lifetime of the Plants (million tons U)
Global-Growth Scenario[(a)] (IEER estimate)	5.04	8.80	4.55	7.95
IAEA (middle scenario)	5.39	---	4.81	---
IAEA (high scenario)	7.58	---	6.44	---
Steady-State Growth[(a)] Scenario (IEER estimate)	8.76	18.17	7.24	15.01

(a) The IEER estimates for uranium requirements assume a constant rate of growth in installed nuclear capacity starting in 2010 reaching 1,000 or 2,500 GW by 2050. The lifetime of the plants is assumed to be 40 years for consistency with the assumptions made in the economic analysis of Chapter Two. The reduced tails assays considered are 0.25 percent U-235 for the global-growth scenario and 0.2 percent for the steady-state growth scenario. Our estimated uranium needs also include an allowance for a total loss of 2 percent of the uranium during the conversion, enrichment, and fuel fabrication stages which is consistent with the assumptions of the MIT analysis.[1062]

Despite the differences in assumptions, we find that our estimates for the uranium requirements of the global-growth and steady-state growth scenarios are similar to those of the two growth scenarios considered by the IAEA. It is important to note that, like the global-growth scenario of the MIT report, the IAEA "middle" scenario is not a projection of a likely growth, but is instead described by the IAEA as "being rather optimistic and challenging."[1063] As a final check, we can compare our estimates to those made by the authors of the MIT study. The MIT analysis of uranium demand assumes a total capacity of 1,500 GW online by 2050, a plant life of 50 years, and a higher level of burnup with a correspondingly greater enrichment of the fuel. Adjusting their values to be consistent with our base case assumptions about plant life and fuel burnup, we find they would estimate a lifetime uranium demand of 11.9 million tons for 1,500 GW of capacity.[1064] Using the same methodology employed in developing Table A.6, we would predict a lifetime demand of 12.0 million tons which is in excellent agreement with the MIT estimate.

Comparing Tables A.1, A.3, and A.6 we find that, even without the additional production from speculative resources or any allowance for an increase in the resource base over today's estimates, that the level of production estimated by the IAEA for uranium prices below $130 per kilogram (5.30 million tons) is sufficient to supply the global-growth scenario with or without a reduction in tail assays (4.55 to 5.04 million tons). In addition, the estimates for all known and unknown conventional resources presently available (11.44 million tons) are also more than sufficient to fuel the full life-cycle needs of the reactors envisioned under the proposed global-growth scenario (7.95 to 8.80 million tons). The inclusion of an increasing resource base with increasing price would add further to the likely cushion between supply and demand in this case.

For the steady-state growth scenario, however, the IAEA projections for production through 2050 would fall well short of the demand even if a significant reduction in tail assays was made (5.30 million tons of production versus 7.24 million tons of demand). A similar shortfall is found between present estimates for conventional resources and total life-cycle demand (11.44 million tons of reserves versus 15.01 million tons of total demand). However, this apparent shortfall is likely to be more than compensated for by the increase in recoverable resources and improvements in extraction technology as discussed in Section A.1. Our rough estimate for the level of production that could be possible by 2050 at less than $130 per kilogram was between 17.5 and 54 million tons of uranium. Even if the actual amount was just half of the lower estimate, there would still be more than enough uranium to fuel the steady-state growth scenario through 2050. Similarly, the estimates for the total amount of uranium recoverable at $130 per kilogram ranged from 34 to 104 million tons, and thus less than half the lower estimate would again be sufficient to supply the life-cycle needs of even the extremely aggressive steady-state growth scenario.

Thus, while the exploration and development of the available uranium resources would require a very significant amount of money, time, and effort and there are important uncertainties remaining with regards to the available supply and the ability of it to be brought to market, it appears reasonable to conclude that the once-through fuel cycle could be adequately sustained through mid-century and beyond at uranium prices below $130 per kilogram. While this is well above the recent price of $30 to $40 per kilogram, such an increase would not add significantly to the cost of generating nuclear power since, in general, the fuel costs are a small percentage of the total generation costs and the uranium costs are a small percentage of the fuel costs.[1065]

Section A.5 – Impacts of Uranium Supply and Demand on Proliferation

Because uranium can be used in both nuclear reactors and nuclear bombs, the fuel-cycle choices made as part of an expansion of nuclear power will have significant impacts on proliferation. As already noted, the choice to lower the tail assays to conserve natural uranium supplies would increase the need for enrichment plants adding to concerns over the spread of advanced centrifuge technology. Connected to these concerns is the question of whether or not it would be possible to directly safeguard uranium resources. Mohamed ElBaradei, the Director General of the IAEA, has noted that

> The technical barriers to mastering the essential steps of uranium enrichment – and to designing weapons – have eroded over time, which inevitably leads to the conclusion that the control of technology, *in and of itself*, is not an adequate barrier against further proliferation."[1066]

This conclusion has lead to a renewed focus on the viability of directly safeguarding uranium mines themselves as a way to mitigate the threat posed by clandestine enrichment facilities. The Additional Protocol (see Section 3.4.1) already added stronger reporting requirements regarding uranium and thorium mines, including the requirement to report "the estimated annual production capacity of uranium mines and concentration plants and thorium concentration plants, and the current annual production of such mines and concentration plants" and, upon request by the IAEA, the countries must report "the current annual production of an individual mine or concentration plant." However, the Additional Protocol also states that "provision of this information does not require detailed *nuclear material* accountancy".[1067]

While it is unclear if a system of extensive safeguards at the mines themselves would be of significant value even under today's operational conditions, the very large expansion of uranium exploration and development and the spread of production centers around the world would likely make such a system of safeguards prohibitively difficult, time consuming, and expensive. In addition to the potential for diversion from known uranium production centers, the large increase in extraction envisioned under either the global-growth or steady-state growth scenarios would also make it increasingly difficult to detect clandestine mines. To give a sense of the level of accuracy needed in actually carrying out such safeguard efforts we note that just under five tons of natural uranium is required to make enough HEU for one nuclear weapon.[1068] For comparison, this amount of uranium would be just 0.0050 percent of the average annual demand for natural uranium to manufacture reactor fuel under the global growth scenario.

The second, and most important, connection between proliferation and uranium resources is the potential for a presumed scarcity of uranium to increase pressures for reprocessing and the use of plutonium as a reactor fuel. As noted, the concerns of some within the nuclear complex over uranium's availability were a major driving force behind the original push for fast breeder reactors. While arguments about long-term waste management are typically more prominent today, the conservation of uranium resources remains a rationale noted by some for pursuing closed fuel-cycle technologies.[1069] Significantly, four of the six Generation IV designs currently being pursued by the U.S. Department of Energy are fast reactors with a closed fuel cycle based on reprocessing and a fifth design is capable of using either a once-through or closed-fuel cycle.[1070]

However, as we have shown, it is likely that uranium resources recoverable at less than $130 per kilogram would be more than sufficient to cover the needs of even the extremely aggressive steady-state growth scenario. Even at this uranium price, reprocessing and the use of MOX fuel would be likely to remain uneconomical. The MIT study estimated that a uranium price of $560 per kilogram would be required before the cost of MOX would breakeven with the cost of the once-through fuel cycle.[1071] Similar results were found by the authors of the Harvard study. Using a range of estimates for the cost of various components of the fuel cycles, the authors estimated that a uranium price of $220 to $450 per kilogram would be required to make MOX economical. Their central estimate was $370 per kilogram.[1072] All of these estimates are well above the $130 per kilogram we have considered in our analysis and, in fact, the central estimates from the MIT and Harvard studies are actually greater than some recent estimates for the cost of recovering uranium from seawater. While there is a great deal of uncertainty in these costs estimates, the OECD and IAEA have cited research indicating that uranium could potentially be recovered from seawater at approximately

$300 per kilogram. The oceans are estimated to contain a total of 4 to 4.5 billion tons of uranium, and thus recovering even a tiny fraction of this amount could potentially add significant quantities of uranium to the market well before the MOX fuel cycle would become economical.[1073]

Finally, we note that there is the potential for improvements in nuclear disarmament and non-proliferation efforts through expanded efforts to down blend highly enriched uranium (HEU) and use it in existing reactors. Since 1993, the U.S. and Russia have agreed to declare 645 tons of highly enriched uranium from their military stockpiles to be surplus (500 tons from Russia and 145 tons from the United States). This material would be mixed with natural, depleted, or slightly enriched uranium to produce material suitable for use as fuel in power reactors. The U.S. is limited to accepting only the equivalent of 30 tons of HEU per year from Russia. This amount of HEU, once down blended, can provide enough LEU to fuel approximately 44 light-water reactors in the U.S. The current amount of declared surplus is only a fraction of the estimated HEU produced by these two countries, and thus the amount available for commercial use could be significantly increased. Specifically, the IAEA base case estimate for future uranium demand is that an additional 305 tons (250 Russian and 55 U.S.) will be declared surplus in the coming years. As a high case, the IAEA estimated that as much as 755 tons (500 Russian and 255 U.S.) could be freed up over the presently agreed upon levels.[1074]

By June 2005, a total of 245 tons of HEU had already been down-blended for commercial use amounting to 7,225 tons of LEU. The remaining surplus under the two IAEA projections (705 to 1,155 tons) would be enough to yield approximately 20,800 to 34,100 tons LEU if it was eventually down-blended. Assuming a continuation of the recent levels of uranium demand, this would be enough low-enriched uranium to fuel both the U.S. and Russian fleets for roughly 6.5 to 11 years.[1075] If a total of 1,100 tons (out of a possible 1,400 tons) was eventually declared surplus, this amount would be sufficient to fuel the entire U.S. and Russian fleets for approximately eight years. If the down blending of this HEU was stretched out over 20 years, it could supply approximately 40 percent of the annual demand. At this level, the surplus HEU would replace the equivalent of 6,650 MTSWU of enrichment services. For comparison, this would be greater than the level of enrichment services that would be provided by both of the new centrifuge plants that are currently seeking licenses in the United States (the National Enrichment

Facility in New Mexico and the American Centrifuge Commercial Plant facility in Ohio).[1076]

The expanded use of down blended HEU would have two advantages from a non-proliferation standpoint. The first is that it would provide a significant amount of low enriched uranium without the need for additional enrichment services. Already Mohamed ElBaradei has called for a temporary moratorium on new enrichment plants due to concerns over the potential proliferation of nuclear weapons.[1077] The available resource from down blended HEU would reinforce the adequacy of existing enrichment capacity and lessen the incentive to build new enrichment plants in the near term. The second advantage would be that the HEU would be converted into a form that is no longer directly usable in nuclear weapons thereby greatly reducing the threat of theft, as well as making the disarmament step represented by declaring this material to be surplus significantly more irreversible. IEER continues to support the expanded use of down blended HEU to augment the capacity of presently operating enrichment facilities in order to fuel the existing fleet of reactors.

References

10 CFR 20 2005	U.S. Code of Federal Regulations, "Title 10 – Energy: Chapter I – Nuclear Regulatory Commission; Part 20 – Standards for Protection Against Radiation", January 1, 2005
ACA 1999	Paul Kerr, press contact, "The IAEA 1997 Additional Safeguards Protocol", Arms Control Association Fact Sheet, September 199
ACEEE Factsheet	American Council for an Energy-Efficient Economy, "Energy Efficiency Progress and Potential", online at http://www.aceee.org/energy/effact.htm
ACEEE R&D Factsheet	American Council for an Energy-Efficient Economy, "DOE Energy Efficiency R&D and Technology Deployment Programs: Critically Needed, Sound Investments in the Nation's Energy Future", online at http://www.aceee.org/energy/rdtech.htm
Acheson and Lilienthal 1946	Dean Acheson et al., "A Report on the International Control of Atomic Energy", March 16, 1946
ACIA 2004	ACIA, *Impacts of a Warming Arctic: Arctic Climate Impact Assessment*, Cambridge University Press, 2004
ACNW 2001	George Hornberger, Chairman of the Advisory Committee on Nuclear Waste, Letter Report to Richard Meserve, Chairman of the U.S. Nuclear Regulatory Committee, September 18, 2001
Adam 2001	David Adam, "The North Sea Bubble", *Nature*, Vol. 411, 518 (2001)
AEC 1948	Atomic Energy Commission, "Report to the U.S. Congress, No. 4", Washington, DC, 1948
Agriculture Factsheet	University of Illinois College of College of Agricultural, Consumer and Environmental Sciences, *A Brief History of Agriculture, Background Information, Sustainability: Considering Lessons from History*, online at http://www.aces.uiuc.edu/~sare/backinfo.html
Albright 1994	David Albright, "South Africa and the Affordable Bomb", *Bulletin of the Atomic Scientist*, Volume 50 Number 4, July/August 1994

Albright, Berkhout, and Walker 1997	David Albright, Frans Berkhout, and William Walker, <u>Plutonium and Highly Enriched Uranium 1996 World Inventories, Capabilities and Policies</u>, Oxford University Press, Oxford (1997)
Alvarez 2002	Robert Alvarez, "What about the spent fuel", *Bulletin of the Atomic Scientists*, Volume 58 Number 1, January/February 2002
Alvarez et al. 2003	Robert Alvarez et al., "Reducing the Hazards from Stored Spent Power-Reactor Fuel in the United States", *Science and Global Security*, Volume 11, accepted January 22, 2003
Amick et al. 2002	Phil Amick et al., "A Large Coal IGCC Power Plant", Nineteenth Annual International Pittsburgh Coal Conference, September 23-27, 2002
ANA 2005	Alliance for Nuclear Accountability, "Fiscal Year 2006 Budget", August 2005. online at http://www.ananuclear.org/fy06budget.pdf
ANA Factsheet	Alliance for Nuclear Accountability, "Nuclear Weapons in the FY04 Energy and Water Subcommittee and Defense Authorization Reports", November 2003, online at http://www.ananuclear.org/FY04budgetcompare.html
Andersson 2000	Björn A. Andersson, "Materials Availability for Large-scale Thin-film Photovoltaics", *Progress in Photovoltaics: Research and Applications*, Vol. 8, 61-76 (2000)
Andra	French National Radioactive Waste Management Agency, "Research> Laboratory> Background", online at http://www.andra.fr/interne.php3?id_article=347&id_rubrique=127
Andrews 2004	Anthony Andrews, "Spent Nuclear Fuel Storage Locations and Inventory", Congressional Research Service, Updated December 21, 2004 (Order Code RS22001)
Anspaugh, Catlin, and Goldman 1988	Lynn Anspaugh, Robert Catlin, and Marvin Goldman, "The Global Impact of the Chernobyl Reactor Accident", *Science*, Volume 242, 1513-1519 (1988)
Archer and Jacobson 2003	Cristina Archer and Mark Jacobson, "Spatial and temporal distributions of U.S. wind power at 80 m derived from measurements", *Journal of Geophysical Research*, Vol. 108, ACL 10, (2003)

Archer and Jacobson 2004	Cristina Archer and Mark Jacobson, "Correction to 'Spatial and temporal distributions of U.S. wind power at 80 m derived from measurements'", *Journal of Geophysical Research*, Vol. 109, D20116, (2004)
Arrhenius 1896	S. Arrhenius, "On the Influence of Carbonic Acid in the Air Upon the Temperature of the Ground", *Philosophical Magazine*, Volume 41 Number 237 (1896) excerpts available online at http://web.lemoyne.edu/~giunta/Arrhenius.html
ASE et al. 1997	Alliance to Save Energy, American Council for an Energy-Efficient Economy, Natural Resources Defense Council, the Tellus Institute, and Union of Concerned Scientists, "Energy Innovations: A Prosperous Path to a Clean Environment", 1997
Associated Press 2005	Associated Press, "Katrina may cost as much as four years of war: Government certain to pay more than $200 billion following hurricane", September 10, 2005
Associated Press 2005b	Associated Press, "Katrina's Death Toll in La. Tops 1,000", October 7, 2005
Associated Press 2005c	Associated Press, "Katrina Evacuees Uncertain of Destinations", October 14, 2005
Associated Press 2006	Associated Press, "Iran Ends Voluntary Cooperation on Nukes", February 6, 2006
AWEA 2004	American Wind Energy Association, "Global Wind Energy Market Report", March 2004, online at http://www.awea.org/pubs/documents/globalmarket2004.pdf
AWEA Factsheet	Thomas O. Gray, "Wind Energy - Views on the Environment: Clean and Green", American Wind Energy Association, online at http://www.americanwindenergy.org/pubs/documents/oppoll.PDF
Ayres, MacRae, and Stogran 2004	Matt Ayres, Morgan MacRae, and Melanie Stogran, "Levelised Unit Electricity Cost Comparison of Alternate Technologies for Baseload Generation in Ontario", Canadian Energy Research Institute, August 2004
Bacher 1949	Robert Bacher, "Research and Development of Atomic Energy", *Science*, Volume 109 Number 2819, 2-7 (1949)

Baltimore-Sun 2004	Del Quentin Wilber, Gus Sentementes, and Alec MacGillis, "'A wall of flame'; Burning Load of Fuel consumes 5 Vehicles; Blast Heard, Smoke Seen Darkening Sky for Miles", *Baltimore Sun*, January 14, 2004
Bamford 2001	James Bamford, <u>Body of Secrets: Anatomy of the Ultra-Secret National Security Agency from the Cold War through the Dawn of a New Century</u>, Doubleday, New York, NY (2001)
Baruch 1946	Bernard Baruch, "Speech before the first session of the United Nations Atomic Energy Commission", June 14, 1946
BAS Press Release 2005	British Antarctic Survey Press Office, "Antarctic glaciers thinning fast", February 2, 2005
BBC 2002	BBC News, "Comparing the worst oil spills", November 19, 2002
BBC 2002b	BBC News, "Antarctic ice shelf breaks apart", March 19, 2002
BBC 2004	BBC News, "Flood death toll rises in Assam", October 11, 2004
Bergeron 2002	Kenneth Bergeron, <u>Tritium on Ice: The Dangerous New Alliance of Nuclear Weapons and Nuclear Power</u>, MIT Press, Cambridge, MA (2002)
Berkhout and Gadekar 1997	Frans Berkhout and Surendra Gadekar, "Reprocessing in India", *Energy & Security*, No. 2, January 1997, online at http://www.ieer.org/ensec/no-2/india-b.html
BMU 2003	Franz Josef Schafhausen and Hildegard Kaiser *ed.*, "Climate Change Policy in Germany", Federal Ministry for the Environment, Nature Conservation and Nuclear Safety (Germany), October 2003
Boston Globe 2003	Gareth Cook, "Napping Reactor Operators Startles MIT", *Boston Globe*, July 29, 2003
Boucher 2004	Richard Boucher, "U.S.-India and U.S.-Pakistan Partnerships -- Close and Productive", U.S. Department of State Office of the Spokesman, March 22, 2004
Brandt et al. 1980	Brandt et al., <u>North-South: A Program for Survival</u>, The Report of the Independent Commission on International Development Issues under the Chairmanship of Willy Brandt, The MIT Press, Cambridge, MA (1980)

Brenner 1981	Michael Brenner, Nuclear Power and Non-Proliferation: The Remaking of U.S. Policy, Cambridge University Press, New York (1981)
Broecker 1997	Wallace S. Broecker, "Thermohaline Circulation, the Achilles Heel of Our Climate System: Will Man-Made CO_2 Upset the Current Balance?", *Science*, Vol. 278, 1582-1588 (1997)
Bromet et al. 2002	Evelyn J. Bromet et al., "Somatic Symptoms in Women 11 Years after the Chernobyl Accident: Prevalence and Risk Factors", *Environmental Health Perspectives*, Vol. 110 Supplement 4, 625-629 (2002)
Brugge and Goble 2002	Doug Brugge and Rob Goble, "The History of Uranium Mining and the Navajo People", *American Journal of Public Health*, Vol. 92 No. 9, 1410-1419 (2002)
Bryant 1997	Edward Bryant, Climate Process & Change, Cambridge University Press, Cambridge (1997)
Bueno and Carta 2005	C. Bueno and J.A. Carta, "Technical-economic analysis of wind-powered pumped hyrdrostorrage systems. Part I: model development", *Solar Energy*, Vol. 78, 382-395 (2005)
Bueno and Carta 2005b	C. Bueno and J.A. Carta, "Technical-economic analysis of wind-powered pumped hyrdrostorrage systems. Part II: model application to the island of El Hierro", *Solar Energy*, Vol. 78, 396-405 (2005)
Bunn et al. 2003	Matthew Bunn, Steve Fetter, John P. Holdren, and Bob van der Zwaan, "The Economics of Reprocessing vs. Direct Disposal of Spent Nuclear Fuel", Project on Managing the Atom, Belfer Center for Science and International Affairs, John F. Kennedy School of Government, December 2003
Bunn et al. 2005	Matthew Bunn, John P. Holdren, Steve Fetter, and Bob Van Der Zwaan, "The Economics of Reprocessing Versus Direct Disposal of Spent Nuclear Fuel", *Nuclear Technology*, Vol. 150, 209-230 (2005)
Burk 2004	Susan Burk, "U.S. – IAEA Additional Protocol", Testimony before the Senate Foreign Relations Committee, January 29, 2004 online at http://www.state.gov/t/isn/rls/rm/29249.htm

Cameco McArthur History	Cameco Corporation, "McArthur River: History", page updated November 9, 2005, online at http://www.cameco.com/operations/uranium/mcarthur_river/history.php
Cameco Reserves	Cameco Corporation, "Uranium Reserves", page updated April 15, 2005, online at http://www.cameco.com/operations/uranium/mcarthur_river/reserves.php
Campbell et al. 2003	Katherine Campbell, "Chlorine-36 data at Yucca Mountain: statistical tests of conceptual models for unsaturated-zone flow", *Journal of Contaminant Hydrology*, Volume 62-63 (2003)
Castronuovo and Lopes 2004	Edgardo D. Castronuovo and João A. Peças Lopes, "Optional operation and hydro storage sizind of a wind-hydro power plant", *Electrical Power and Energy Systems*, Vol. 26, 771-778 (2004)
Castronuovo and Lopes 2004b	Edgardo D. Castronuovo and J.A. Peças Lopes, "On the Optimization of the Daily Operation of a Wind-Hydro Power Plant", *IEEE Transactions On Power Systems*, Vol. 19, 1599-1606 (2004)
CBO 2003	Congressional Budget Office, "Cost Estimate: S.14 Energy Policy Act of 2003", May 7, 2003
CBO 2005	Congressional Budget Office, "Cost Estimate: S.10 Energy Policy Act of 2005", June 9, 2005
CDC/NCI 2001	Jeffrey P. Koplan, "Progress Report: A Feasibility Study of the Health Consequences to the American Population of Nuclear Weapons Tests Conducted by the United States and Other Nations", A Report Prepared for the U.S. Congress by the Department of Health and Human Services Centers for Disease Control and Prevention and the National Cancer Institute, August 2001
Census 2004	U.S. Census Bureau, "Statistical Abstract of the United States 2004-2005: The National Databook", 124th Edition, 2004
Chaffee 2004	Devon Chaffee, "Freedom of Force on the High Seas? Arms Interdiction and International Law", *Science for Democratic Action*, Vol. 12 No. 3, June 2004
Charpin, Dessus, and Haut 2000	Jean-Michel Charpin, Benjamin Dessus, and Rene Pellat Haut, "Étude économique prospective de la filière électrique nucléaire", A Report for the Prime Minister, July 2000

Cheney et al. 2001	Dick Cheney et al., "Reliable, Affordable, and Environmentally Sound Energy for America's Future: Report of the National Energy Policy Development Group", May 2001
Chiesa et al. 2005	Paolo Chiesa, Stefano Consonni, Thomas Kreutz, and Robert Williams, "Co-production of hydrogen, electricity and CO_2 from coal with commercially ready Technology. Part A: Performance and emissions", *International Journal of Hydrogen Energy*, Vol. 30, 747-767 (2005)
Chirac 2006	Speech by Jacques Chirac, President of the French Republic, during his visit to The Strategic Air and Maritime Forces at Landivisiau / L'Ile Longue, January 19, 2006. online at http://www.elysee.fr/elysee/anglais/speeches_and_documents/2006/speech_by_jacques_chirac_president_of_the_french_republic_during_his_visit_to_the_stategic_forces.38447.html
Chomsky 1992	Noam Chomsky, <u>Deterring Democracy</u>, Hill and Wang, New York, NY (1992)
Chomsky 1993	Noam Chomsky, <u>Year 501: The Conquest Continues</u>, South End Press, Cambridge, MA (1993)
Christian Science Monitor 2006	Mark Clayton, "Terror risks of nuclear fuel", *Christian Science Monitor*, March 16, 2006
CIA World Factbook	U.S. Central Intelligence Agency, "The World Factbook", online at http://www.cia.gov/cia/publications/factbook/
CISAC and PSGS 2005	Christopher Chyba, Harold Feiveson, and Frank von Hippel (co-directors), "Preventing Nuclear Proliferation and Nuclear Terrorism: Essential steps to reduce the availability of nuclear-explosive materials", Center for International Security and Cooperation (Stanford University) and Program on Science and Global Security (Princeton University), March 2005
Clary 2004	Christopher Clary, "Dr. Khan's Nuclear Walmart", *Disarmament Diplomacy*, Vol. 76, 31-36 (March/April 2004)
Clemmer et al. 2001	Steven Clemmer, Deborah Donovan, Alan Nogee, and Jeff Deyette, "Clean Energy Blueprint: A Smarter National Energy Policy for Today and the Future", Union of Concerned Scientists with American Council For An Energy-Efficient Economy and the Tellus Institute, October 2001

CNN 2000	"Japan's Nuclear Power Accident Claims Second Fatality", *CNN*, April 28, 2000, online at http://www1.cnn.com/2000/ASIANOW/east/04/27/japan.nuclear.accident
CNN 2002	"Child's Body Recovered from Bridge Collapse", *CNN*, May, 29, 2002, online at http://www.cnn.com/2002/US/05/29/bridge.collapse
CNN 2006	"2 hurt in Japan nuclear plant fire", *CNN*, March 22, 2006, online at http://www.cnn.com/2006/WORLD/asiapcf/03/22/japan.fire.ap/index.html
Cohen and Graham 2004	Avner Cohen and Thomas Graham Jr., "An NPT for non-members", *Bulletin of the Atomic Scientists*, Volume 60 Number 3, May/June 2004
Cole 1953	Sterling Cole, Letter to Congressman John Phillips, May 20, 1953, with cover not from AEC secretary Roy Snapp, July 9. 1953, DOE Archives, Box 1290, Folder 2
Committee on Appropriations 2005	U.S. Congress, House of Representatives, Committee on Appropriations, "Report, Energy and Water Development Appropriations Bill, 2006", May 18, 2005
Committee on Energy and Commerce 2005	U.S. Congress, House of Representatives, Committee on Energy and Commerce, *Discussion Draft of the Energy Policy Act of 2005*, available online at http://energycommerce.house.gov/108/energy_pdfs.htm (last viewed February 22, 2005)
Court of Appeals 2004	United States District Court of Appeals for the District of Columbia Circuit, "On Petitions for Review of Orders of the Environmental Protection Agency, the Department of Energy, and the Nuclear Regulatory Commission", Decided July 9, 2004
Craig 2004	Paul Craig, "Rush to Judgment at Yucca Mountain", *Science for Democratic Action*, Vol. 12 No. 3, June 2004
Daily Yomiuri 2005	"Gravitational force on 2nd car similar to that of air crash", *The Daily Yomiuri (Tokyo)*, May 8, 2005
Davis 1994	Mary Byrd Davis, "The French Mess Nucléaire", *Bulletin of the Atomic Scientist*, Volume 50 Number 4, July/August 1994
Davis 2001	Mary Byrd Davis. "La France nucléaire: matières et sites 2002", Paris: Wise-Paris, 2001.online at http://www.francenuc.org

Davis-Besse Factsheet	U.S. Department of Energy, "U.S. Nuclear Plants: Davis-Besse, Ohio", Energy Information Administration, page last modified on Fri Mar 18 2005 online at http://www.eia.doe.gov/cneaf/nuclear/page/at_a_glance/reactors/davisbesse.html
Deller, Makhijani, and Burroughs 2003	Nicole Deller, Arjun Makhijani, and John Burroughs, Rule of Power or Rule of Law?: An Assessment of the U.S. Polices and Actions Regarding Security-Related Treaties, Apex Press, New York (2003)
deMenocal 2001	Peter deMenocal, "Cultural Responses to Climate Change During the Late Holocene", *Science*, Vol. 292, 667-673 (2001)
Deseret Morning News 2006	Suzanne Struglinski, "Cedar Mountain OK dents nuclear plans", *Deseret Morning News*, January 7, 2006
DOD 2001	U.S. Department of Defense, "Nuclear Posture Review", submitted to Congress December 31, 2001, excerpts online at http://www.globalsecurity.org/wmd/library/policy/dod/npr.htm
DOE 1991	U.S. Department of Energy Office of New Production Reactors, "Draft Environmental Impact Statement for the Siting, Construction, and Operation of New Production Reactor Capacity", Vol. 1, April 1991 (DOE/EIS-0144D)
DOE 1994	U.S. Department of Energy, "Additional Information Concerning Underground Nuclear Weapon Test of Reactor-Grade Plutonium", June 27, 1994 online at http://www.osti.gov/html/osti/opennet/document/press/pc29.html
DOE 1996	U.S. Department of Energy, "Plutonium: The First 50 Years, United States plutonium production, acquisition, and utilization from 1944 to 1994", February 1996
DOE 1997	U.S. Department of Energy, "Nonproliferation and Arms Control Assessment of Weapons-Usable Fissile Material Storage and Excess Plutonium Disposition Alternatives", January 1997 (DOE/NN-0007)
DOE 1997b	U.S. Department of Energy, "Integrated Data Base Report – 1996: U.S. Spent Nuclear Fuel and Radioactive Waste, Inventories, Projections, and Characteristics", Office of Environmental Management, December 1997 (DOE/RW-0006, Rev. 13)

DOE 1999	U.S. Department of Energy, "Final Programmatic Environmental Impact Statement for Alternative Strategies for the Long-Term Management and Use Of Depleted Uranium Hexafluoride", Volume 1: Main Text, April 1999 (DOE/EIS-0269), online at http://web.ead.anl.gov/uranium/documents/nepacomp/peis/parts/maintext.cfm
DOE 2001	U.S. Department of Energy, "A Roadmap to Deploy New Nuclear Power Plants in the United States by 2010: Volume I Summary Report", October 31, 2001
DOE 2001b	U.S. Department of Energy, "Generation-IV Roadmap: Report of the Fuel Cycle Crosscut Group", March 18, 2001
DOE 2002c	U.S. Department of Energy Nuclear Energy Research Advisory Committee and the Generation IV International Forum, "A Technology Roadmap for Generation IV Nuclear Energy Systems", December 2002 (GIF-002-00)
DOE 2002e	U.S. Department of Energy, "Final Environmental Impact Statement for a Geologic Repository for the Disposal of Spent Nuclear Fuel and High-Level Radioactive Waste at Yucca Mountain, Nye County, Nevada", Volume II Appendixes A through O, Office of Civilian Radioactive Waste Management, February 2002 (DOE/EIS-0250)
DOE 2002f	U.S. Department of Energy, "National Transmission Grid Study", May 2002, online at http://www.pi.energy.gov/pdf/library/TransmissionGrid.pdf
DOE 2003	U.S. Department of Energy Office of Fossil Energy, "FutureGen – A Sequestration and Hydrogen Research Initiative", February 2003
DOE 2003b	U.S. Department of Energy, "The U.S. Generation IV Implementation Strategy: Preparing for Tomorrow's Energy Needs", September 2003 (03-GA50438-06)
DOE 2004b	U.S. Department of Energy, "How Coal Gasification Power Plants Work", Office of Fossil Energy, October 13, 2004, online at http://www.fe.doe.gov/programs/powersystems/gasification/howgasificationworks.html
DOE 2004d	U.S. Department of Energy, "Secretary Abraham Announces $36 Million for Minnesota Clean Coal", Office of Public Affairs, October 26, 2004

DOE 2005	U.S. Department of Energy, "Monthly Summary of Program Financial and Budget Information As of February 28, 2005", Office of Civilian Radioactive Waste Management, Office of Program Management
DOE 2005b	U.S. Department of Energy, "FutureGen Project Launched", Office of Public Affairs, December 6, 2005
DOE 2006	U.S. Department of Energy, "DOE Awards $235 Million to Southern Company to Build Clean Coal Plant", Office of Public Affairs, February 22, 2006
DTI 2003	Department of Trade and Industry, "Energy Whitepaper: Our Energy Future - Creating a Low Carbon Economy", February 2003 (Cm 5761)
DTI 2003b	Department of Trade and Industry, "Review of the Feasibility of Carbon Dioxide Capture and Storage in the UK", September 2003 (DTI/Pub URN 03/1261)
DTI 2003c	Mott MacDonald, "The Carbon Trust & DTI Renewables Network Impact Study Annex 4: Intermittency Literature Survey & Roadmap", The Carbon Trust and Department of Trade and Industry, November 2003
Dubin & Rothwell 1990	J.A. Dubin and G.S. Rothwell, "Subsidy to Nuclear Power Through Price-Anderson Liability Limit", *Contemporary Policy Issues*, Volume 8 Number 3, July 1990
Dulvy, Sadovy, & Reynolds 2003	N.K. Dulvy, Y. Sadovy, and J.D. Reynolds, "Extinction vulnerability in marine populations", *Fish and Fisheries*, Vol. 4, 25-64 (2003)
Dumas 1999	Lloyd Dumas, Lethal Arrogance: Human Fallibility and Dangerous Technologies, St. Martins Press, New York, NY (1999)
EEI 2003	Edison Electric Institute, "2002 Financial Review Plus 2003 Developments", Annual Report of the Shareholder-Owned Electric Utility Industry, 2003
EIA 1986	U.S. Department of Energy, "Analysis of Nuclear Power Plant Construction Costs", Energy Information Administration Office of Coal, Nuclear, Electric and Alternative Fuels, 1986 (DOE/EIA-0485)
EIA 1992	U.S. Department of Energy, "Federal Energy Subsidies: Direct and Indirect Interventions in Energy Markets", Energy Information Administration Office of Energy Markets and End Use, 1992 (SR/EMEU/92-02)

EIA 1999	U.S. Department of Energy, "Federal Financial Interventions and Subsidies in Energy Markets 1999: Primary Energy", Energy Information Administration Office of Integrated Analysis and Forecasting, September 1999 (SR/OIAF/99-03)
EIA 2001	Ronald Hagen, John Moens, and Zdenek Nikodem, "Impact of U.S. Nuclear Generation on Greenhouse Gas Emissions", U.S. Department of Energy, Energy Information Administration, November 2001
EIA 2002	U.S. Department of Energy, "U.S. Nuclear Reactor List – Shutdown", Energy Information Administration, last modified on July 12, 2002, online at http://eia.doe.gov/cneaf/nuclear/page/nuc_reactors/shutdown.html
EIA 2002b	U.S. Department of Energy, "Electric Power Annual 2000 Volume II", Energy Information Administration Office of Coal, Nuclear, Electric and Alternate Fuels, November 2002 (DOE/EIA-0348(00)/2)
EIA 2003	U.S. Department of Energy, "Emissions of Greenhouse Gases in the United States 2002", Energy Information Administration Office of Integrated Analysis Forecasting, October 2003 (DOE/EIA-0573)
EIA 2003c	U.S. Department of Energy, "Electric Power Annual 2002", Energy Information Administration Office of Coal, Nuclear, Electric and Alternate Fuels, December 2003 (DOE/EIA-0348(2002))
EIA 2003d	U.S. Department of Energy, "The Global Liquefied Natural Gas Market: Status & Outlook", Energy Information Administration, December 2003 (DOE/EIA-0637 (2003))
EIA 2004c	U.S. Department of Energy, "Electric Power Monthly September 2004 With Data for June 2004", Energy Information Administration, Office of Coal, Nuclear, Electric and Alternate Fuels, September 2004 (DOE/EIA-0226 (2004/09))
EIA 2004d	Damien Gaul, "U.S. Natural Gas Imports and Exports: Issues and Trends 2003", U.S. Department of Energy, Energy Information Administration, Office of Oil and Gas, August 2004

EIA 2004e	U.S. Department of Energy, "Annual Energy Review 2003", Energy Information Administration Office of Energy Markets and End Use, September 2004 (DOE/EIA-0384(2003)), Table 8.2a online at http://www.eia.doe.gov/emeu/aer/txt/ptb0802a.html and Table 12.3 online at http://www.eia.doe.gov/emeu/aer/txt/ptb1203.html
EIA 2004f	U.S. Department of Energy, "Analysis of Five Selected Tax Provisions of the Conference Energy Bill of 2003", Energy Information Administration Office of Integrated Analysis and Forecasting, February 2004 (SR/OIAF/2004-01)
EIA 2005	U.S. Department of Energy, "Annual Energy Outlook 2005 With Projections to 2025", Energy Information Administration, Office of Integrated Analysis and Forecasting, February 2005 DOE/EIA-0383(2005), additional information available online at http://www.eia.doe.gov/oiaf/archive/aeo05/index.html
EIA 2006	U.S. Department of Energy, "Natural Gas Monthly April 2006", Energy Information Administration, Office of Oil and Gas, 2006, DOE/EIA-0130(2006/04)
EIA Natural Gas Factsheet	U.S. Department of Energy, "Natural Gas Prices", Energy Information Administration, updated April 27. 2006, online at http://tonto.eia.doe.gov/dnav/ng/ng_pri_sum_dcu_nus_a.htm
Eisenhower 1953	Dwight Eisenhower, "Address Before the General Assembly of the United Nations on the Peaceful Uses of Atomic Energy, New York City", December 8, 1953, online at: http://www.presidency.ucsb.edu/site/docs/ppus.hph?admin=034&year=1953&id=256#
ElBaradei 2004	Mohamed ElBaradei, "Nuclear Non-Proliferation: Global Security in a Rapidly Changing World", Statement of the Director General of the International Atomic Energy Agency, June 21, 2004
ElBaradei 2004b	Mohamed ElBaradei, "Transcript of the Director General's Interview with Al-Ahram News", Statement of the Director General of the International Atomic Energy Agency, July 27, 2004

ElBaradei 2004c	Mohamed ElBaradei, "In Search of Security: Finding an Alternative to Nuclear Deterrence", Statement of the Director General of the International Atomic Energy Agency, November 4, 2004
ElBaradei 2005	Mohamed ElBaradei, "Seven Steps to Raise World Security", Statement of the Director General of the International Atomic Energy Agency, February 2, 2005
ElBaradei 2005b	Mohamed ElBaradei, "Treaty on the Non-Proliferation of Nuclear Weapons", Statement of the Director General of the International Atomic Energy Agency, May 2, 2005
EPA 1999	U.S. Environmental Protection Agency, "Cancer Risk Coefficients for Environmental Exposure to Radionuclides", Federal Guidance Report No. 13, September 1999 (EPA 402-R-99-001)
EPA 2001b	U.S. Environmental Protection Agency, "40 CFR Part 197: Public Health and Environmental Radiation Protection Standards for Yucca Mountain, NV; Final Rule", *Federal Register*, Vol. 66 No. 114, 32074-32135 (June 13, 2001)
EPA 2004	U.S. Environmental Protection Agency, "Toxic Release Inventory Program", 2002 Data Updated as of August 2, 2004 online at http://www.epa.gov/triexplorer/
EPA 2005	United States. Environmental Protection Agency. "40 CFR Part 197: Public Health and Environmental Radiation Protection Standards for Yucca Mountain, Nevada: Proposed Rule." Federal Register, v.70, no.161, August 22, 2005, pages 49014-49065.
Eskom 2005	Eskom News, "SA researcher makes Solar breakthrough", June 10, 2005, online at http://www.eskom.co.za/live/content.php?Item_ID=702
FAA 2001	Federal Aviation Administration, "Press Release: FAA Restricts All Private Aircraft Flying Over Nuclear Facilities", October 30, 2001, available online at http://www.faa.gov/apa/pr/pr.cfm?id=1451
Faltermayer 1988	Edmund Faltermayer, "Taking Fear Out of Nuclear Power", *Fortune*, August 1, 1988
Fan and Chen 2000	Chin-Feng Fan and Wen-Hou Chen, "Accident sequence analysis of human-computer interface design", *Reliability Engineering and System Safety*, Volume 67(1), 2000

FEMA 2001	Hilary Styron, "CSX Tunnel Fire Baltimore, MD July 2001", United States Fire Administration, Federal Emergency Management Agency, 2001 (USFA-TR-140)
FENOC 2002	FirstEnergy Nuclear Operating Company, "Davis-Besse Nuclear Power Station: Management and Human Performance Root Causes", August 15, 2002, online at http://www.nrc.gov/reactors/operating/ops-experience/vessel-head-degradation/vessel-head-degradation-files/nrc08-15-final-presentation.pdf
Feynman 1988	Richard Feynman, "What Do You Care What Other People Think?": Further Adventures of a Curious Character, W.W. Norton & Company, New York, NY (1988)
Forbes 1985	James Cook, "Nuclear Follies", *Forbes*, February 11, 1985
Ford 1982	Daniel Ford, The Cult of the Atom: The Secret Papers of the Atomic Energy Commission, Simon and Schuster, New York, NY (1982)
Frame et al. 2005	D.J. Frame *et al.*, "Constraining climate forecasts: The role of prior assumptions", *Geophysical Research Letters*, Vol. 32, L09792, (2005)
Freund 2003	P. Freund, "Making deep reduction in CO_2 emissions from coal-fired power plant using capture and storage of CO_2", Proceedings of the Institution of Mechanical Engineers. Part A: Power and Energy, Vol. 217 No. 1, 1-7 (2003)
Fthenakis 2000	Vasilis M. Fthenakis, "End-of-life management and recycling of PV modules", *Energy Policy*, Vol. 28, 1051-1058 (2000)
Fthenakis and Moskowitz 2000	V.M. Fthenakis and P.D. Moskowitz, "Photovoltaics: Environmental, Health and Safety Issues and Perspectives", *Progress in Photovoltaics: Research and Applications*, Vol. 8, 27-38 (2000)
Fuller 1975	John Fuller, We Almost Lost Detroit, Reader's Digest Press, New York, NY (1975)
FXHistory	OANDA Corporation, "FXHistory: historical currency exchange rates", online at http://www.oanda.com/convert/fxhistory
Gaddis 1972	John Lewis Gaddis, The United States and the Origins of the Cold War, 1941-1947, Columbia University Press, New York, NY (1972)

Gale 2004	John Gale, "Geological storage of CO_2: What do we know, where are the gaps and what more needs to be done?", *Energy*, Vol. 29, 1329-1338 (2004)
GAO 1986	General Account Office, "Financial Consequences of a Nuclear Power Plant Accident", July 1986 (GAO/RCED-86-193BR)
GAO 1997	General Accounting Office, "Preventing Problem Plants Requires More Effective NRC Action", March 1997 (GAO/RCED-91-145)
GAO 1999	General Accounting Office, "Strategy Needed to Regulate Safety Using Information on Risk", March 1999 (GAO/RCED-99-95)
GAO 2001	General Accounting Office, "Nuclear Waste: Technical, Schedule, and Cost Uncertainties of the Yucca Mountain Repository Project", December 21, 2001 (GAO-02-191)
GAO 2004	General Accounting Office, "Nuclear Regulation: NRC Needs to More Aggressively and Comprehensively Resolve Issues Related to the Davis-Besse Nuclear Power Plant's Shutdown", May 2004 (GAO-04-415)
GAO 2004b	General Accounting Office, "Yucca Mountain: Persistent Quality Assurance Problems Could Delay Repository Licensing and Operation", April 2004 (GAO-04-460)
GAO 2005	Government Accountability Office, "Nuclear Regulatory Commission: NRC Needs to Do More to Ensure that Power Plants are Effectively controlling Spent Nuclear Fuel", April 2005 (GAO-05-339)
GAO 2005b	Government Accountability Office, "Nuclear Regulatory Commission: Challenges Facing NRC in Effectively Carrying Out Its Mission", Testimony Before the Subcommittee on Clean Air, Climate Change, and Nuclear Safety, Committee on Environment and Public Works, U.S. Senate, May 26, 2005 (GAO-05-754T)
Garwin 1998	Richard L. Garwin, "Reactor-Grade Plutonium Can be Used to Make Powerful and Reliable Nuclear Weapons: Separated plutonium in the fuel cycle must be protected as if it were nuclear weapons", Draft as of August 26, 1998, online at http://www.fas.org/rlg/980826-pu.htm
Gehman et al. 2003	Harold Gehman et al., "Report of the Columbia Accident Investigation Board Volume I", August 2003

General Electric 2004	General Electric Company, "GE Energy, Bechtel Announce Alliance for Cleaner Coal Projects", October 4, 2004 online at http://www.gepower.com/about/press/en/2004_press/100404.htm
Gibb 2000	Fergus Gibb, "A new scheme for the very deep geological disposal of high-level radioactive waste", *Journal of the Geological Society, London*, Vol. 157, 27-36 (2000)
Global Policy Forum	Celine Nahory, Giji Gya, and Misaki Watanabe, "Subjects of UN Security Council Vetoes", Global Policy Forum, online at http://www.globalpolicy.org/security/membship/veto/vetosubj.htm
Global Security	GlobalSecurity.org, "Nuclear Weapons Program", online at http://www.globalsecurity.org/wmd/world/japan/nuke.htm
GNEP Factsheet	U.S. Department of Energy, Global Nuclear Energy Partnership Factsheet, online at http://www.gnep.energy.gov/pdfs/06-GA50035b.pdf
Goldberg 2000	Marshall Goldberg, "Federal Energy Subsidies: Not all Technologies are Created Equal", Renewable Energy Policy Project Research Report, July 2000
Green 2000	M.A. Green, "Photovoltaics: technology overview", *Energy Policy*, Vol. 28, 989-998 (2000)
Greenpeace 2002	Greenpeace USA, "Losing the Clean Energy Race: How the U.S. Can Retake the Lead and Solve Global Warming", Two Case Studies" Wind Power and Solar. 2002, online at http://www.greenpeace.org/raw/content/usa/press/reports/losing-the-clean-energy-race.pdf
Greenwire 2004	Darren Samuelsohn and Brian Stempeck, "Coal: IGCC Leads Clean Technologies, But Will It Pass Utility Muster?", *Greenwire*, August 11, 2004
Guardian 2003	Paul Brown, "Sellafield shutdown ends the nuclear dream; £1.8bn Thorp plant that promised limitless electricity to close by 2010", *Guardian (London)*, August 26, 2003
Guardian 2004	Patrick Wintour and Paul Brown, "Blair reignites nuclear debate", *Guardian (London)*, July 7, 2004

Guardian 2004b	Duncan Campbell, "Vanunu under house arrest: Soldiers snatch nuclear whistleblower from church sanctuary as Israel accuses him of passing on more secrets", *Guardian (London)*, November 12, 2004
Hagen 1980	E.W. Hagen, "Common-Mode/Common Cause Failure: A Review", *Nuclear Safety*, Volume 21 Number 2, March-April 1980
Hanford Factsheet	GlobalSecurity.org, "Weapons of Mass Destruction (WMD): Hanford", online at http://www.globalsecurity.org/wmd/facility/hanford.htm
Hansen 2005	James Hansen, "A Slippery Slope: How Much Global Warming Constitutes 'Dangerous Anthropogenic Interference'?", Climatic Change, Vol. 68, 269-279 (2005)
Hansen and Lebedeff 1987	J.E. Hansen and S. Lebedeff, "Global trends of measured surface air temperatures", *Journal of Geophysical Research*, Volume 92 Issue D11, 1987
Harrison 2000	Tim Harrison, "Very deep borehole: Deutag's opinion on boring, canister emplacement and retriveability", SKB Rapport R-00-35 (May 2000)
Harwood 1996	William Harwood, "NASA Officials Calculate Shuttle Risks", *Space News Business Report*, January 29, 1996
Hatch 2005	Orrin Hatch, "Major PFS Partners Backing Out of Skull Valley Plan", News Release, December 8, 2005
Hatch 2005b	Orrin Hatch, "Majority of PFS Shareholders Now Out: Florida Power & Light Joins PFS Exodus", News Release, December 14, 2005
Hewlett and Anderson 1990	Richard G. Hewlett and Oscar E. Anderson, Jr., The New World: A History of the United States Atomic Energy Commission, Volume I 1939-1946, University of California Press, Berkeley, CA (1990)
Heyes & Liston-Heyes 1998	A.G. Heyes and C. Liston-Heyes, "Subsidy to Nuclear Power Through Price-Anderson Liability Limit: Comment", *Contemporary Economic Policy*, Volume 16 Number 1, January 1998

Heyes & Liston-Heyes 1998b	Anthony Heyes and Catherine Liston-Heyes, "Liability Capping and Financial Subsidy in North American Nuclear Power: Some Financial Results Based on Insurance Data", Paper Presented at the Conference on Liability, Economics, and Insurance, Odense, Denmark, October 22-24, 1998 online at http://www.akf.dk/som/pdf/som32/5heyes.pdf
Heyes 2002	Anthony Heyes, "Determining the Price of Price-Anderson", *Regulation*, Winter 2002-2003
High-level Panel 2004	Report of the Secretary-General's High-level Panel on Threats, Challenges, and Change, "A More Secure World: Our Shared Responsibility", United Nations, 2004
Hodell, Curtis, Brenner 1995	David Hodell, Jason Curtis, and Mark Brenner, "Possible role of climate in the collapse of the Classic Maya civilization", *Nature*, Vol. 375, 391-394 (1995)
Holloway 1994	David Holloway, Stalin and the Bomb, Yale University Press, New York, NY (1994)
Holt 2005	Mark Holt, "Civilian Nuclear Waste Disposal", Congressional Research Service, Updated March 8, 2005 (Order Code IB92059)
Holt 2005b	Mark Holt, "Nuclear Energy Policy", Congressional Research Service, Updated March 17, 2005 (Order Code IB88090)
Holt 2005c	Mark Holt, "Nuclear Energy Policy", Congressional Research Service, Updated June 14, 2005 (Order Code IB88090)
Hominid Timeline	Washington State University, "Hominid Species Timeline", online at http://www.wsu.edu/gened/learn-modules/top_longfor/timeline/timeline.html
Hotchkiss 2003	R. Hotchkiss, "Coal gasification technologies", Proceedings of the Institution of Mechanical Engineers. Part A: Power and Energy, Vol. 217 No. 1, 27-33 (2003)
Hu, Makhijani, & Yih 1992	Howard Hu, Arjun Makhijani, and Katherine Yih, ed., Plutonium: Deadly Gold of the Nuclear Age, International Physicians Press, Cambridge, MA (1992)
Huybrechts and de Wolde 1999	Philippe Huybrechts and Jan de Wolde, "The Dynamic Response of the Greenland and Antartic Ice Sheets to Multiple-Century Climatic Warming", *Journal of Climate*, Vol. 12, 2169-2188 (1999)

IAEA 1997	International Atomic Energy Agency, "Model Protocol Additional to the Agreement(s) Between State(s) and the International Atomic Energy Agency for the Application of Safeguards", September 1997 (INFCIRC/540 (Corrected))
IAEA 1999	International Atomic Energy Agency, "Report on the preliminary fact finding mission following the accident at the nuclear fuel processing facility in Tokimura, Japan", November 1999
IAEA 2001	International Atomic Energy Agency, "Analysis of Uranium Supply to 2050", May 2001 (STI/PUB/1104)
IAEA 2001b	International Atomic Energy Agency Press Release, "Calculating the New Global Nuclear Terrorism Threat", November 1, 2001, online at http://www.iaea.org/NewsCenter/PressReleases/2001/nt_pressrelease.shtml
IAEA 2003b	International Atomic Energy Agency, " Scientific and Technical Basis for the Geological Disposal of Radioactive Wastes", Technical Reports Series No. 413, February 2003
IAEA 2003c	International Atomic Energy Agency, "Implementation of the NPT safeguards agreement in the Islamic Republic of Iran: Report by the Director General", June 6, 2003 (GOV/2003/40)
IAEA 2003d	International Atomic Energy Agency, "Implementation of the NPT Safeguards Agreement in the Islamic Republic of Iran: Report by the Director General", August 26, 2003 (GOV/2003/63)
IAEA 2005	International Atomic Energy Agency, "Multilateral Approaches to the Nuclear Fuel Cycle: Expert Group Report submitted to the Director General of the International Atomic Energy Agency", Information Circular, February 22, 2005 (INFCIRC/640)
IAEA 2005b	International Atomic Energy Agency, "Non-Proliferation of Nuclear Weapons & Nuclear Security: IAEA Safeguards Agreements and Additional Protocols", May 2005, online at http://www.iaea.org/Publications/Booklets/engl_nuke.pdf

IAEA 2005c	International Atomic Energy Agency, "Strengthened Safeguards System: Status of Additional Protocols", information as of June 8, 2005, online at http://www.iaea.org/OurWork/SV/Safeguards/sg_protocol.html
IAEA 2005d	International Atomic Energy Agency, "Country Nuclear Fuel Cycle Profiles, Second Edition", Technical Reports Series No. 425, 2005
IAEA 2005e	International Atomic Energy Agency, "Implementation of the NPT Safeguards Agreement in the Islamic Republic of Iran", Resolution Adopted September 24, 2005 (GOV/2005/77)
IAEA 2005f	International Atomic Energy Agency, "Global Public Opinion on Nuclear Issues and the IAEA: Final Report from 18 Countries", Prepared for the International Atomic Energy Agency by GlobeScan Incorporated, October 2005, online at http://www.iaea.org/Publications/Reports/gponi_report2005.pdf
IAEA 2006	International Atomic Energy Agency, "Implementation of the NPT Safeguards Agreement in the Islamic Republic of Iran", Resolution adopted on 4 February 2006 (GOV/2006/14)
IAEA 2006b	International Atomic Energy Agency, "Implementation of the NPT Safeguards Agreement in the Islamic Republic of Iran", Report by the Director General, February 27, 2006 (GOV/2006/15)
IAEA/NEA 2001	A Joint Report by the OECD Nuclear Energy Agency and the International Atomic Energy Agency, "Management of Depleted Uranium", 2001
ICMA-RC 2004	ICMA Retirement Corporation, "10-Year Treasury Note Yield", chart posted March 26, 2004
IEA 2000	International Energy Agency, "World Energy Outlook: 2000", Organization for Economic Cooperation and Development, 2000
IEA 2001	International Energy Agency, "World Energy Outlook: Assessing Today's Supplies to Fuel Tomorrow's Growth, 2001 Insights", Organization for Economic Cooperation and Development, 2001

IEA 2001b	International Energy Agency, "Nuclear Power in the OECD", Organization for Economic Cooperation and Development, 2001
IEA 2002	International Energy Agency, "World Energy Outlook 2002", Organization for Economic Cooperation and Development, 2002
IEA 2003	International Energy Agency, "Renewables For Power Generation: Status & Prospects", Organization for Economic Cooperation and Development, 2003
IEA 2003b	Claude Mandil. "World Energy Investment Outlook: North American Energy Investment Challenges: 2003 Insights." International Energy Agency. Online at http://www.csis.org/energy/031112_mandil.pdf
IEA 2005	International Energy Agency, "Variability of Wind Power and Other renewable: Management options and strategies", Organization for Economic Cooperation and Development, June 2005
IEA 2005b	International Energy Agency, "Key World Energy Statistics: 2005", Organization for Economic Cooperation and Development, 2005
IEA 2005c	International Energy Agency, "Projected Costs of Generating Electricity: 2005 Update", Organization for Economic Cooperation and Development Nuclear Energy Agency, 2005
IEER 2002	Arjun Makhijani, "Independent Institute Recommends Alternative Nuclear Waste Plan: Safer and more environmentally sound than the proposed Yucca Mountain repository", Press Release, June 4, 2002, online at http://www.ieer.org/comments/waste/yuccaalt.html
IEER 2005	Institute for Energy and Environmental Research, "Iodine-131 Releases from the July 1959 Accident at the Atomics International Sodium Reactor Experiment", January 13, 2005
ILWG 2000	Interlaboratory Working Group on Energy-Efficient and Clean Energy Technologies, "Scenarios for a Clean Energy Future", November 2000 (ORNL/CON-476 and LBNL-44029)
Independent 1992	Wolfgang Achtner, "Bombers murder Italy's leading anti-Mafia judge", *The Independent (London)*, May 24, 1992

IPCC 2001	T. Morita et al., "Greenhouse Gas Emission Mitigation Scenarios and Implications" in *Climate Change 2001: Mitigation. Contributions of Working Group III of the Intergovernmental Panel on Climate Change* [Bert Metz et al. Ed.], Cambridge University Press, New York (2001)
IPCC 2001b	W.R. Moomaw et al., "Technological and Economic Potential of Greenhouse Gas Emissions Reduction" in *Climate Change 2001: Mitigation. Contributions of Working Group III of the Intergovernmental Panel on Climate Change* [Bert Metz et al. Ed.], Cambridge University Press, New York (2001)
IPCC 2001c	Tariq Banuri et al., "Technical Summary" in *Climate Change 2001: Mitigation. Contributions of Working Group III of the Intergovernmental Panel on Climate Change*, Cambridge University Press, New York (2001)
IPCC 2001d	U. Cubasch et al., "Projections of Future Climate Change" in *Climate Change 2001: The Scientific Basis. Contributions of Working Group I of the Intergovernmental Panel on Climate Change* [J.T. Houghton et al. Ed.], Cambridge University Press, New York (2001)
IPCC 2001e	J.A. Church et al., "Changes in Sea Level" in *Climate Change 2001: The Scientific Basis. Contributions of Working Group I of the Intergovernmental Panel on Climate Change* [J.T. Houghton et al. Ed.], Cambridge University Press, New York (2001)
IPCC 2001f	J.T. Houghton et al. Ed., *Climate Change 2001: The Scientific Basis. Contributions of Working Group I of the Intergovernmental Panel on Climate Change*, Cambridge University Press, New York (2001)
IPCC 2001g	R. Akhtar et al., "Human Health" in *Climate Change 2001: Impacts, Adaptation, and Vulnerability. Contributions of Working Group II of the Intergovernmental Panel on Climate Change* [James J. McCarthy et al. Ed.], Cambridge University Press, New York (2001)
IPCC 2001i	J.B. Smith et al., "Vulnerability to Climate Change and Reasons for Concern: A Synthesis" in *Climate Change 2001: Impacts, Adaptation, and Vulnerability. Contributions of Working Group II of the Intergovernmental Panel on Climate Change* [James J. McCarthy et al. Ed.], Cambridge University Press, New York (2001)

IPCC 2002	H. Gitay *et al.*, "Climate Change and Biodiversity" in *Climate Change and Biodiversity: IPCC Technical Paper V* [H. Gitay et al. Ed.], paper requested by the United Nations Convention on Biological Diversity and prepared under the auspices of the IPCC Chair, Dr. Robert T. Watson, April 2002
IPCC 2005	Bert Metz et al., "IPCC Special Report on Carbon Dioxide Capture and Storage", Prepared for Working Group III of the Intergovernmental Panel on Climate Change, Cambridge University Press, 2005
Iraq Casualties	Iraq Coalition Casualty Count, last viewed March 7, 2006, online at http://icasualties.org/oif/US_chart.aspx and http://icasualties.org/oif/IED.aspx
Ishimatsu et al. 2004	Atsushi Ishimatsu et al., "Effects of CO_2 on Marine Fish: Larvae and Adults", *Journal of Oceanography*, Vol. 60, 731-741 (2004)
Jackson and Oliver 2000	Tim Jackson and Mark Oliver, "The viability of solar photovoltaics", *Energy Policy*, Vol. 28, 983-988 (2000)
Jacobson 2005	Mark Jacobson, "Studying ocean acidification with conservative, stable numerical schemes for nonequilibrium air-ocean exchange and ocean equilibrium chemistry", *Journal of Geophysical Research*, Vol. 110, D07302, (2005)
Jacobson and Masters 2001	Mark Jacobson and Gilbert Masters, "Exploiting Wind Versus Coal", *Science*, Vol. 293, 1438 (2001)
Jager-Waldau 2004	Arnulf Jäger-Waldau, "Status of thin film solar cells in research, production and the market", *Solar Energy*, Vol. 77, 667-678 (2004)
Jedicke 1989	Peter Jedicke, "The NRX Incident", Fanshawe College, London, Ontario, 1989, online at http://www.cns-snc.ca/history/nrx.html
Jensen 2003	James T. Jensen, "The LNG Revolution", *The Energy Journal*, Vol. 24 No. 2, 1-45 (2003)
Jerusalem Post 2003	Arieh O'Sullivan, "Strike while the ions are hot", *Jerusalem Post*, August 29, 2003
Jones, Rosa, and Johnson 2004	Steven G. Jones, Doug R. Rosa, Johnny E. Johnson, "Acid-gas injection design requires numerous considerations", *Oil & Gas Journal*, Vol. 102 No. 10, 45-51 (2004)

Juhlin and Sandstedt 1989	Christopher Juhlin and Håkan Sandstedt, "Storage of nuclear waste in very deep boreholes: Feasibility study and assessment of economic potential", Parts I and II, SKB Technical Report 89-39, Stockholm: Svensk Kärnbränslehantering. (December 1989)
Kang and von Hippel 2005	Jungmin Kang and Frank von Hippel, "Limited Proliferation-Resistance Benefits from Recyling Unseparated Transuranics and Lanthanides from Light-Water Reactor Spent Fuel", *Science and Global Security*, Vol. 13, 169-181 (2005)
Kemeny Commission 1979	John Kemeny et al., "Report of the President's Commission on the Accident at Three Mile Island", October 1979
Kendall et al. 1977	Henry Kendall (study director) et al., The Risks of Nuclear Power Reactors: A Review of the NRC Reactor Safety Study WASH-1400 (NUREG-75/014), Union of Concerned Scientists, Cambridge, MA (August 1977)
Kennedy et al. 2006	John Kennedy, Andreas Zsiga, Laurie Conheady, and Paul Lund, "Credit Aspects of North American and European Nuclear Power", Standard & Poor's, January 9, 2006
Kersting et al. 1999	A.B. Kersting et al., "Migration of plutonium in ground water at the Nevada Test Site", *Nature*, Volume 397 (January 199)
Knief 1992	Ronald Allen Knief, Nuclear Engineering: Theory and Technology of Commercial Nuclear Power, Second Edition, Taylor & Francis, Washington, DC (1992)
Knutson and Tuleya 2004	T.R. Knutson and R.E. Tuleya, "Impact of CO_2-Induced Warming on Simulated Hurricane Intensity and Precipitation: Sensitivity to the Choice of Climate Model and Convective Parameterization", *Journal of Climate*, Vol. 17 No. 18, September 15, 2004
Knutson et al. 2001	Thomas R. Knutson et al., "Impact of CO_2-Induced Warming on Hurricane Intensities as Simulated in a Hurricane Model with Ocean Coupling", *Journal of Climate*, Vol. 14 No. 11, June 1, 2001
Knutson, Tuleya, & Kurihara 1998	Thomas R. Knutson, Robert E. Tuleya, and Yoshio Kurihara, "Simulated Increase of Hurricane Intensities in a CO_2-Warmed Climate", *Science*, Vol. 279, February 13, 1998
Komanoff 1981	Charles Komanoff, Power Plant Cost Escalation: Nuclear and Coal Capital Costs, Regulation, and Economics, Van Nostrand Reinhold Company Inc., New York, 1981

Komanoff and Roelofs 1992	Charles Komanoff and Cora Roelofs, "Fiscal Fission: The Economic Failure of Nuclear Power", Komanoff Energy Associates for Greenpeace USA, December 1992
Krabill et al. 2004	W. Krabill et al., "Greenland Ice Sheet: Increased costal thinning", *Geophysical Research Letters*, Vol. 31, L24402 (2004)
Kreutz et al. 2005	Thomas Kreutz, Robert Williams, Stefano Consonni, and Paolo Chiesa, "Co-production of hydrogen, electricity and CO_2 from coal with commercially ready Technology. Part B: Economic analysis", *International Journal of Hydrogen Energy*, Vol. 30, 747-767 (2005)
Kurihara, Shimode, and Shirayama 2004	Haruko Kurihara, Shinji Shimode, and Yoshihisa Shirayama, "Sub-Lethal Effects of Elevated Concentrations of CO_2 on Planktonic Copepods and Sea Urchins", *Journal of Oceanography*, Vol. 60, 743-750 (2004)
Kyoto Status 2005	United Nations Framework Convention on Climate Change, "Kyoto Protocol Status of Ratification", last modified on February 6, 2006, online at http://unfccc.int/files/essential_background/kyoto_protocol/application/pdf/kpstats.pdf
Lamb and Resnikoff 2000	Matthew Lamb and Marvin Resnikoff, "Review of NUREG/CR-6672 Reexamination of Spent Fuel Shipment Risk Estimates", Radioactive Waste Management Associates, October 2000
LANL Biography	Los Alamos National Laboratory, "Staff Biographies: Robert Fox Bacher", online at http://www.lanl.gov/history/people/R_Bacher.shtml
Lansdell 1958	Norman Lansdell, The Atom and the Energy Revolution, Penguin Books, Harmondsworth, Middlesex (1958)
Lawler 2001	Andrew Lawler, "Writing Gets a Rewrite", *Science*, Volume 292 Number 5526, June 2001
Le Monde 2006	Hervé Morin, "Bévues à la centrale nucléaire de Civaux", *Le Monde*, April 4, 2006
Leveson 1994	Nancy Leveson, "High-Pressure Steam Engines and Computer Software", *Computer*, Volume 27(10), 1994
Leveson 2004	Nancy Leveson, " Role of Software in Spacecraft Accidents", *Journal of Spacecraft and Rockets*, Vol. 41 No. 4, 564-575 (2004)

Lewis 1978	H.W. Lewis *et al.*, "Risk Assessment Review Group Report to the U.S. Nuclear Regulatory Commission", U.S. Nuclear Regulatory Commission, NUREG/CR-0400, September 1978
Lilienthal et al. 1947	David Lilienthal et al., "First Report of the U.S. Atomic Energy Commission", *Science*, Volume 105 Number 2721, February 1947
Lions 1996	Jacques-Louis Lions et al., "Ariane 5 Flight 501 Failure: Report of the Inquiry Board", July 19, 1996
LLW Compact Factsheet	Nuclear Regulatory Comission, "Low-Level Waste Compacts", Last revised November 30, 2004, online at http://www.nrc.gov/waste/llw-disposal/compacts.html
Lochbaum 1998	David Lochbaum, "The Good, the Bad, and the Ugly: A Report on Safety in America's Nuclear Power Industry", *Union of Concerned Scientists*, June 1998
Lochbaum 2000	David Lochbaum, "Nuclear Plant Risk Studies: Failing the Grade", Union of Concerned Scientists, August 2000
London Times 1981	Christopher Walker, "Iraqi nuclear reactor bombed in raid by Israeli jets", *Times (London)*, June 9, 1981
London Times 2004	The Sunday Times Insight Team, "Mordechai Vanunu: The Sunday Times articles", *Times (London)*, April 21, 2004
Long and Ewing 2004	Jane C.S. Long and Rodney C. Ewing, "Yucca Mountain: Earth-Science Issues at a Geologic Repository for High-Level Nuclear Waste", *Annual Review of Earth and Planetary Sciences*, Vol. 32, 363-401 (2004)
Los Angeles Times 1998	Frank Clifford, "Plutonium From Bombs May be Used in Reactors", *Los Angeles Times*, November 15, 1998
Los Angeles Times 2003	Bennett Ramberg, "Iran May Hide its Nuclear Ambitions from Some, but Not Israel", *Los Angeles Times*, December 10, 2003
Los Angeles Times 2005	Elise Castelli, "Probe of Possible Fabricated Reports Stalls Nuclear Dump; Yucca Mountain license process hinges on the findings of a water-safety survey inquiry", *Los Angeles Times*, April 6, 2005
Lu et al. 2003	N. Lu et al., "Sorption kinetics and impact of temperature, ionic strength and colloid concentration on the adsorption of plutonium-239 by inorganic colloids", *Radiochimica Acta*, Volume 91 (2003)

Lyman 2001	Edwin Lyman, "Public Health Risks of Substituting Mixed-Oxide for Uranium Fuel in Pressurized Water Reactors", *Science and Global Security*, Volume 9, 33-79 (2001)
Macfarlane 2003	Allison Macfarlane, "Underlying Yucca Mountain: The Interplay of Geology and Policy in Nuclear Waste Disposal", *Social Studies of Science*, Vol. 33 No. 5, 783-807 (October 2003)
Makhijani 1998	Arjun Makhijani, "Achieving Enduring Nuclear Disarmament", *Science for Democratic Action*, Vol. 6 No. 4 and Vol. 7 No. 1, October 1998
Makhijani 2001	Arjun Makhijani, "Plutonium End Game: Managing Global Stocks of Separated Weapons-Usable Commercial and Surplus Nuclear Weapons Plutonium", January 22, 2001
Makhijani 2001b	Arjun Makhijani, "Securing the Energy Future of the United States: Oil, Nuclear, and Electricity Vulnerabilities and a post-September 11, 2001 Roadmap for Action", November 2001
Makhijani 2003	Arjun Makhijani, "Yucca Mountain: An Example Not to Follow", Paper on Nuclear Waste Management for presentation to a Greenpeace Briefing Quale Futuro per le Scorie Nucleari Italiene?, December 2, 2003, online at http://www.ieer.org/comments/waste/yuccaitaly.html
Makhijani 2005	Arjun Makhijani, "Bad to the Bone: Analysis of the Federal Maximum Contaminant Levels for Plutonium-239 and Other Alpha-Emitting Transuranic Radionuclides in Drinking Water", June 2005
Makhijani and Deller 2003	Arjun Makhijani and Nicole Deller, "NATO and Nuclear Disarmament: An Analysis of the Obligations of the NATO Allies of the United States under the Nuclear Non-Proliferation Treaty and the Comprehensive Test Ban Treaty", October 2003
Makhijani and Makhijani 1995	Arjun Makhijani and Annie Makhijani, <u>Fissile Materials in a Glass, Darkly: Technical and Policy Aspects of the Disposition of Plutonium and Highly Enriched Uranium</u>, IEER Press, Takoma Park, MD (1995)

Makhijani and Makhijani 2006	Arjun Makhijani and Annie Makhijani, Disposal of Long-Lived Highly Radioactive Wastes in France: An IEER Evaluation, *Science for Democratic Action*, Volume 13 Number 4, January 2006, online at http://www.ieer.org/sdafiles/13-4.pdf
Makhijani and Makhijani 2006b	Annie Makhijani and Arjun Makhijani, "Low-Carbon Diet without Nukes in France: An Energy Technology and Policy Case Study on Simultaneous Reduction of Climate Change and Proliferation Risks", May 4, 2006, online at http://www.ieer.org/reports/energy/france/lowcarbonreport.pdf
Makhijani and Poole 1975	Arjun Makhijani with Alan Poole, Energy and Agriculture in the Third World, Ballinger Publishing Company, Cambridge, MA (1975)
Makhijani and Saleska 1992	Arjun Makhijani and Scott Saleska, High-Level Dollars Low-Level Sense: A Critique of Present Policy for the Management of Long-Lived Radioactive Waste and Discussion of an Alternative Approach, A Report of the Institute for Energy and Environmental Research, Apex Press, New York, NY (1992)
Makhijani and Saleska 1999	Arjun Makhijani and Scott Saleska, The Nuclear Power Deception: U.S. Nuclear Mythology from Electricity "Too Cheap to Meter" to "Inherently Safe" Reactors, Apex Press, New York, NY (1999)
Makhijani and Smith 2004	Arjun Makhijani and Brice Smith, "Costs and Risks of Management and Disposal of Depleted Uranium from the National Enrichment Facility Proposed to be Built in Lea County New Mexico by LES", report prepared for Nuclear Information and Resource Service / Public Citizen who are interveners in the Louisiana Energy Services case before the Atomic Safety and Licensing Board, November 24, 2004
Makhijani and Smith 2005	Arjun Makhijani and Brice Smith, "Update to *Costs and Risks of Management and Disposal of Depleted Uranium from the National Enrichment Facility Proposed to be Built in Lea County New Mexico by LES* by Arjun Makhijani, PhD. and Brice Smith, Ph.D. based on information obtained since November 2004", report prepared for Nuclear Information and Resource Service / Public Citizen who are interveners in the Louisiana Energy Services case before the Atomic Safety and Licensing Board, July 5, 2005

Makhijani and Smith 2005b	Arjun Makhijani and Brice Smith, "Comments on the U.S. Environmental Protection Agency's Proposed Rule for the Public Health and Environmental Radiation Protection Standards for Yucca Mountain, Nevada Submitted on Behalf of the Institute for Energy and Environmental Research", November 21, 2005, online at http://www.ieer.org/comments/waste/yuccaepa.html
Makhijani and Smith 2006b	Arjun Makhijani and Brice Smith, "Dangerous Discrepancies: Missing Weapons Plutonium in Los Alamos National Laboratory Waste Accounts", November 29, 2005 (reissued April 21, 2006), online at http://www.ieer.org/reports/lanl/weaponspureport.pdf
Makhijani et al. 2004	Arjun Makhijani, Peter Bickel, Aiyou Chen, and Brice Smith, "Cash Crop on the Wind Farm: A New Mexico Case Study of the Cost, Price, and Value of Wind-Generated Electricity", Prepared for presentation at the North American Energy Summit Western Governors' Association, Albuquerque, New Mexico, April 15-16, 2004
Makhijani, Chalmers, and Smith 2004	Arjun Makhijani, Lois Chalmers, and Brice Smith, "Uranium Enrichment: Just Plain Facts to Fuel an Informed Debate on Nuclear Proliferation and Nuclear Power", A report prepared by Institute for Energy and Environmental Research for the Nuclear Policy Research Institute, October 15, 2004
Makhijani, Hu, and Yih 2000	Arjun Makhijani, Howard Hu, and Katherine Yih editors, Nuclear Wastelands: A Global Guide to Nuclear Weapons Productions and Its Health and Environmental Effects, MIT Press, Cambridge, MA (2000)
Marland, Boden, and Andres 2003	G. Marland, T.A. Boden, and R.J. Andres, "Global Regional, and National CO_2 Emissions In Trends: A Compendium of Data on Global Change", Carbon Dioxide Information Analysis Center, Oak Ridge National Laboratory, 2003
Mayors 2005	Office of the Mayor of Seattle, "U.S. Mayors' Climate Protection Agreement", online at http://www.ci.seattle.wa.us/mayor/climate/ (last viewed January 10, 2006)
Medvedev 1990	Zhores Medvedev, The Legacy of Chernobyl, W.W. Norton & Company, New York, NY (1990)

Meier 2002	Paul J. Meier, "Life-Cycle Assessment of Electricity Generation Systems and Applications for Climate Change Policy Analysis", Ph.D. Dissertation, University of Wisconsin-Madison, August 2002, online at http://fti.neep.wisc.edu/pdf/fdm1181.pdf
Meserve 2002	Richard Meserve, "Safety Culture: An NRC Perspective", Remarks at the 2002 INPO CEO Conference in Atlanta Georgia, November 8, 2002
Millennium Assessment 2005	Walter V. Reid *et al.*, "Ecosystems and Human Well-being: Synthesis", A Report of the Millennium Ecosystem Assessment, 2005
MIT 2003	John Deutch and Ernest J. Moniz (co-chairs) et al., "The Future of Nuclear Power", An Interdisciplinary MIT Study, 2003, online at http://web.mit.edu/nuclearpower/pdf/nuclearpower-full.pdf
Monju Accident Factsheet	MONJU Construction Office, "The MONJU Sodium Leak", online at http://www.jaea.go.jp/jnc/zmonju/mjweb/NaL.htm
Monju Reactor Factsheet	MONJU Construction Office, "From site to power", online at http://www.jaea.go.jp/jnc/zmonju/mjweb/fromsite.htm
MSNBC 2005	"Vice President Cheney on inauguration day: Don Imus interviews the candidate he once poked fun at – and gets pork chops as a parting gift from the VP", *MSNBC*, January 20, 2005
Murphy et al. 2004	James Murphy *et al.*, "Quantification of modeling uncertainties in a large ensemble of climate change simulations", *Nature*, Vol. 430, 768-772 (2004)
Murray 1953	Thomas E. Murry, Memorandum to Lewis L. Strauss, September 16, 1953, DOE Archives, Record Group 326, Box 1290, Folder 2
Murray and Teare 1993	Daniel Murray and Scott Teare, "Probability of a tossed coin landing on edge", Physical Review E, Volume 48 (4) 2547-2552, October 1993
Narula, Wen, and Himes 2002	Ram Narula, Harvey Wen, and Kenneth Himes, "Technical and Economic Comparison of CO_2 Reducing Technologies for Power Plants", Becthel Power, November 5, 2002

NAS 1995	John P. Holdren et al., <u>Management and Disposition of Excess Weapons Plutonium: Reactor-Related Options</u>, Committee on International Security and Arms Control, National Academy Press, Washington, DC (1995)
NAS/NRC 1992	John Ahearne et al., <u>Nuclear Power: Technical and Institutional Options for the Future</u>, National Academy Press, Washington, DC (1992)
NAS/NRC 1995	Robert Fri et al., <u>Technical Bases for Yucca Mountain Standards</u>, National Academy Press, Washington, DC (1995)
NAS/NRC 1996	National Research Council. Committee on Decontamination and Decommissioning of Uranium Enrichment Facilities. <u>Affordable Cleanup? Opportunities for cost reduction in the decontamination and decommissioning of the nation's uranium enrichment facilities</u>. Washington, DC: National Academies Press, 1996
NAS/NRC 1997	Douglas Chapin et al., <u>Digital Instrumentation and Control Systems in Nuclear Power Plants: Safety and Reliability Issues</u>, National Academy Press, Washington, DC (1997)
NAS/NRC 2000b	Gregory R. Choppin et al., <u>Electrometallurgical Techniques for DOE Spent Fuel Treatment: Final Report</u>, National Academy Press, Washington, DC (2000)
NAS/NRC 2001	D. Warner North et al., <u>Disposition of High-Level Waste and Spent Nuclear Fuel: The Continuing Societal and Technical Challenges</u>, National Academy Press, Washington, DC (2001)
NAS/NRC 2001b	Robert W. Fri et al., <u>Energy Research at DOE Was It Worth It? Energy Efficiency and Fossil Fuel Energy Research 1978 to 2000</u>, National Academy Press, Washington, DC (2001)
NAS/NRC 2002	Richard Alley et al., <u>Abrupt Climate Change: Inevitable Surprises</u>, National Academy Press, Washington, DC (2002)
NAS/NRC 2002b	Lewis M. Branscomb et al., <u>Making the Nation Safer: The Role of Science and Technology in Countering Terrorism</u>, National Academy Press, Washington, DC (2002)
NAS/NRC 2004	Michael P. Rampage *et al.*, <u>The Hydrogen Economy: Opportunities, Costs, Barriers, and R&D Needs</u>, National Academy Press, Washington, DC (2004)

NAS/NRC 2006c	Richard R. Monson (Chair) *et al.*, "Health Risks from Exposure to Low Levels of Ionizing Radiation: BEIR VII – Phase 2", Committee to Assess Health Risks from Exposure to Low Levels of Ionizing Radiation, Board on Radiation Effects Research, National Academies Press, Washington, DC (2006)
NAS/NRC 2005c	Frank L. Parker *et al.*, <u>Tank Wastes Planned for On-Site Disposal at Three Department of Energy Sites: The Savannah River Site - Interim Report</u>, Committee on the Management of Certain Radioactive Waste Streams Stored in Tanks at Three Department of Energy Sites, National Academies Press, Washington, DC (2005)
NAS/NRC 2006	Louis J. Lanzerotti *et al.*, <u>Safety and Security of Commercial Spent Nuclear Fuel Storage</u>, National Academy Press, Washington, DC (2006)
NAS/NRC 2006b	Neal F. Lane et al., <u>Going the Distance? The Safe Transport of Spent Nuclear Fuel and High-Level Radioactive Waste in the United States</u>, National Academy Press, Washington, DC (2006)
NATO 2001	North Atlantic Treaty Organization, "NATO Handbook", 2001 online at http://www.nato.int/docu/handbook/2001/pdf/handbook.pdf
NEI 2004	"Two years on, Davis-Besse reached 100% again", *Nuclear Engineering International*, Volume 49 Number 597, April 2004
NEI 2004b	"MOX gets the go-ahead", *Nuclear Engineering International*, Volume 49 Number 597, April 2004
NEI 2006	Nuclear Energy Institute, online at http://www.nei.org (last viewed on January 10, 2006)
New Scientist 2005	Celeste Biever, "Melting permafrost pulls plug on Arctic lakes", *New Scientist*, June 11, 2005
New Scientist 2005b	Fred Pearce, "Climate warning as Siberia melts", *New Scientist*, August 11, 2005
New York Times 1954	Elie Abel, "Hydrogen Blast Astonished Scientists, Eisenhower Says", *New York Times*, March 25, 1954
New York Times 1954b	Lewis Strauss, "Text of Statement and Comments by Strauss on Hydrogen Bomb Tests in the Pacific", *New York Times*, April 1, 1954

New York Times 1954c	"Rites for H-Bomb Victim Held", *New York Times*, October 10, 1954
New York Times 1978	Anthony J. Parisi, "Shell Oil Forcasts Slackening Imports", *New York Times*, February 11, 1978
New York Times 1981	Bernard Nossiter, "Israelis Condemned by Security Council for Attacks on Iraq", *New York Times*, June 20, 1981
New York Times 1982	Joseph Lelyveld, "Bombs Damage Atom Plant Site in South Africa", *New York Times*, December 20, 1982
New York Times 1986	William Robbins, "Untested Process Was in Use at Time of Fatal Leak", *New York Times*, January 6, 1986
New York Times 1986b	Ben Franklin, "Uranium Plant Fined in Fatal Blast", *New York Times*, October 16, 1986
New York Times 2000	Andre Revkin, "Con Ed Says that Customers Will Absorb $600,000-a-Day Cost Caused by Leak at A-Plant", *New York Times*, March 4, 2000
New York Times 2002	Ross Milloy, "Confirmed Deaths Rise to 7 After Collapse of Bridge on Interstate in Oklahoma", *New York Times*, May 28, 2002
New York Times 2004	Mohamed El Baradei, "Saving Ourselves from Self-Destruction", *New York Times*, February 12, 2004
New York Times 2004b	William Broad el. al., "A Tale of Nuclear Proliferation: How Pakistani Built his Network", *New York Times*, February 12, 2004
New York Times 2004c	Craig Smith, "Roots of Pakistan Atomic Scandal Traced to Europe", *New York Times*, February 19, 2004
New York Times 2004d	Robert McFadden and Alison Leigh Cowan, "Bumper to Bumper, Travelers Ride Out An I-95 Nightmare", *New York Times*, March 27, 2004
New York Times 2005	Jad Mouawad, "Iran Offers Europe 'Guarantees' on Its Nuclear Program", *New York Times*, March 17, 2005
New York Times 2005c	Matthew Wald, "Fraud is Seen in Nuclear-Waste Site Study", *New York Times*, April 2, 2005
New York Times 2005d	Matthew Wald, "Interest in Building Reactors, but Industry Is Still Cautious", *New York Times*, May 2, 2005
New York Times 2005e	Salman Masood and David Rhode, "Pakistan Now Says Scientist Did Send Koreans Nuclear Gear", *New York Times*, August 25, 2005

New York Times 2006	Matthew Wald, "Big Question Marks on Nuclear Waste Facility", *New York Times*, February 14, 2006
Nirex 2004	United Kingdom Nirex Limited, "Nirex Report: A Review of the Deep Borehole Disposal Concept for Radioactive Waste", Nirex Report no. N/108, June 2004
NIST 2002	Kevin McGrattan and Anthony Hamins, "Numerical Simulation of the Howard Street Tunnel Fire, Baltimore, Maryland, July 2001", National Institute of Standards and Technology, August 2002 (NISTIR 6902)
NRC 1979	U.S. Nuclear Regulatory Commission, "NRC Statement on Risk Assessment and the Reactor SafetyStudy Report (WASH-1400) in Light of the Risk Assessment Review Group Report", January 18, 1979
NRC 1983	U.S. Nuclear Regulatory Commission, "The Price-Anderson Act - The Third Decade", December 1983 (NUREG-0957)
NRC 1987	Nuclear Regulatory Commission, "Report on the Accident at the Chernobyl Nuclear Power Station", January 1987 (NUREG-1250)
NRC 1996	Nuclear Regulatory Commission, "Regulatory Concerns From Precursor Events at Fuel Cycle Facilities", SECY-96-007, Attachment 2 Enclosure 3 (April 16, 1996)
NRC 1996b	Nuclear Regulatory Commission, "Individual Plant Examination Program: Perspectives on Reactor Safety and Plant Performance Summary Report Draft Report for Comment", NUREG-1560 Vol. 1, Part 1 (October 1996)
NRC 1996c	Nuclear Regulatory Commission, "Generic Environmental Impact Statement for License Renewal of Nuclear Plants", May 1996 (NUREG-1437 Vol. 1), online at http://www.nrc.gov/reading-rm/doc-collections/nuregs/staff/sr1437/v1/
NRC 1998	Nuclear Regulatory Commission, "The Price-Anderson Act - Crossing the Bridge to the Next Century: A Report to Congress", October 1998 (NUREG/CR-6617)
NRC 1999	Nuclear Regulatory Commission, "NRC Certifies Westinghouse Electric Company's AP600 Reactor Design", NRC News, December 15, 1999
NRC 2000	U.S. Nuclear Regulatory Commission, "Reexamination of Spent Fuel Shipment Risk Estimates", March 2000 (NUREG/CR-6672 SAND2000-0234)

NRC 2000b	U.S. Nuclear Regulatory Commission, Advisory Committee on Reactor Safeguards, "Meeting: Human Factors", March 15, 2000 online at: http://www.nrc.gov/reading-rm/doc-collection/acrs/tr/subcommittee/2000/hf000315.html
NRC 2001	U.S. Nuclear Regulatory Commission, "Review of Finding for Human Error Contribution to Risk in Operating Events", August 2001 (INEEL/EXT-01-01166)
NRC 2002	Nuclear Regulatory Commission, "NRC Issues 'Lessons Learned' Task Force Report on Agency's Handling of Davis-Besse Reactor Vessel Head Damage", October 9, 2002 (No. 02-119)
NRC 2003a	Nuclear Regulatory Commission, "NRC Grants TVA License Amendment to Produce Tritium at Watts Bar Nuclear Power Plant for Energy Department", NRC News, September 24, 2002
NRC 2003b	Nuclear Regulatory Commission, "NRC Grants TVA License Amendment to Produce Tritium at Sequoyah Nuclear Power Plant fin Tennessee", NRC News, October 1, 2002
NRC 2003c	Nuclear Regulatory Commission, "Evaluation of the Effects of the Baltimore Tunnel Fire on Rail Transportation of Spent Nuclear Fuel", SECY-03-0002, January 6, 2003
NRC 2004	U.S. Nuclear Regulatory Commission, "U.S. Nuclear Regulatory Commission Staff Evaluation of U.S. Department of Energy Analysis Model Reports, Process Controls, And Corrective Actions", April 7, 2004
NRC 2004b	U.S. Nuclear Regulatory Commission, "List of Power Reactor Units", last revised September 21, 2004, online at http://www.nrc.gov/reactors/operating/list-power-reactor-units.html
NRC 2005	U.S. Nuclear Regulatory Commission, "Information Digest 2004-2005 Edition", NUREG-1350, Volume 16, Rev. 1, Date Published: July 2004, Date Revised: February 2005
NRC 2005b	NRC News, "$5,450,000 Fine for Davis-Besse Reactor Vessel Head Violations", U.S. Nuclear Regulatory Commission Office of Public Affairs, April 21, 2005
NRC 2005c	U.S. Nuclear Regulatory Commission, "Event Notification Report for October 12, 2005", Event Number 42050, October 12, 2005

NRC 2005d	U.S. Nuclear Regulatory Commission, "Preliminary Notification of Event or Unusual Occurrence: Technical Specification Required Shutdown of Palo Verde Units 2 and 3", PNO-IV-05-016, October 12, 2005
NRC 2005e	U.S. Nuclear Regulatory Commission, "Status of License Renewal Applications and Industry Activities", last revised November 30, 2005, online at http://www.nrc.gov/reactors/operating/licensing/renewal/applications.html
NRC 2005f	Nuclear Regulatory Commission, "NRC Denies Utah's Final Appeals, Authorizes Staff to Issue License for PFS Facility", NRC News, September 9, 2005
NRC 2006	U.S. Nuclear Regulatory Commission, "Status of License Renewal Applications and Industry Activities", Last revised Tuesday, January 31, 2006, online at http://www.nrc.gov/reactors/operating/licensing/renewal/applications.html
NRC 2006b	Nuclear Regulatory Commission, "NRC Publishes Certification Rule for Westinghouse's AP1000 Advanced Reactor Design", NRC News, January 31, 2006
NRC 2006c	Nuclear Regulatory Commission, "NRC Issues License to Private Fuel Storage for Spent Nuclear Fuel Storage Facility in Utah", NRC News, February 22, 2006
NRDC 2001	Daniel Lashof, Patricio Silva, et al., "A Responsible Energy Policy for the 21st Century", Natural Resources Defense Council, March 2001
NREL 2005	Complied by J. Aabakken, "Power technologies Energy Data Book: Third Edition", National Renewable Energy Laboratory, U.S. Department of Energy Office of Energy Efficiency and Renewable Energy, April 2005 (NREL/TP-620-37930)
NSDM 1975	Henry A. Kissinger, "U.S. – Iran Nuclear Cooperation", National Security Decision Memorandum 292, April 22, 1975, online at http://www.ford.utexas.edu/library/document/nsdmnssm/nsdm292a.htm
NSDM 1976	Brent Scowcroft, "Negotiations of a Nuclear Agreement with Iran", National Security Decision Memorandum 324, April 20, 1976, online at http://www.ford.utexas.edu/library/document/nsdmnssm/nsdm324a.htm

NSSM 1975	Henry A. Kissinger (memorandum signed by Brent Scowcroft), "U.S. – Iran Agreement on Cooperation in Civil Uses of Atomic Energy", National Security Study Memorandum 219, March 14, 1975, online at http://www.ford.utexas.edu/library/document/nsdmnssm/nssm219a.htm
NTI 2004	Nikolai Sokov, "Russia's Nuclear Doctrine", Produced for the Nuclear Threat Initiative by the Center for Nonproliferation Studies at the Monterey Institute of International Studies, online at http://www.nti.org/e_research/e3_55a.html
NVSR 2002	Centers for Disease Control, "Deaths: Final Data for 2000", *National Vital Statistics Reports*, Vol. 50 No. 15, September 16, 2002
NVSR 2003	Centers for Disease Control, "Deaths: Final Data for 2001", *National Vital Statistics Reports*, Vol. 52 No. 3, September 18, 2003
NVSR 2004	Centers for Disease Control, "Deaths: Final Data for 2002", *National Vital Statistics Reports*, Vol. 53 No. 5, October 12, 2004
NWTRB 2002	Nuclear Waste Technical Review Board, Letter Report to Congress and the Department of Energy, January 24, 2002
NWTRB 2003	Nuclear Waste Technical Review Board, "An Evaluation of Key Elements in the U.S. Department of Energy's Proposed System for Isolating and Containing Radioactive Waste" as attached to a Letter Report to Margaret Chu, Director of the Office of Civilian Radioactive Waste Management, November 25, 2003
NWTRB 2004	Nuclear Waste Technical Review Board, Letter Report to Congress and the Department of Energy, December 30, 2004
NWTRB 2005	Nuclear Waste Technical Review Board, Board letter to Paul Golan following November Board Meeting, December 19, 2005 online at http://www.nwtrb.gov/corr/bjg046.pdf
NWTRB 2005b	Nuclear Waste Technical Review Board, "Report to Congress and the Secretary of Energy January 1, 2004 to December 31, 2004", May 2005

OECD 1986	Organisation for Economic Co-Operation and Development Nuclear Energy Agency, "Projected Costs of Generating Electricity from Nuclear and Coal Fired Power Stations for Commissioning in 1995", OECD/NEA/6/P94, 1986
OECD 1996	E.G. Kudriavtsev, "Tomsk-7 Nuclear Event: Causes, Consequences and Lessons Learned", in Organisation for Economic Co-Operation and Development Committee on the Safety of Nuclear Installations, *Topical Meeting: Safety of the Nuclear Fuel Cycle Cadarche, France 20-21 September 1994*, OCDE/GD(96)18, 1996
OECD 1998	Organisation for Economic Co-Operation and Development Nuclear Energy Agency, "Projected Costs of Generating Electricity Update 1998", 1998
OECD/IAEA 2002	A Joint Report by the Organisation for Economic Co-Operation and Development Nuclear Energy Agency and the International Atomic Energy Agency, "Uranium 2001: Resources, Production and Demand", 2002
OECD/IAEA 2004	A Joint Report by the Organisation for Economic Co-Operation and Development Nuclear Energy Agency and the International Atomic Energy Agency, "Uranium 2003: Resources, Production and Demand", 2004
OIG 2002	Office of the Inspector General, "NRC's Regulation of Davis-Besse Regarding Damage to the Reactor Vessel Head", U.S. Nuclear Regulatory Commission, December 30, 2002
Ortmeyer and Makhijani 1997	Pat Ortmeyer and Arjun Makhijani, "Worse Than We Knew", *Bulletin of the Atomic Scientists*, November/December 1997
OTA 1977	U.S. Congress, Office of Technology Assessment, *Nuclear Proliferation and Safeguards*, June 1977 (NTIS order #PB-275843)
OTA 1993	U.S. Congress, Office of Technology Assessment, *Energy Efficiency: Challenges and Opportunities for Electric Utilities*, OTA-E-561 (Washington, DC: U.S. Government Printing Office, September 1993)
OTA 1993b	U.S. Congress, Office of Technology Assessment, *Proliferation of Weapons of Mass Destruction: Assessing the Risks*, OTA-ISC-559 (Washington, DC: U.S. Government Printing Office, August 1993)

OTA 1995	U.S. Congress, Office of Technology Assessment, *Nuclear Safeguards and the International Atomic Energy Agency*, OTA-ISS-615 (Washington, DC: U.S. Government Printing Office, June 1995).
Palm Beach Post 2005	Palm Beach Post, "Hurricane Special: Florida's summer of storms", online at http://www.palmbeachpost.com/storm/content/weather/special/storm/2004/atlantic/ (last viewed February 21, 2005)
Palo Verde Factsheet	U.S. Department of Energy, "U.S. Nuclear Plants: Palo Verde, Arizona", Energy Information Administration, page last modified on Fri Mar 18 2005 online at http://www.eia.doe.gov/cneaf/nuclear/page/at_a_glance/reactors/palo_verde.html
Pediatrics 2003	American Academy of Pediatrics, "Policy Statement: Radiation Disasters and Children", *Pediatrics*, Vol. 111 No. 6, 1455-1466 (2003)
Perrow 1999	Charles Perrow, Normal Accidents: Living with High-Risk Technologies, Princeton University Press, Princeton, NJ (1999 first published in 1984)
Peterson 2003	Per F. Peterson, "Will the United States Need a Second Geologic Repository?", *The Bridge*, National Academy of Engineering, Fall 2003
PIU 2002	U.K. Cabinet Office Performance and Innovation Unit, "The Energy Review: A Performance and Innovation Unit Report", February 2002
Portner, Langenbuch, and Reipschlager 2004	Hans O. Portner, Martina Langenbuch, and Anke Reipschlager, "Biological Impact of Elevated Ocean CO_2 Concentrations: Lessons from Animal Physiology and Earth History", *Journal of Oceanography*, Vol. 60, 705-718 (2004)
Rahmstorf 2000	Stefan Rahmstrof, "The Thermohaline Ocean Circulation: A System with Dangerous Thresholds?", *Climactic Change*, Vol. 46, 247-256 (2000)
Rahmstorf and Ganopolski 1999	Stefan Rahmstrof and Andrey Ganopolski, "Long-Term Global Warming Scenarios Computed with an Efficient Coupled Climate Model", *Climactic Change*, Vol. 43, 353-367 (1999)
Rahmstorf and Zickfeld 2005	Stefan Rahmstrof and Kristen Zickfeld, "Thermohaline Circulation Changes: A Question of Risk Assessment", *Climactic Change*, Vol. 68, 241-247 (2005)

Ramana, D'Sa, and Reddy 2005	M.V. Ramana, Antonette D'Sa, Amulya K.N. Reddy, "Economics of Nuclear Power from Heavy Water Reactors", *Economic and Political Weekly*, April 23, 2005
Rampton and Stauber 2000	Sheldon Rampton and John Stauber, <u>Trust Us Were Experts: How Industry Manipluates Science and Gambles With Your Future</u>, Penguin Putnam, New York, NY (2000)
Règle N° III.2.f	Règle N° III.2.f (10 juin 1991) Règles fondamentales de sûreté relatives aux installations nucléaires de base autres que reacteurs Tome III: production, contrôle et traitement des effluents et déchets. Chapitre 2: Déchets solides. [France]
REPP 1996	Barbara Farhar, "Energy and the Environment: The Public View", Renewable Energy Policy Project Issue Brief, November 12, 1996, online at http://crest.org/repp_pubs/pdf/issuebr3.pdf
Riccio 2001	Jim Riccio, "Risky Business: The Probability and Consequences of a Nuclear Accident", A Study for Greenpeace USA, 2001, online at http://www.greenpeace.org/raw/content/usa/press/reports/risky-business-the-probabilit.pdf
Rignot et al. 2004	E. Rignot et al., "Accelerated ice discharge from the Antarctic Peninsula following the collapse of Larsen B ice shelf", *Geophysical Research Letters*, Vol. 31, L18401 (2004)
Robbins, Makhijani, & Yih 1991	Anthony Robbins, Arjun Makhijani, and Katherine Yih, <u>Radioactive Heaven and Earth: The health and environmental effects of nuclear weapons testing in, on, and above the earth</u>, A report of the IPPNW International Commission to Investigate the Health and Environmental Effects of Nuclear Weapons Production and the Institute for Energy and Environmental Research, The Apex Press, New York, NY (1991)
Rosen and Meier 1999	Karen Rosen and Alan Meier, "Energy Use of Televisions and Videocassette Recorders in the U.S.", Environmental Energy Technologies Division Lawrence Berkeley National Laboratory, March 1999 (LBNL-42393)

Royal Society 2005	The Royal Society, "Ocean acidification due to increasing atmospheric carbon dioxide", June 2005
Ruckelshaus 1984	William D. Ruckelshaus, "Risk in a Free Society", *Risk Analysis*, Vol. 4 No. 3, 157-162 (1984)
Runde et al. 2002	W. Runde et al., "Solubility and sorption of redox-sensitive radionuclides (Np, Pu) in J-13 water from the Yucca Mountain site: comparison between experiment and theory", *Applied Geochemistry*, Volume 17 (2002)
Sabine et al. 2004	Christopher L. Sabine et al., "The Ocean Sink for Anthropogenic CO_2", *Science*, Vol. 305, 367-371 (2004)
Sahr 2004	Robert Sahr, "Consumer Price Index (CPI) Conversion Factors 1800 to estimated 2014 to Convert to Dollars of 2004 (estimated)", Political Science Department, Oregon State University, March 1, 2004, online at http://www.orst.edu/Dept/pol_sci/fac/sahr/cf16652014.pdf
Sanders 2005	Jackie Wolcott Sanders, "How to Strengthen the NPT", *Foreign Policy Agenda* (U.S. State Department electronic journal), March 2005
Sawai 2001	Masako Sawai, "Rokkasho: A Troubled Nuclear Fuel Cycle Complex", *Science for Democratic Action*, Vol. 9 No. 4, August 2001
Scambos et al. 2004	T.A. Scambos et al., "Glacier acceleration and thinning after ice shelf collapse in the Larsen B embayment, Antarctica", *Geophysical Research Letters*, Vol. 31, L18402 (2004)
Schäfer et al. 2003	T. Schäfer et al., "Colloid-Borne Americium Migration in Gorleben Groundwater: Significance of Iron Secondary Phase Transformation", *Environmental Science and Technology*, Volume 37 (2003)
Schiermeier 2004	Quirin Schiermeier, "A rising tide", *Nature*, Vol. 428, 114-115 (2004)
Schwartz and Randall 2003	Peter Schwartz and Doug Randall, "An Abrupt Climate Change Scenario and Its Implications for United States National Security", October 2003
Schwarzenegger 2005	Arnold Schwarzenegger, "California Leading the Way on Fight Against Global Warming", Press Room, Update from the Governor's Office, June 6, 2005

SDA 1999	*Science for Democratic Action*, Volume 7 Number 3, May 1999, online at http://www.ieer.org/sdafiles/vol_7/7-3/index.html
Seaborg 1967	Glenn Seaborg, "The Promise of the International Atomic Energy Agency", *Science*, Volume 158 Number 3798, 226-230 (1967)
Sharpe 2003	Saxon Sharpe, "Future Climate Analysis - 10,000 Years to 1,000,000 Years After Present", MOD-01-001 Rev 01, March 17, 2003
Sharpe 2003b	Saxon Sharpe, "Climate: Past, Present, and Future", TR 03-001 Rev 00, October 2003
Shell 2001	Shell International, "Energy Needs, Choices and Possibilities: Scenarios to 2050", Global Business Environment, 2001, online at http://www.shell.com/static/media-en/downloads/51852.pdf
Shepherd et al. 2003	Andrew Shepherd et al., "Larsen Ice Shelf Has Progressively Thinned", *Science*, Vol. 302, 856-859 (2003)
Sissine 2004	Fred Sissine, "Renewable Energy: Tax Credit, Budget, and Electricity Production Issues", Congressional Research Service, Updated March 12, 2004 (Order Code IB10041)
Smith 2003	Brice Smith, "The 'Usable' Nuke Strikes Back", *Science for Democratic Action*, Vol. 11 No. 4, September 2003
Smith 2004	Brice Smith, "What the DOE Knows it Doesn't Know about Grout: Serious Doubts Remain About the Durability of Concrete Proposed to Immobilize High-Level Nuclear Waste in the Tank Farms at the Savannah River Site and other DOE Sites", updated October 18, 2004, online at http://www.ieer.org/reports/srs/grout.pdf
Smith et al. 2005	L.C. Smith et al., "Disappearing Artic Lakes", *Science*, Vol. 308 No. 5727, 1429 (2005)
Sokolski 2005	Henry Sokolski, "After Iran: Keeping Nuclear Energy Peaceful", *Foreign Policy Agenda* (U.S. State Department electronic journal), March 2005
SOTW 1999	Lester Brown et al., State of the World 1999: A Worldwatch Institute Report on Progress Toward a Sustainable Society, W.W. Norton & Company, New York, NY (1999)

Stacy 2000	Susan Stacy, "Proving the Principle", Idaho Operations Office of the U.S. Department of Energy, 2000 (DOE/ID-10799)
Stainforth et al. 2005	D.A. Stainforth et al., "Uncertainty in predictions of the climate response to rising levels of greenhouse gases", *Nature*, Vol. 433, January 27, 2005
State Complaint 2003	Thomas F. Reilly, Richard Blumenthal, and G. Steven Rowe. *Commonwealth of Massachusetts, State of Connecticut, and State of Main, Plaintiffs v. Christine Todd Whitman, in her capacity as Administrator of the United States Environmental Protection Agency, Defendant.* United States District Court District of Connecticut, June 4, 2003, online at http://www.ct.gov/ag/lib/ag/press_releases/2003/enviss/epa-mass.pdf
State Complaint 2004	Richard Blumenthal, Eliot Spitzer, Bill Lockyer, Thomas J. Miller, Peter C. Harvey, Patrick C. Lynch, William H. Sorrell, Peg Lautenschlager, and Michael A. Cardozo. *State of Connecticut, State of New York, People of the State of, California Ex Rel. Attorney General Bill Lockyer, State of Iowa, State of New Jersey, State of Rhode Island, State of Vermont, State of Wisconsin, City of New York, Plaintiffs, v. American Electric Power Company, Inc., American Electric Power Service Corporation, The Southern Company, Tennessee Valley Authority, Xcel Energy Inc., Cinergy Corporation, Defendants.* United States District Court for the Southern District of New York, July 21, 2004, online at http://www.oag.state.ny.us/press/2004/jul/jul21a_04_attach.pdf
State Dept 2004	U.S. Department of State, "The United States and the Republic of the Marshall Islands Proliferation Security Initiative Shipboarding Agreement", August 14, 2004 online at http://www.state.gov/r/pa/prs/ps/2004/35236.htm
State Dept 2005	U.S. Department of State, "The United States and Belize Proliferation Security Initiative Ship Boarding Agreement", August 5, 2005 online at http://www.state.gov/r/pa/prs/ps/2005/50787.htm
Stipp 2004	David Stipp, "Climate Collapse: The Pentagon's Weather Nightmare", *Fortune*, January 26, 2004

Stobaugh and Yergin 1979	Robert Stobaugh and Daniel Yergin (editors) et al., <u>Energy Future: Report of the Energy Project at the Harvard Business School</u>, Random House, New York, NY (1979)
Stocker and Schmittner 1997	Thomas F. Stocker and Andreas Schmittner, "Influence of CO_2 emission rates on the stability of the thermohaline circulation", *Nature*, Vol. 388, 862-865 (1997)
Strauss 1954	Lewis Strauss, "Remarks Prepared for Delivery at the Founders' Day Dinner, National Association of Science Writers", September 16, 1954
Suits 1951	C.G. Suits, "Power from the Atom – An Appraisal", *Nucleonics*, Vol. 8 No. 2, February 1951
Sutcliffe 1997	W. G. Sutcliffe, "Proliferation Concern With Nuclear Power", in <u>The Enduring Nuclear Fuel Cycle</u>, Carl E. Walter (editor), Proceedings of a Panel Discussion, Carl E. Walter and Robert A. Krakowski (co-chairs), American Nuclear Society Winter Meeting, Albuquerque, NM, November 18, 1997 (CONF-971125)
Takagi et al. 1997	J. Takagi et al., "Comprehensive Social Impact Assessment of MOX Use In Light Water Reactors", Final Report of the International MOX Assessment, November 1997
Thomas et al. 2004	Chris D. Thomas et al., "Extinction risk from climate change", *Nature*, Vol. 427, January 8, 2004
Thomas et al. 2004b	R. Thomas et al., "Accelerated Sea-Level Rise from West Antarctica", *Science*, Vol. 306, 255-258 (2004)
Toronto Star 2003b	Reuters, "Can't exclude strike on North Korea, Perle says", *Toronto Star*, June 12, 2003
Toulhoat 2002	P. Toulhoat, "Confinement and migration of radionuclides in a nuclear waste deep repository", *Comptes Rendus Physique*, Volume 3 (2002)
Torp and Gale 2004	Tore A. Torp and John Gale, "Demonstrating storage of CO_2 in geological reservoirs: The Sleipner and SACS projects", *Energy*, Vol. 29, 1361-1369 (2004)
Truman, Attlee, and King 1945	Harry S. Truman, C.R. Attlee, and W.L. Mackenzie King, "Declaration on Atomic Bomb by President Truman and Prime Ministers Attlee and King", November 15, 1945
TVA Factsheet	Tennessee Valley Authority, "Tritium Production at TVA: Factsheet", online at http://www.tva.gov/news/tritium.htm
TVA Watts Bar	Tennessee Valley Authority, "Watts Bar Nuclear Plant", online at http://www.tva.gov/power/nuclear/wattsbar.htm

U Chicago 2004	"The Economic Future of Nuclear Power", A Study Conducted at The University of Chicago, August 2004
U.S. Congress 1985	U.S. Congress, "NRC Authorization for Fiscal Years 1986-87: Hearing before the Subcommittee on Energy Conservation and Power of the Committee on Energy and Commerce House of Representatives", April 17, 1985 (Serial No. 99-22)
U.S. Congress 2005	U.S. Congress, "H.R. 6 – Energy Policy Act of 2005", 109th Congress First Session, signed into law on August 8, 2005
U.S. Senate 1945	United States Senate, "Report of Proceedings Hearing Held before U.S. Congress, Senate, Special Committee on Atomic Energy S. Res. 179 Volume 2", November 28, 1945
UCS 1990	Union of Concerned Scientists, "Advanced Reactor Study", prepared by MHB Technical Associates, July 1990
UCS 2003	Union of Concern Scientists, "Davis-Besse: One Year Later", March 3, 2003
UCS 2004	Union of Concerned Scientists, "Nuclear Weapon Budget Request for FY 2005", February 10, 2004, online at http://www.ucsusa.org/global_security/nuclear_weapons/page.cfm?pageID=1384#2
UIC 2002	Uranium Information Center Ltd., "Supply of Uranium", UIC Nuclear Issues Briefing Paper #75, August 2002, online at http://www.uic.com/nip75.htm
UN Disarmament	United Nations, "Treaty on the Non-Proliferation of Nuclear Weapons", online at http://disarmament2.un.org/wmd/npt/npttext.html
UNDP 2000	United Nations Development Programme, United Nations Department of Economic and Social Affairs, and the World Energy Council, "World Energy Assessment: Energy and the Challenge of Sustainability", 2000
UNHCR 2001	United Nations High Commissioner for Human Rights, "Concluding Observations of the Committee on the Elimination of Racial Discrimination: United States of America", August 14, 2001
UNICEF 2003	United Nations Children's Fund, "Hot topics; Overview: Children and Women in Iraq", March 2003 available online at http://www.unicef.org/media/media_9779.html

United Nations 2005	World Health Organization, International Atomic Energy Agency, and the United Nations Development Programme, "Chernobyl: the true scale of the accident", Joint News Release, September 5, 2005, online at http://www.who.int/mediacentre/news/releases/2005/pr38/en/index.html
USBR 2004	U.S. Bureau of Reclamation, "The Role of Hydropower Development in the U.S. Energy Equation", Revised October 12, 2004 online at http://www.usbr.gov/power/edu/hydrole.html
USEC 2003	U.S. Enrichment Corporation, "Securities and Exchange Commission Form 10-K, Transition Report Pursuant to Section 13 OR 15 (d) of the Securities Exchange Act of 1934 for the six-month period ended December 31, 2002", March 3, 2003 online at http://media.corporate-ir.net/media_files/NYS/USU/reports/10k_1202_FINAL1.pdf
USEC 2005	United States Enrichment Corporation, "Progress Report: US-Russian Megatons to Megawatts Program, Recycling Nuclear Warheads into Electricity", as of June 30, 2005, online at http://www.usec.com/v2001_02/HTML/Megatons_status.asp (last viewed on August 15, 2005)
Ux Weekly 2004	Ux Consulting, "Ux Weekly: Third Quarter Spot Conversion & SWU Review", Vol. 18 No. 41, October 11, 2004, online at http://www.uxc.com/products/UxW18-41.pdf
Ux Weekly 2005	Ux Consulting, "Ux Weekly: First Quarter Spot U_3O_8 Review", Vol. 19 No. 15, April 11, 2005, online at http://www.uxc.com/products/UxW19-15.pdf
van der Zwaan and Rabl 2003	Bob van der Zwaan and Ari Rabl, "Prospects for PV: a learning curve analysis", Solar Energy, Vol. 74, 19-31 (2003)
Vandegrift et al. 2004	George F. Vandegrift et al., "Designing and Demonstration of the UREX+ Process Using Spent Nuclear Fuel", Final Paper for Presentation at: ATATLANTE 2004, Advances for Future Nuclear Fuel Cycles, International Conference, Nimes, France. June 21-24, 2004
Walther et al. 2002	Gian-Reto Walther, et al., "Ecological responses to recent climate change", *Nature*, Vol. 416, March 28, 2002
Washington Post 1979a	Associated Press, "Hundreds Arrested at Plant Sites", *Washington Post*, June 3, 1979

Washington Post 1979b	Karlyn Barker, "Nearly 1,000 Rally at Vepco Plant", *Washington Post*, June 3, 1979
Washington Post 1981a	Joanne Omang, "Calif. A-Plant Faces License Suspension", *Washington Post*, November 18, 1981
Washington Post 1981b	Michael Berlin, "U.S., Iraqis Agree on U.N. Resolution Condemning Israel", *Washington Post*, June 19, 1981
Washington Post 1981c	News Services and Staff Reporters, "Record Number of Arrests at A-Plant", *Washington Post*, September 23, 1981
Washington Post 1982	Associated Press, "South African Reactor Damaged by 4 Explosions; Sabotage Cited", *Washington Post*, December 20, 1982
Washington Post 1988	Associated Press, "State Reaches Tentative Deal to Tear Down Shoreham Plant", *Washington Post*, May 27, 1988
Washington Post 1999	John Deutch, Henry Kissinger, and Brent Scowcroft, "Test-Ban Treaty: Let's Wait a While", *Washington Post*, October 6, 1999
Washington Post 2002	Aluf Benn, "Where First Strikes Are Far From the Last Resort", *Washington Post*, November 10, 2002
Washington Post 2002b	Lois Romano and Ellen Nakashima, "Barge Hits Bridge on I-40 in Okla; Rescuers Search River; Several Feared Dead", *Washington Post*, May 27, 2002
Washington Post 2003	Jim Hoagland, "Israel's Red Flag on Iran", *Washington Post*, August 13, 2003
Washington Post 2004a	Peter Slevin, "Brazil Shielding Uranium Facility: Nation Seeks to Keep its Proprietary Data from U.N. Inspectors", *Washington Post*, April 4, 2004
Washington Post 2004b	Macela Sanchez, "Nuclear Spat Stokes Brazil's Resentment", *Washington Post*, April 8, 2004
Washington Post 2004d	Michael Amon and Jamie Stockwell, "At Crash Site, Exploring Why 4 Didn't Come Home", *Washington Post*, January 18, 2004
Washington Post 2004e	John Ward Anderson, "Nuclear Spy, Icon Release in Israel; Vanunu Exposed Weapons Program", *Washington Post*, April 22, 2004
Washington Post 2005b	Robin Wright, "U.S. Wants Guarantees on Iran Effort; Support for U.N. Action Sought if Theran Does Not Abandon Nuclear Program", *Washington Post*, March 4, 2005

Washington Post 2006	Peter Finn, "Russia Cuts Off Gas to Ukraine In Controversy Over Pricing, Move Raises Concerns About Energy Supply for Other European Countries", Washington Post, January 2, 2006
Washington Post 2006b	Peter Finn, "Russia and Ukraine Reach Deal on Gas, Ending Dispute", Washington Post, January 5, 2006
Washington Post 2006c	John Deutch and Ernest J. Moniz, "A Plan for Nuclear Waste", Washington Post, January 29, 2006
Washington Post 2006d	Peter Baker, Dafna Linzer, and Thomas E. Ricks, "U.S. is Studying Military Strike Options on Iran", Washington Post, April 9, 2006
Washington State	Sam Reed, "Washington State 2004 General Elections", Washington Secretary of State, online at http://vote.wa.gov/general/measures.aspx?a=297#map
Watts Bar Factsheet	U.S. Department of Energy, "U.S. Nuclear Plants: Watts Bar, Tennessee", Energy Information Administration, page last modified on Fri Mar 18 2005 online at http://www.eia.doe.gov/cneaf/nuclear/page/at_a_glance/reactors/wattsbar.html
Weinberg 1994	Alvin Weinberg, The First Nuclear Era: The Life and Times of a Technology Fixer, AIP Press, New York (1994)
Weiss and Bradley 2001	Harvey Weiss and Raymond Bradley, "What Drives Societal Collapse?", *Science*, Vol. 291 No. 5504, p. 609-610 (2001)
Whatley 2000	Laurie Whatley, "Acid-gas injection proves economic for West Texas gas plant", *Oil & Gas Journal*, Vol. 98 No. 21, 58-61 (2000)
White House 1993	"President Clinton Fact Sheet on Nonproliferation And Export Control Policy", Office of the Press Secretary, September 27, 1993, online at http://www.rertr.anl.gov/REFDOCS/PRES93NP.html
White House 2002	Office of the Press Secretary, "Fact Sheet: President Bush Announces Clear Skies & Global Climate Change Initiatives", February 14, 2002
White House 2002b	George W. Bush, "The National Security Strategy of the United States of America", September 2002

White House 2002c	George W. Bush, "National Strategy to Combat Weapons of Mass Destruction", December 2002
White House 2003b	White House Office of the Press Secretary, "Proliferation Security Initiative: Statement of Interdiction Principles", September 4, 2003 online at http://www.state.goc/t/np/rls/fs/23764.htm
White House 2004	George W. Bush, "President Announces New Measures to Counter the Threat of WMD", Remarks at the National Defense University, February 11, 2004
White House 2005	George W. Bush, "President Discusses Energy at National Small Business Conference", April 27, 2005
White House 2005b	"Joint Statement Between President George W. Bush and Prime Minister Manmohan Singh", July 18, 2005, online at http://www.whitehouse.gov/news/releases/2005/07/print/20050718-6.html
White House 2006	George W. Bush, "The National Security Strategy of the United States of America", March 2006
WHO 2003	World Health Organization, "Climate Change and Human Health – Risks and Responses: Summary", 2003
Wichert and Royan 1997	Edward Wichert and Tom Royan, "Acid gas injection eliminates sulfur recovery expense", *Oil & Gas Journal*, Vol. 95 No. 17, 67-72 (1997)
Wilson et al. 1985	R. Wilson et al., "Report to the American Physical Society of the study group on radionuclide release from severe accidents at nuclear power plants", *Reviews of Modern Physics*, Volume 57 Number 3 Part II, July 1985
WIPP Factsheet	Waste Isolation Pilot Plant, "WIPP Milestones", U.S. Department of Energy Carlsbad Field Office, online at http://www.wipp.energy.gov/fctshts/Milestones.pdf
Wise 1999	M. Wise, "Management of UKAEA Graphite Liabilities". IAEA Technical Committee Meeting on "Nuclear Graphite Waste Management", held from 18-20 October 1999 in Manchester, United Kingdom, International Atomic Energy Agency, Vienna (Austria), online at http://www.iaea.org/inis/aws/htgr/fulltext/manchester_16.pdf

WISE 2001	WISE International, "Annual data on 'civilian' plutonium per country", online at http://www.wise-paris.org/english/ourfigures/011231stock_pu_iaea_en.htm
World Bank 1991	World Bank Environment Department, "Work Bank Technical Paper Number 154: Environmental Assessment Sourcebook Volume III Guidelines for Environmental Assessment of Energy and Industry Projects", 1991
Zaidi 2005	Ammar Zaidi, "Iran piped gas to be cheaper than shipping LPG", *rediff.com*, June 8, 2005, online at http://in.rediff.com/money/2005/jun/08gas.htm
Zerriffi and Makhijani 2000	Hisham Zerriffi and Annie Makhijani, "The Nuclear Alchemy Gamble: An Assessment of Transmutation as a Nuclear Waste Management Strategy", Institute for Energy and Environmental Research, August 25, 2000

Endnotes

[1] Strauss 1954

[2] The most important greenhouse gases are: carbon dioxide, methane, nitrous oxide, hydrofluorcarbons, perfluorocarbons, and sulfur hexafluoride. Typically, emissions of greenhouse gases are reported in terms of "carbon equivalent" which is a way of relating how effective the various chemicals are at trapping heat relative to carbon dioxide. Carbon dioxide was chosen as the standard because is the most important greenhouse gas and was responsible for more than 84% of all effective emissions in 2000. The next highest contributor was methane which was responsible for just over 9% of all effective emissions in 2000. [EIA 2003 in Tables 4, 5, 13, 23, and 30]

[3] Meier 2002 p. i-iii and 25-27

[4] As of February 6, 2006 the Kyoto Protocol has been ratified by 160 countries or regional economic organizations (including Russia and all of the European Union). These countries account for more than three-fifths of global carbon emissions. The United States, along with Australia, Monaco, Croatia, Kazakhstan, and Zambia, have signed, but not ratified the treaty. [Kyoto Status 2005] Significantly, the Bush administration has publicly declared its hostility to the treaty and has decided to pursue its own unilateral strategies regarding carbon emissions. This is particularly important given the fact that the United States alone accounts for more than half of all carbon emissions from countries that are not bound by the Kyoto Protocol. For a further discussion see *Rule of Power or Rule of Law?: An Assessment of U.S. Polices and Actions Regarding Security-Related Treaties* [Deller, Makhijani, and Burroughs 2003 p. 101-112].

[5] [IPCC 2001c p. 65] The Intergovernmental Panel on Climate Change is an organization of the world's leading climate scientists chartered through the World Meteorological Organization and the United Nations Environment Program in 1988.

[6] In this work we will use MW to stand for megawatt-electric (i.e. the rated electric power of a generating station) and MWt to stand for megawatt-thermal (i.e. the rated thermal power of a generating station). For typical nuclear plants in operation today, the electric generating power is approximately one-third of the reactor's thermal power.

[7] Strauss 1954

[8] as quoted in Ford 1982 p. 30

[9] as quoted in Ford 1982 p. 30

[10] as quoted and described in Ford 1982 p. 23-24

[11] For a further discussion see Chapter Three of *Nuclear Power Deception* by Arjun Makhijani and Scott Saleska. This work was the first to discover the information presented on the internal view of nuclear power by those within Congress and the nuclear complex. [Makhijani and Saleska 1999 p. 53-69]

[12] Robbins, Makhijani, & Yih 1991 p. 75 and New York Times 1954b

[13] New York Times 1954

[14] Robbins, Makhijani, & Yih 1991 p. 75-76

[15] New York Times 1954c

[16] New York Times 1954b and Robbins, Makhijani, & Yih 1991 p. 78
[17] For a further discussion see [Makhijani and Saleska 1999 p. 53-69]
[18] Eisenhower 1953
[19] Eisenhower 1953
[20] Eisenhower 1953
[21] New York Times 1954b
[22] [Holloway 1994 p. 352-353] In understanding the evolution of nuclear power it is important to note that some of those involved in its earliest stages viewed the risks posed by radiation in a very different way than we do today. For example, one British representative to the Geneva Conference on the Peaceful Uses of Atomic Energy said that, in light of the growing world population, "diminished fertility and shortened life-span [due to the potential radiation risks associated with nuclear power] might not be altogether to be deplored." [as quoted in Lansdell 1958 p. 177]
[23] For a further discussion see [Makhijani and Saleska 1999 p. 53-69].
[24] Holloway 1994 p. 346-347
[25] Makhijani and Saleska 1999 p. 85-86
[26] Makhijani and Saleska 1999 p. 87
[27] The contracts were known as "turnkey" contracts because the utilities agreed to pay a fixed price for the power plant no matter what the actual construction cost was, and once the contractor had completed the plant all the utility would have to do is turn the key and start generating electricity. [Makhijani and Saleska 1999 p. 87-88]
[28] According to the National Research Council report, the least expensive nuclear plant examined by the Committee that came online between 1985 and 1988 cost $1,342 per kW (in 1988 dollars) while the most expensive plant that became operational between 1971 and 1976 cost $1,284 per kW (in 1988 dollars). [NAS/NRC 1992 p. 31]
[29] NAS/NRC 1992 p. 31
[30] NAS/NRC 1992 p. 33
[31] U.S. Senate 1945 p. 122-123
[32] Acheson and Lilienthal 1946 p. 4
[33] Lilienthal et al. 1947 p. 203
[34] The treaty was opened for signatures on July 1, 1968, and came into force on March 5, 1970. [UN Disarmament] It currently has 189 signatories which is the largest number of parties to any international agreement outside of the United Nations Charter. Only India, Pakistan, and Israel have never signed onto the treaty. In January 2003, North Korea withdrew from the NPT without providing the required three month notification.
[35] Bergeron 2002 p. 29-30, and 32
[36] London Times 1981
[37] Washington Post 1981b and New York Times 1981
[38] London Times 2004
[39] NAS/NRC 1992 p. 33
[40] Komanoff and Roelofs 1992 p. 23

[41] CBO 2003 p. 11 and Komanoff and Roelofs 1992 p. iii
[42] Watts Bar Factsheet and TVA Watts Bar
[43] Forbes 1985
[44] Hansen and Lebedeff 1987
[45] Rampton and Stauber 2000 p. 270-283 and Faltermayer 1988 p. 105
[46] World Bank 1991 p. 84
[47] NEI 2006
[48] DOE 2001 p. iv
[49] DOE 2001 p. 1
[50] Holt 2005b p. CRS-4
[51] Guardian 2004
[52] Cheney et al. 2001 p. xi, xiii, and 1-6
[53] Deller, Makhijani, and Burroughs 2003 p. 106-110
[54] Arrhenius 1896
[55] For example, between 1900 and 1997 the world's consumption of coal increased by more than four fold while the use of oil increased by more than 160 fold, and natural gas usage increased by more than 240 fold. Taken together, these three fossil fuels accounted for approximately three-fourths of all energy used in 1997 as compared to less than three-fifths in 1900. [SOTW 1999 p. 23]
[56] For example, the proposal put forward by the Bush administration in February 2002 would lead to an increase in greenhouse gas emissions by 2012 of nearly 24 percent over their 1990 levels assuming 3 percent annual growth in GDP. For comparison, the Kyoto Protocol, which is a very small first step towards addressing the threat of climate change, would have required the U.S. to achieve a reduction in emissions by the 2008-2012 period of 7 percent over its 1990 level. [White House 2002, EIA 2004e p. 335, and Deller, Makhijani, and Burroughs 2003 p. 104]
[57] State Complaint 2003
[58] The five companies named in the suit are: American Electric Power Co., Cinergy Corp., Southern Company, the Tennessee Valley Authority, and Xcel Energy Inc. Together they are the largest carbon emitters in the U.S. electricity sector. [State Complaint 2004]
[59] Mayors 2005
[60] Schwarzenegger 2005
[61] IPCC 2001f p. 2,10
[62] IPCC 2001f p. 7 and 12 and IPCC 2001d p. 527
[63] IPCC 2001f p. 13
[64] IPCC 2001e p. 642
[65] Stainforth et al. 2005 p. 403 and 404
[66] See for example Murphy et al. 2004 and Frame et al. 2005.
[67] Bryant 1997 p. 90 and BBC 2004
[68] IPCC 2001i p. 958-959
[69] See for example Knutson, Tuleya, & Kurihara 1998, Knutson et al. 2001, and Knutson and Tuleya 2004
[70] Palm Beach Post 2005

[71] Associated Press 2005, Associated Press 2005b, and Associated Press 2005c

[72] "The Millennium Ecosystem Assessment was called for by United Nations Secretary-General Kofi Annan in 2000… The MEA was conducted under the auspices of the United Nations, with the secretariat coordinated by the United Nations Environment Programme, and it was governed by a multistakeholder board that included representatives of international institutions, governments, business, NGOs, and indigenous peoples. The objective of the MA was to assess the consequences of ecosystem change for human well-being and to establish the scientific basis for actions needed to enhance the conservation and sustainable use of ecosystems and their contributions to human well-being." [Millennium Assessment 2005 p. ii]

[73] Millennium Assessment 2005 p. 1

[74] Millennium Assessment 2005 p. 17

[75] IPCC 2002 p. 22

[76] New Scientist 2005, New Scientist 2005b, and Smith et al. 2005

[77] ACIA 2004 in Preface

[78] ACIA 2004 p. 59 and 61

[79] ACIA 2004 p. 58

[80] ACIA 2004 p. 70-73

[81] IPCC 2002 p. 24

[82] Thomas et al. 2004 p. 145 and 147

[83] Walther et al. 2002 p. 393-394 and Thomas et al. 2004 p. 147

[84] Thomas et al. 2004 p. 147

[85] Sabine et al. 2004 p. 367 and 370

[86] Portner, Langenbuch, and Reipschlager 2004 p. 706 and Jacobson 2005 p. 1 and 12

[87] [Sabine et al. 2004 p. 367-368]. The Thermohaline Circulation (THC), also known as the Gulf Stream or North Atlantic "Conveyor," is a complex series of ocean currents that transport heat from the equatorial latitudes northwards in the Atlantic ocean. It is this heat transport that makes northern and western Europe far warmer than it would otherwise be given its latitude, and it has a major influence on global weather patterns. Although its Atlantic branch is the one most often focused on, the THC is a complex system that extends to Antarctica and into the Indian and Pacific Oceans as well. [see for example Broecker 1997]

[88] Kurihara, Shimode, and Shirayama 2004 p. 743

[89] Jacobson 2005 p. 12 and Royal Society 2005 p. 14

[90] Portner, Langenbuch, and Reipschlager 2004 p. 705 and 711, Ishimatsu et al. 2004 p. 731 and 737-739, and Kurihara, Shimode, and Shirayama 2004 p. 743-744 and 746-748

[91] Ishimatsu et al. 2004 p. 732-733

[92] Kurihara, Shimode, and Shirayama 2004 p. 748

[93] Ishimatsu et al. 2004 p. 733-734

[94] Portner, Langenbuch, and Reipschlager 2004 p. 715 and Kurihara, Shimode, and Shirayama 2004 p. 748

[95] Dulvy, Sadovy, & Reynolds 2003 p. 25

[96] Millennium Assessment 2005 p. 12, 27, and 32

[97] More than 130 local, regional, or global extinctions of marine populations have been identified with more than 90 percent attributable to human exploitation or habitat loss. For example, only eight individual white abalone are now known to exist in an area of the ocean that two decades ago was home to a population of 16,000 to 820,000. The last known successful breeding event of this species occurred more than 35 years ago. [Dulvy, Sadovy, & Reynolds 2003 p. 25 and 47]

[98] Kurihara, Shimode, and Shirayama 2004 p. 748

[99] Walther et al. 2002 p. 394, Royal Society 2005 p. 25-27, IPCC 2002 p. 20-22, and Dulvy, Sadovy, & Reynolds 2003 p. 37-38

[100] Royal Society 2005 p. 23

[101] IPCC 2001g p. 453

[102] WHO 2003 p. 7 and 17

[103] WHO 2003 p. 19

[104] [IPCC 2001g p. 477 and IPCC 2002 p. 25-27] The terms Global North and Global South refer to the countries that are typically labeled as the "developed" world and the "developing" world respectively. The countries of the Global South, which contain the majority of the world's population, are also sometimes referred to as the "Third World." The concept of a Global North and South was put forth in 1980 in a report by the Independent Commission on International Development Issues entitled *North - South: A Program for Survival* which called for a re-assessment of the concept of development as well as for more equitable economic relations between countries. [Brandt et al. 1980]

[105] See for example Hodell, Curtis, and Brenner 1995, deMenocal 2001, and Weiss and Bradley 2001.

[106] NAS/NRC 2002 p. 1

[107] DTI 2003 p. 7-8

[108] See for example Broecker 1997, Stocker and Schmittner 1997, Rahmstorf and Ganopolski 1999, Rahmstorf 2000, and Rahmstorf and Zickfeld 2005

[109] IPCC 2001d p. 562-563

[110] See for example BBC 2002b, Shepherd et al. 2003, Thomas et al. 2004b, Rignot et al. 2004, and Scambos et al. 2004

[111] BAS Press Release 2005

[112] See for example Huybrechts and de Wolde 1999, Schiermeier 2004, Krabill et al. 2004, and Hansen 2005

[113] Bryant 1997 p. 90

[114] Bryant 1997 p. 94

[115] Bryant 1997 p. 157

[116] Bryant 1997 p. 157-158

[117] Stipp 2004

[118] Schwartz and Randall 2003 p. 2, 8, 13

[119] Schwartz and Randall 2003 p. 2

[120] Schwartz and Randall 2003 p. 17-18

[121] Schwartz and Randall 2003 p. 19

[122] Schwartz and Randall 2003 p. 19

[123] The report was prepared by an interdisciplinary committee consisting of nine professors from eight departments at the Massachusetts Institute of Technology. The committee was co-chaired by Professors John Deutch (chemistry) and Ernest Moniz (physics), both of whom are former Under Secretaries in the U.S. Department of Energy. The complete text of the report is available online at http://web.mit.edu/nuclearpower. [MIT 2003 p. iii-iv]

[124] MIT 2003 p. 1

[125] In the year 2000, the total amount of electricity generated worldwide by nuclear power was 2,230 billion kWh. This is equivalent to the electricity that would be produced by approximately 300 GW of nuclear power operating at an 85 percent capacity factor. [MIT 2003 p. 115]

[126] MIT 2003 p. 3 and 49

[127] MIT 2003 p. 26

[128] MIT 2003 p. 112 to 115

[129] The MIT study does not include an estimate for future CO_2 emissions from electricity production either globally or for the United States. The DOE's Energy Information Administration predicts that the average annual growth rate of carbon emission for the whole United States through 2025 will be more than 1.4 percent. [EIA 2005 p. 55] Historically, we note that between 1990 and 2000, the CO_2 emissions from the electricity sector rose an average of 2.4 percent per year with a minimum annual change of -0.2 percent (1990-1991) and a maximum increase of +5 percent (1999-2000). The MIT study predicts that U.S. electricity consumption will increase at a rate of 1.68 percent per year between 2000 and 2050, which is broadly consistent with the EIA's estimate. [MIT 2003 p. 112] Using a growth rate for CO_2 emissions of 1.4 percent as a guide, we estimated that, in a business-as-usual scenario, the 2050 emissions from the U.S. electricity sector would be approximately 1,240 million metric tons of carbon as compared to 619 million tons in 2000. [EIA 2003 in Tables 4, 5, 13, 23, and 30] This is consistent with the MIT study's assumption that CO_2 emissions from all sources globally are expected to more than double by 2050. [MIT 2003 p. 3] If 300 GW of new nuclear power was built in the U.S., it would reduce the projected emissions in 2050 to 700 to 1,000 million metric tons. The range of emissions depends upon whether the new nuclear power plants displaced coal or natural gas fired capacity. If a portion of the new nuclear plants were built in lieu of renewable energy sources such as wind or solar, the projected increase in carbon emissions would be proportionally higher.

[130] [MIT 2003 p. 115]. The rate of electricity growth assumed in the MIT analysis is well below the 3.8 percent average annual growth rate in electricity consumption experienced between 1971 and 2000. A more rapid growth in electricity demand would lead to an even greater number of nuclear reactors being required. [IEA 2002 p. 411]

[131] The information about current levels of generation and the relative levels of CO_2 emissions per kWh used in this calculation were taken from [IEA 2002 p. 411 and 413 and MIT 2003 p. 3 and 115]

[132] MIT 2003 p. 40-41
[133] Forbes 1985
[134] NRC 1999, Makhijani and Saleska 1999 p. 138-139, Holt 2005b p. CRS-5, and NRC 2006b
[135] This assumes 200 light-water reactors of one gigawatt capacity each and an additional 286 to 800 HTGRs with capacities of 125 to 350 MW each. [MIT 2003 p. 49]
[136] MIT 2003 p. 93
[137] Weinberg 1994 p. 139-141
[138] [Makhijani and Saleska 1999 p. 140-141 and 152 and EIA 2002] The capacity factor of a power plant is the ratio of the average amount of electricity generated to the amount of electricity that could have been generated over that same time if the plant had run at its maximum power.
[139] DOE 2002c p. 17
[140] For example, 50 years after disposal, the heat output of spent MOX fuel with a burnup of 43 megawatt-days per ton is more than three times that of spent LEU fuel with the same burnup. [Bunn et al. 2003 p. 39]. For a further discussion see Section 5.4.2 and [Makhijani 2001].
[141] MIT 2003 p. 4-5 (emphasis in original)
[142] MIT 2003 p. 146 and 148 and NAS 1995 p. 302
[143] Takagi et al. 1997 p. 163 and 166
[144] Bunn et al. 2003 p. 45-51, MIT 2003 p. 151, and Takagi et al. 1997 p. 164 and 166
[145] MIT 2003 p. 42 and 148, U Chicago 2004 p. 9-5 and A5-8, Charpin, Dessus, and Haut 2000 p. 34, 56, and 219-221, and Takagi et al. 1997 p. 166 and 169
[146] Takagi et al. 1997 p. 164 and 166
[147] MIT 2003 p. 40 (emphasis added)
[148] MIT 2003 p. 40 and 42-43
[149] EIA 2006 p. 65 and 67
[150] MIT 2003 p. 40-41 (emphasis added)
[151] U Chicago 2004 p. 5-17, 5-24, 5-25
[152] U Chicago 2004 p. 5-24 to 5-25 and 9-5 to 9-6
[153] U Chicago 2004 p. 5-25
[154] See for example EIA 1992, EIA 1999, Goldberg 2000, Komanoff and Roelofs 1992, Sissine 2004, Dubin & Rothwell 1990, Heyes & Liston-Heyes 1998, Heyes & Liston-Heyes 1998b, and Heyes 2002
[155] Ramana, D'Sa, and Reddy 2005 p. 1764 and 1770
[156] Ayres, MacRae, and Stogran 2004 p. 15
[157] U Chicago 2004 p. 3-2 to 3-3, 5-17, 9-1, 9-10, 9-14, and 9-18
[158] MIT 2003 p. 43 and U Chicago 2004 p. 3-16 and 9-5 to 9-6
[159] EIA 1986 p. ix
[160] EIA 1986 p. ix
[161] NAS/NRC 1992 p. 31
[162] MIT 2003 p. 40
[163] NAS/NRC 1992 p. 33

[164] NAS/NRC 1992 p. 34
[165] The plants which began construction after 1993 considered by the IAEA were built in Japan, South Korea, China, and India. The shortest construction time among these reactors was 4.5 years, while the longest was 7.2 years. [U Chicago 2004 p. 2-13]
[166] U Chicago 2004 p. 5-17
[167] EIA 1986 p. xv
[168] EIA 1986 p. xvi
[169] CBO 2003 p. 11
[170] IEA 2005c p. 29-30 and 49-50
[171] MIT 2003 p. 141-142
[172] [MIT 2003 p. 45 and 139] The euro to dollar conversion rate has varied a great deal since its introduction from a low of 0.829 on October 26, 2000 to a high of 1.36 on December 31, 2004. The conversion rate used in our estimate was 1.20, which is equal to the average rate between January 1, 2003 and February 1, 2006. [FXHistory] The MIT study used a lower conversion rate of 1.0 in converting the Finnish estimate to U.S. dollars; however, our estimate of 1.20 is more likely to be an underestimate than an overestimate of the actual long-term exchange rate. [Makhijani and Smith 2004 p. 39-40]
[173] The twelve countries examined by the OECD were: Belgium, Canada (Central and Eastern), Finland, France, West Germany, Italy, Japan, Netherlands, Norway, Spain, the United Kingdom, and the United States (Central, Eastern, Rocky Mountain regions). For countries with multiple estimates of nuclear power construction costs we considered the average in comparing it to the estimated cost of coal fired plants in those countries. [OECD 1986 p. 22-23]
[174] OECD 1986 p. 22-23
[175] The six countries that were kept from the 1986 OECD study were: Canada, Finland, France, Japan, Spain, and the United States, while the six new countries considered in the 1998 update were: Brazil, China, India, Russia, South Korea, and Turkey. [OECD 1998 p. 54-57]
[176] OECD 1998 p. 54-57
[177] U Chicago 2004 p. 9-5 to 9-6
[178] MIT 2003 p. 38
[179] NAS/NRC 1992 p. 33
[180] Komanoff and Roelofs 1992 p. 23
[181] See for example [Komanoff 1981 p. 271]
[182] CBO 2003 p. 11 and Komanoff and Roelofs 1992 p. iii
[183] U Chicago 2004 p. 5-6
[184] [New York Times 2005d] Standard & Poor's and Moody's are two of the most influential credit rating, research, and risk analysis agencies in the United States.
[185] U Chicago 2004 p. 5-19
[186] [MIT 2003 p. 45 and 139-140]. At recent treasury rates (March 1998 through March 2004) this would be equivalent to a real interest rate of just 4.1 to 7.6 percent. [ICMA-RC 2004]

[187] MIT 2003 p. 43, 132, and 135 and U Chicago 2004 p. 5-17, 5-21, and 5-23 to 5-24
[188] U Chicago 2004 p. 5-21 (emphasis added)
[189] For example, a 50 percent loan guarantee and a risk free rate of 5 percent, would reduce the effective interest rate for new nuclear plants to just 8.25 to 8.75 percent under the MIT and U Chicago assumptions respectively. These rates would both be below the financing rates assumed for new fossil fuel plants.
[190] CBO 2003 p. 10-11
[191] CBO 2003 p. 12
[192] CBO 2003 p. 12
[193] CBO 2005 p. 7-9 and U.S. Congress 2005 Sections 1702 and 1703
[194] Makhijani and Saleska 1999 p. 88
[195] MIT 2003 p. 8 and 42
[196] Committee on Energy and Commerce 2005 p. Title XIII – 58 to 61
[197] The calculation assumes the reactors can be built for $2,000 per kW as assumed in the MIT base-case analysis.
[198] This would not be the full length of time over which the subsidies could be paid out, and thus the actual total could be even higher. [EIA 2004f p. 7-8]
[199] Holt 2005c p. CRS-3 and U.S. Congress 2005 Section 1306
[200] Kennedy et al. 2006
[201] U Chicago 2004 p. 9-1, 9-6, and 9-9
[202] U Chicago 2004 p. 9-10
[203] White House 2005
[204] U.S. Congress 2005 Section 638
[205] U.S. Congress 2005 Section 638
[206] Kennedy et al. 2006
[207] MIT 2003 p. 41 (emphasis in the original)
[208] U Chicago 2004 p. 9-6 to 9-8
[209] U Chicago 2004 p. 9-10
[210] U Chicago 2004 p. 9-11 to 9-13 and 9-15
[211] EIA 2005 p. 93
[212] MIT 2003 p. 8
[213] U.S. Congress 2005 Section 1301
[214] In 2003, there was an estimated 763.7 billion kWh of nuclear electricity generated. In 2050, with 300 GW of capacity running at 85 percent effective capacity, there would be a total of 2235.3 billion kWh of electricity generated from nuclear power. [EIA 2004e p. 224]
[215] Goldberg 2000 p. 7
[216] EEI 2003 p. 60
[217] MIT 2003 p. 40 and 43
[218] IPCC 2001b p. 242
[219] PIU 2002 p. 108
[220] EIA 2003 in Table 5
[221] EIA 2002b p. 6
[222] EIA 2003 in Table 5

[223] EIA 2004c p. 22, 24, 30, and 36
[224] This last point was noted by the authors of the MIT study. Specifically, they noted that: "With carbon taxes at these high levels, it could become economical to deploy a generating technology involving the gasification of coal, its combustion in a CCGT (IGCC), and the sequestration of carbon dioxide produced in the process." [MIT 2003 p. 42-43]
[225] MIT 2003 p. 42, 112
[226] BMU 2003 p. 6 and 8, DTI 2003 p. 8, and Schwarzenegger 2005
[227] In the United States, the electricity sector is the dominant source of carbon emissions, however, the releases from other areas of the economy are far from insignificant. In 2000, the transportation sector in the U.S. emitted 504 million metric tons of carbon (32 percent of the country's total emissions) compared to approximately 619 million tons from the electricity sector (39 percent of the total). If the U.S. transportation sector was a country, it would have been the third largest emitter in the world for 2000, behind only the rest of the U.S. economy, with approximately 1,077 million tons of carbon emissions, and China, with approximately 762 million tons of carbon emissions. In that year, greenhouse gas emissions from the U.S. transportation sector alone were 28 percent more than from all fossil fuel consumption in Russia, 73 percent more than from India, and more than 5 times the emissions from Brazil. [EIA 2003 in Table 5 and Marland, Boden, and Andres 2003 in fact sheets on USA, China, Russia, India, and Brazil]
[228] For example see ASE et al. 1997, UNDP 2000 p. 174-217, NRDC 2001, Makhijani 2001b, and Clemmer et al. 2001
[229] Makhijani and Poole 1975
[230] as quoted in [Stobaugh and Yergin 1979 p. 143]
[231] New York Times 1978
[232] Stobaugh and Yergin 1979 p. 137
[233] OTA 1993 p. 5
[234] OTA 1993 p. 9
[235] as quoted in ACEEE R&D Factsheet
[236] ACEEE R&D Factsheet
[237] NAS/NRC 2001b p. 36-37
[238] NAS/NRC 2001b p. 4-6, 29, and 37
[239] EIA 2003c p. 7, and 52-54
[240] EIA 2003c p. 7, and 52-54
[241] In 2000, the total electricity consumption of Papua New Guinea, Ghana, Ivory Coast, Cameroon, Kenya, Myanmar, Senegal, Tanzania, Yemen, Sudan, Nepal, Democratic Republic of the Congo, Angola, Uganda, and Ethiopia was equal to 41.4 billion kWh, while the combined population of these 15 countries was 400 million people. [MIT 2003 p. 114]
[242] ACEEE Factsheet
[243] PIU 2002 p. 98
[244] EIA 2004e p. 149, 151, 189, and 205
[245] NREL 2005 p. 57-59 and 64

[246] Rosen and Meier 1999 p. ii
[247] This analysis assumes that the energy replaced either by nuclear power or through a reduction in demand would have been generated by combined cycle natural gas fired plants. [PIU 2002 p. 108]
[248] DTI 2003 p. 16
[249] IPCC 2001b p. 246, 248
[250] In order from largest potential to smallest, the 12 top states are: North Dakota (1,210 billion kWh), Texas (1,190 billion kWh), Kansas (1,070 billion kWh), South Dakota (1,030 billion kWh), Montana (1,020 billion kWh), Nebraska (868 billion kWh), Wyoming (747 billion kWh), Oklahoma (725 billion kWh), Minnesota (657 billion kWh), Iowa (551 billion kWh), Colorado (481 billion kWh), and New Mexico (435 billion kWh). [Makhijani et al. 2004 p. 14]
[251] Makhijani et al. 2004 p. 14
[252] [Archer and Jacobson 2003 p. 10-6 and Archer and Jacobson 2004 p. 2] Class 3 winds average between 6.9 and 7.5 meters per second while Class 4 winds average between 7.5 and 8.1 meters per second. Class 3 sites are considered economically suitable for wind development at current costs, while Class 4 sites or higher are considered to be highly favorable with higher average capacity factors and lower costs.
[253] Archer and Jacobson 2003 p. 10-7 and Archer and Jacobson 2004 p. 2
[254] EIA 2004e p. 224
[255] NREL 2005 p. 40
[256] CIA World Factbook, MIT 2003 p. 112, and AWEA 2004 p. 2
[257] AWEA 2004 p. 6
[258] NREL 2005 p. 40
[259] IPCC 2001b p. 245-247
[260] NRC 1996c Table 8.2
[261] NREL 2005 p. 43
[262] NRC 1996c Table 8.2
[263] Shell 2001 p. 9, 38
[264] DTI 2003 p. 12 and 55 and Greenpeace 2002 p. iii
[265] The U.S. grid actually consists of three grids that are weakly linked together and stretch outside the continental United States at points. These three grids are known as the Eastern Interconnected System, the Western Interconnected System, and the Electric Reliability Council of Texas. [DOE 2002f p. 2-4]
[266] NREL 2005 p. 39
[267] NAS/NRC 2004 p. 228
[268] For a discussion of grid integration costs associated with wind power see [Makhijani et al. 2004 p. 16-18]
[269] Makhijani et al. 2004 p. 18-20
[270] PIU 2002 p. 121
[271] Makhijani et al. 2004 p. 18
[272] Jacobson and Masters 2001 p. 1438, IPCC 2001b p. 256 and 259, IEA 2003 p. 152 and 165, Makhijani et al. 2004 p. 19-21, NREL 2005 p. 37 and 39, and EIA 2005 p. 59-60 and 127

[273] NREL 2005 p. 43
[274] IEA 2003 p. 152 and 165
[275] NREL 2005 p. 39 and 43
[276] IEA 2003b
[277] MIT 2003 p. 132
[278] REPP 1996 p. 5
[279] AWEA Factsheet p. [4]
[280] MIT 2003 p. 168
[281] The countries included in the survey were: Argentina, Australia, Britain, Cameroon, Canada, France, Germany, Hungary, India, Indonesia, Japan, Jordan, Mexico, Morocco, Russia, Saudi Arabia, South Korea, and the United States. [IAEA 2005f p. 18-20]
[282] see for example Makhijani et al. 2004
[283] DTI 2003 p. 11
[284] The difference in emissions for coal and natural gas fired plants depends on such factors as the heat rate and efficiency of the power plants. Estimates for the ratio of the amount of CO_2 emitted per unit of generation for pulverized coal compared to natural gas typically range from about 1.5 to 3.1 [Makhijani and Saleska 1999 p. 197, IPCC 2001b p. 256 and 259, MIT 2003 p. 3 and 42, and U Chicago 2004 p. 8-10]. For consistency with recent utility experience in the U.S., we used the estimated ratio from the MIT study in our calculations in which coal fired plants emit 2.25 times more greenhouse gases than natural gas fired plants [MIT 2003 p. 3 and 42 and EIA 2004e p. 224 and 339].
[285] MIT 2003 p. 1, ILWG 2000 p. 7.1, and IPCC 2001 p. 156-157
[286] EIA 2004d from Table SR2 and EIA 2004e p. 195
[287] EIA 2004e p. 195
[288] [EIA 2004e p. 195] For natural gas, one thousand cubic feet is approximately equal to one million BTU (MMBtu).
[289] EIA 2004e p. 135, 151, 185, and 189
[290] Jensen 2003 p. 18
[291] Jensen 2003 p. 9
[292] EIA 2003d p. 43
[293] Jensen 2003 p. 21-22
[294] EIA 2003d p. 27, EIA 2005 p. 8 and 96, and Jensen 2003 p. 16-17, 36, and 40
[295] EIA 2005 p. 95-96 and 117
[296] EIA Natural Gas Factsheet
[297] [EIA 2005 p. 3 and 97] The EIA projections have the price of natural gas starting at $4.98 per thousand cubic feet in 2003, falling to $3.64 in 2010, and rising to $4.79 by 2025.
[298] EIA 2003d p. 44-45 and Jensen 2003 p. 7
[299] Jensen 2003 p. 31 and EIA 2003d p. 43
[300] EIA 2003d p. 42-46
[301] MIT 2003 p. 43
[302] [EIA 2005 p. 3 and 97, EIA 2003d p. 35, Zaidi 2005, and Jensen 2003 p. 20 and 27] The growth of LNG exports from the Middle East in the longer term is

expected to help stabilize the gas prices between the Atlantic and the Pacific importers given that its transport costs are similar to both areas and thus Middle East gas could provide a balancing capacity between importers in different parts of the world.

[303] MIT 2003 p. 42-43 and U Chicago 2004 p. 6-10

[304] U Chicago 2004 p. 6-10

[305] MIT 2003 p. 40 and 42-43

[306] For a review of the recent literature concerning the ways in which geographic distribution, improvements to the transmission grid, and improved meteorology can reduce the impacts from wind's intermittency see [DTI 2003c p. 7-19] and [IEA 2005 p. 3 and 19-25]

[307] Archer and Jacobson 2003 p. 10-17

[308] Makhijani et al. 2004 p. 42-44

[309] Archer and Jacobson 2003 p. 10-19

[310] Archer and Jacobson 2003 p. 10-19 and Archer and Jacobson 2004 p. 11

[311] With pumped hydro power, electricity can be used to run the system in reverse allowing water to be pumped from the lower reservoir to the upper reservoir where it can be used to regenerate electricity at a later time.

[312] The top ten states by installed pumped hydro capacity are: California (3352.6 MW), Virginia (2347.9 MW), South Carolina (2188 MW), Michigan (1978.8 MW), Tennessee (1530 MW), Massachusetts (1506.5 MW), Georgia (1397.8 MW), Pennsylvania (1269 MW), New York (1240 MW), and Missouri (600.4 MW). Together, these ten states account for 90 percent of the pumped hydro capacity in the United States as of the end of 2002. [IEA 2002 p. 416, 420, 432, and 440 and USBR 2004]

[313] Castronuovo and Lopes 2004b p. 1602 and Bueno and Carta 2005b p. 399

[314] NREL 2005 p. 51-54 and 56

[315] The calculation assumes a baseline cost for wind power of 3 to 5 cents per kWh, a round-trip efficiency of 75 to 80 percent for the pumped system, an operations and maintenance cost for regenerating the hydro power of 0.7 cents per kWh, and a transmission efficiency of 95 percent.

[316] see Castronuovo and Lopes 2004 and Castronuovo and Lopes 2004b

[317] Castronuovo and Lopes 2004b p. 1603-1604

[318] see Bueno and Carta 2005 and Bueno and Carta 2005b

[319] Bueno and Carta 2005b p. 399 and 404

[320] For example, in the MIT analysis roughly 65 to 75 percent of the total cost of generating electricity from natural gas fired plants is attributable to the cost of fuel under the moderate and high gas price scenarios. Therefore, going from a capacity factor of 85 percent down to 75 percent only adds 0.1 cents per kWh to the cost of electricity from natural gas. This can be compared to an increase of 0.4 cents per kWh for coal and 0.8 cents per kWh for nuclear for the same change in capacity factors. [MIT 2003 p. 42-43]

[321] IEA 2005 p. 48

[322] IEA 2005 p. 14-15

[323] This estimate assumes one percent of currently "unused land" would be covered with photovoltaics. [IPCC 2001b p. 247-248 and IEA 2005b p. 6 and 24]
[324] NREL 2005 p. 26
[325] Jackson and Oliver 2000 p. 984
[326] EIA 2004e p. 224
[327] Jager-Waldau 2004 p. 667
[328] [Green 2000 p. 996-997] Costs for electricity from photovoltaics were between 24 and 30 cents per kWh in 2000. [NREL 2005 p. 28]
[329] van der Zwaan and Rabl 2003 p. 19-20 and 30
[330] Green 2000 p. 993
[331] Green 2000 p. 993 and IPCC 2001b p. 247-248
[332] Eskom 2005
[333] NREL 2005 p. 28
[334] NREL 2005 p. 28
[335] Andersson 2000
[336] See for example Fthenakis 2000 and Fthenakis and Moskowitz 2000
[337] NREL 2005 p. 9
[338] EIA 2005 p. 60
[339] IPCC 2005 p. 155
[340] U Chicago 2004 p. 7-1
[341] DOE 2004b
[342] Amick et al. 2002 p. 1-2
[343] IPCC 2001b p. 239, 256, and 259, IEA 2002 p. 361, PIU 2002 p. 198, DOE 2004b, and U Chicago 2004 p. 6-4
[344] See for example Chiesa et al. 2005 and Kreutz et al. 2005
[345] NAS/NRC 2004 p. 94
[346] NAS/NRC 2004 p. 93
[347] NAS/NRC 2004 p. 142-144
[348] NAS/NRC 2004 p. 49
[349] Hotchkiss 2003 p. 29 and 31
[350] Greenwire 2004, U Chicago 2004 p. 5-24 and 6-4 to 6-6, Hotchkiss 2003 p. 30, Freund 2003 p. 4, and Narula, Wen, and Himes 2002
[351] General Electric 2004
[352] DOE 2004d and DOE 2006
[353] DTI 2003 p. 91
[354] IGCC technology is likely to be able to begin making a significant contribution by 2015 to 2020, with the International Energy Agency predicting that "coal gasification will have become the rule in new coal-fired plants" by 2030. [IEA 2002 p. 84 and 361]
[355] U Chicago 2004 p. 2-9 to 2-10, 5-24 to 5-25, and 9-5 to 9-6, IPCC 2001b p. 256 and 259, Narula, Wen, and Himes 2002 p. [4], MIT 2003 p. 42, and EIA 2005 p. 89
[356] IPCC 2005 p. 347, U Chicago 2004 p. 2-9, IPCC 2001b p. 256 and 259, and PIU 2002 p. 103, 108, and 198
[357] DTI 2003 p. 90, PIU 2002 p. 198, and DOE 2004b

[358] DOE 2003
[359] According to the DOE: "The FutureGen Industrial Alliance will contribute $250 million to the project. Current Alliance members are: American Electric Power (Columbus, Ohio); BHP Billiton (Melbourne, Australia); CONSOL Energy Inc. (Pittsburgh, Pa.); Foundation Coal (Linthicum Heights, Md.); China Huaneng Group (Beijing, China); Kennecott Energy (Gillette, Wyo.); Peabody Energy (St. Louis, Mo.); and Southern Company (Atlanta, Ga.)." [DOE 2005b]
[360] ILWG 2000 p. 7.1, IPCC 2001 p. 157, and DTI 2003b
[361] Freund 2003 p. 2
[362] NAS/NRC 2004 p. 86, U Chicago 2004 p. 8-5, and DTI 2003b p. 9
[363] Freund 2003 p. 2
[364] See for example Wichert and Royan 1997, Whatley 2000, and Jones, Rosa, and Johnson 2004
[365] IPCC 2005 p. 201 and 212
[366] NAS/NRC 2004 p. 86, Adam 2001, Torp and Gale 2004 p. 1361 and 1365-1366, and U Chicago 2004 p. 8-5
[367] Torp and Gale 2004 p. 1361-1366
[368] IPCC 2005 p. 201 and 203
[369] Adam 2001, and U Chicago 2004 p. 8-5
[370] DTI 2003b p. 11, Gale 2004 p. 1330, and IPCC 2001c p. 39
[371] IPCC 2001c p. 39
[372] IPCC 2005 p. 5-33 to 5-35
[373] IPCC 2005 p. 5-35
[374] Gale 2004 p. 1330-1331
[375] Narula, Wen, and Himes 2002 and U Chicago 2004 p. 8-7
[376] DTI 2003b p. 9
[377] DTI 2003b p. 14
[378] NAS/NRC 2004 p. 89
[379] U Chicago 2004 p. 8-6 to 8-8 and NAS/NRC 2004 p. 87
[380] NAS/NRC 2004 p. 88-89
[381] U Chicago 2004 p. 8-6 to 8-8, Narula, Wen, and Himes 2002, IPCC 2001b p. 256 and 259, and IPCC 2005 p. 347
[382] [Freund 2003 p. 3-4] This analysis also estimated a baseline cost of electricity from the IGCC plant of 4.8 cents per kWh which is consistent with the upper end of the range of costs we are considering in the current analysis.
[383] DTI 2003b p. ii and 16-17
[384] While greatly reduced, the emissions from fossil fuel plants with carbon sequestration are not zero. A $50 per ton carbon tax would add an approximately 0.07 cents per kWh to the generation cost of a natural gas fired plant, 0.13 cents per kWh to the cost of an IGCC plant, and 0.17 cents per kWh to the generation cost of a pulverized coal fired plant. Thus, a carbon tax at this level would not affect our conclusions regarding the economics of carbon sequestration compared to new nuclear power.
[385] NRC 2005 p. 135
[386] Forbes 1985

[387] MIT 2003 p. 69
[388] Acheson and Lilienthal 1946 p. 4
[389] High-level Panel 2004 p. 40
[390] CDC/NCI 2001 p. 6
[391] For a further discussion of this history see [Ortmeyer and Makhijani 1997]
[392] Fissile materials are those that are capable of being fissioned by low energy neutrons and of sustaining a chain reaction. The two most important fissile materials for both nuclear weapons and nuclear power production are uranium-235, which is naturally occurring, and plutonium-239, which is produced in nuclear reactors from naturally occurring uranium-238. An additional fissile material that is sometimes considered for nuclear fuel is uranium-233, which can be produced in reactors from naturally occurring thorium-232. The bomb that destroyed Hiroshima used 60 to 65 kilograms of highly-enriched uranium (80 to 90 percent U-235) while the bomb that destroyed Nagasaki used 6.1 kilograms of Pu-239. Modern weapons designs use approximately 20 to 25 kilograms of HEU or just three to four kilograms of Pu-239.
[393] Currently the United States, Russia, Britain, China, France, India, Pakistan, Israel, and North Korea all possess or are believed to possess nuclear weapons.
[394] as quoted in [Makhijani 1998 p. 14]
[395] Sutcliffe 1997 p. 47
[396] High-level Panel 2004 p. 39
[397] ElBaradei 2004c
[398] In 2003, light-water reactors made up 87 percent of the installed capacity in the world while heavy water reactors, which were the next largest contributor, made up just 6.4 percent of the capacity. [NRC 2005 p. 137]
[399] NRC 2005 p. 137
[400] For a description of uranium enrichment technologies and the current status of enrichment programs around the world see [Makhijani, Chalmers, and Smith 2004]
[401] The three enrichment plants that have been built in the U.S are located at Paducah, Kentucky, Portsmouth, Ohio, and Oak Ridge, Tennessee. Currently only the Paducah plant is in operation, while the Oak Ridge plant has been shut down and the Portsmouth plant is in a standby state. The Paducah plant was design to produce only low-enriched uranium and therefore did not contribute directly to the U.S. stockpile of weapons usable HEU. [NAS/NRC 1996 p. 17]
[402] Makhijani, Chalmers, and Smith 2004 p. 7
[403] USEC 2003 p. 8 and EPA 2004
[404] OTA 1977 p. 14
[405] MIT 2003 p. 88
[406] New York Times 2004b and New York Times 2004c
[407] For a further discussion of the current policy regarding the interdiction of ships carrying missile technology or suspected nuclear, chemical, or biological weapons related materials see Section 3.4.3.
[408] White House 2004 and New York Times 2005e
[409] Boucher 2004

[410] ElBaradei 2004
[411] MIT 2003 p. 88
[412] OTA 1993b p. 33 and Makhijani, Chalmers, and Smith 2004 p. 14
[413] For an overview of the history of the South African nuclear weapons program see [Albright 1994].
[414] OTA 1993b p. 36 (emphasis in the original)
[415] The capacity of enrichment plants is given in units of kilogram separative work units of SWUs (pronounced "swooze"). Typical production scale facilities have a capacity that is on the order of a few hundred to a few thousand metric ton SWU (1 MTSWU = 1,000 SWU). The number of SWUs is related to the amount of effort required to achieve a given level of enrichment and is directly proportional to the amount of energy the plant consumes.
[416] Makhijani, Chalmers, and Smith 2004 p. 18-29
[417] MIT 2003 p. 30 and 145-146
[418] IAEA 2005d p. 18
[419] Makhijani, Chalmers, and Smith 2004 p. 18-20 and 26-29
[420] IAEA 2005 p. 49
[421] These estimates assume an enrichment of the depleted uranium tails of 0.2 to 0.3 percent U-235. [Makhijani, Chalmers, and Smith 2004 p. 4, 20, and 37]
[422] Bergeron 2002 p. 30
[423] [Davis 2001 p. 290]. For a discussion of the history of the plutonium economy and its failure on economic, safety, and proliferation grounds see [Makhijani 2001].
[424] Makhijani and Saleska 1999 p. 121-122
[425] In the early days of the nuclear era, it was often claimed that plutonium recovered from commercial spent fuel was incapable of being used in nuclear weapons. Plutonium separated from fuel that has been irradiated for longer times, as is done in typical power plants, have higher concentrations of Pu-240 than in plutonium typically used for nuclear weapons. This impurity makes it more difficult to manufacture a bomb that will not "fizzle" (i.e. explode with a much smaller yield than intended). However, as noted by the Department of Energy itself

> At the lowest level of sophistication, a potential proliferating state or subnational group using designs and technologies no more sophisticated that those used in first-generation nuclear weapons could build a nuclear weapon from reactor-grade plutonium that would have an assured, reliable yield of one or a few kilotons (and a probable yield significantly higher than that). At the other end of the spectrum, advanced nuclear weapons states such as the United States and Russia, using modern designs, could produce weapons from reactor-grade plutonium having reliable explosive yields, weight, and other characteristics generally comparable to those of weapons made from weapons-grade plutonium. [DOE 1997 p. 38-39]

Beyond theoretical work showing that such a bomb is possible, in 1994 the DOE released additional information concerning a successful test that had been conducted by the United States in 1962 of a nuclear weapon made with so-called "reactor grade" plutonium. This test was conducted well before the detonation of the Indian nuclear device in 1974. [DOE 1994]

[426] Bergeron 2002 p. 32

[427] OTA 1977 p. 12

[428] [White House 1993] Currently France, Britain, Russia, India, and Japan all have active commercial reprocessing programs. The British thermal oxide reprocessing plant (Thorp) located at the Sellafield complex is set to end commercial reprocessing by 2010 due to poor sales and difficulties encountered with the vitrification of the plant's high-level waste. British Nuclear Fuels (BNFL), which had declared bankruptcy and transferred an estimated £41 billion ($67 billion) in waste liabilities to the government, will continue to operate the facility for waste management activities beyond 2010. The older British reprocessing plant which handles the spent fuel from the Magnox reactors is set to close in 2012. [Guardian 2003]

[429] Cheney et al. 2001 p. 5-17

[430] Committee on Appropriations 2005 p. 125

[431] U.S. Congress 2005 p. 291-293

[432] GNEP Factsheet

[433] WISE 2001

[434] DOE 1996 p. 25 and NAS/NRC 2002b p. 40

[435] Due to the larger critical mass of reactor grade plutonium compared to weapons grade plutonium, approximately 30 percent more is needed to make a nuclear weapon with a similar yield. The "Fat Man" bomb dropped on Nagasaki used 6.1 kilograms of weapons grade plutonium which would be equivalent to the use of 8 kilograms of reactor grade plutonium. [Garwin 1998] The IAEA currently uses the estimate of eight kilograms of reactor grade plutonium as the amount of fissile material required for one nuclear weapon. [IAEA 2001b]

[436] Only Russia, India, and Japan have plans to commercialize fast-breeder reactors over the coming decades, while only Russia currently operates a commercial scale breeder reactor. [Bunn et al. 2003 p. 67]

[437] NEI 2004b p. 3

[438] Makhijani 2001 p. 37-46

[439] Makhijani 2001 p. 17-18

[440] This calculation assumes an over loss of 5 percent in the chemical processing steps required to separate the plutonium from the MOX fuel and to convert it into a metal.

[441] MIT 2003 p. 122

[442] Sokolski 2005 p. 25

[443] Berkhout and Gadekar 1997, Sawai 2001, Bunn et al. 2003 p. 26 and 29, and IAEA 2005d p. 20, 50, 52, and 70

[444] [MIT 2003 p. 121-123] This calculation assumes 1,000 GW of nuclear power online by 2050 with a capacity factor of 85 percent.

[445] MIT 2003 p. 123
[446] GNEP Factsheet
[447] Zerriffi and Makhijani 2000 p. 41-43, NAS/NRC 2000b p. 17-20, and Kang and von Hippel 2005
[448] Vandegrift et al. 2004 p. 2 and 6 and Kang and von Hippel 2005 p. 171-174
[449] Christian Science Monitor 2006
[450] Kang and von Hippel 2005 p. 171-174
[451] Kang and von Hippel 2005 p. 177
[452] CISAC and PSGS 2005 p. 55-56
[453] CISAC and PSGS 2005 p. 55
[454] Global Security
[455] WISE 2001
[456] MIT 2003 p. 4-5 (emphasis in the original)
[457] MIT 2003 p. 5
[458] For a discussion of plutonium disposition strategies see [Makhijani and Makhijani 1995 p. 19-70].
[459] Makhijani and Smith 2006b
[460] DOE 1991 p. S-1
[461] On February 27, 1999, the Watts Bar reactor irradiated 32 lithium rods in the reactor as a test of the tritium production technology. This number of rods is less than 1.4 percent of the amount allowed to be placed in the reactor under the 2002 license amendment and, therefore, October 2003 marked the commencement of the first production scale run. [TVA Factsheet, NRC 2003a, and NRC 2003b]
[462] DOE 1991 p. S-5
[463] Bergeron 2002 p. 138
[464] Los Angeles Times 1998
[465] The N Reactor at the Hanford reservation in Washington State, completed in 1963 and operated until 1987, was a plutonium production reactor that sold high-temperature steam as a byproduct to a local utility for the production of electricity. The N Reactor was ordered permanently deactivated in 1991. [Hanford Factsheet] While this arrangement was also a blurring of the line between civilian and military uses of nuclear energy, it was not as significant as the conversion of commercial reactors to produce materials for direct use in the U.S. nuclear arsenal while the reactors remain under civilian control.
[466] Bergeron 2002 p. 87, 139
[467] MIT 2003 p. 65
[468] MIT 2003 p. 66
[469] Schwartz and Randall 2003 p. 19
[470] [MIT 2003 p. 112 - 114] The high and low growth scenarios listed in this table were derived from the expected electricity consumption of each country in 2050 and were not the result of further specific analysis on the likelihood of nuclear development. As described by the authors,

> This is certainly not a prediction of rapid growth in nuclear power. Rather, it is an attempt to understand what the distribution of nuclear

power deployment would be if robust growth were realized, perhaps driven by a broad commitment to reducing greenhouse gas emissions and a concurrent resolution of the various challenges confronting nuclear power's acceptance in various countries. [MIT 2003 p. 110-111]

It is interesting to note the increase in North Korea's nuclear power sector predicted in the MIT report despite the fact that the study was published after North Korea's January 2003 withdrawal from the NPT and their announced resumption of nuclear weapons production.

[471] Albright 1994 p. 37-38
[472] MIT 2003 p. 113
[473] MIT 2003 p. 67
[474] NSSM 1975, NSDM 1975, and NSDM 1976
[475] ElBaradei 2004c
[476] U Chicago 2004 p. A7-7
[477] MIT 2003 p. 67
[478] Truman, Attlee, and King 1945
[479] U.S. Senate 1945 p. 122-123
[480] Acheson and Lilienthal 1946 p. 4
[481] Acheson and Lilienthal 1946 p. 4-5
[482] This summary of expanded inspection powers granted by the Additional Protocol was adapted from an Arms Control Association fact sheet. [ACA 1999] The full text of the generic Model Additional Protocol is available from the International Atomic Energy Agency. [IAEA 1997]
[483] Burk 2004
[484] IAEA 2003c p. 1-3 and 5-6
[485] IAEA 2003d p. 6-7
[486] IAEA 2005e, IAEA 2006, and IAEA 2006b
[487] IAEA 2005c
[488] IAEA 2005c
[489] Associated Press 2006
[490] MIT 2003 p. 90
[491] IAEA 2005b p. 5 and 13-14, and IAEA 2005c
[492] IAEA 2005b p. 13-14 and IAEA 2005c
[493] UIC 2002
[494] Makhijani, Chalmers, and Smith 2004 p. 24, Washington Post 2004a, and Washington Post 2004b
[495] as quoted in Deller, Makhijani, and Burroughs 2003 p. 21
[496] Burk 2004
[497] Burk 2004
[498] Acheson and Lilienthal 1946 p. 30
[499] Acheson and Lilienthal 1946 p. 23
[500] Acheson and Lilienthal 1946 p. 34-35
[501] Baruch 1946

[502] Gaddis 1972 p. 332-335, Hewlett and Anderson 1990 p. 580-619, and Holloway 1994 p. 161-166
[503] Brenner 1981 p. 15-16 and 23, Albright 1994, and IAEA 2005d p. 11
[504] IAEA 2005 p. 25
[505] White House 2004
[506] White House 2004
[507] GNEP Factsheet
[508] UN Disarmament
[509] Sanders 2005 p. 10-11
[510] [Makhijani, Chalmers, and Smith 2004] As of February 2004, China operated uranium enrichment plants for making LEU and planed to bring additional capacity online, however, they had no international contracts to supply enrichment services for commercial fuel. In addition, Brazil was preparing to open a commercial uranium enrichment plant, however, the plant was not yet operational when President Bush made his original proposal. It is unlikely that the U.S. could oppose its operation now that Brazil has reportedly reached an agreement with the IAEA over inspections, but its status under the President's proposed monopoly is not yet completely clear.
[511] For a detailed examination of the history of the attempts to privatize the U.S. uranium enrichment services and its impact on nonproliferation efforts see *Nuclear Power and Non-Proliferation: The Remaking of U.S. Policy* by Michael Brenner. [Brenner 1981]
[512] Brenner 1981 p. 34-35
[513] Brenner 1981 p. 36 and 52
[514] Brenner 1981 p. 15-16, 28, and 52-55
[515] MIT 2003 p. 114
[516] Washington Post 2006 and Washington Post 2006b
[517] OTA 1995 p. 70
[518] ElBaradei 2004
[519] ElBaradei 2005
[520] MIT 2003 p. 13 and 89
[521] IAEA 2005 p. 26
[522] IAEA 2005 p. 101
[523] This calculation assumes that each 1000 MW plant requires 100 to 120 MTSWU per year of enrichment services and that those enrichment services are provided at a cost of $100 to $120 per SWU.
[524] OTA 1995 p. 109-110
[525] White House 2005b
[526] White House 2005b
[527] White House 2005b (emphasis added)
[528] Sutcliffe 1997 p. 47
[529] ElBaradei 2004
[530] Baruch 1946
[531] MIT 2003 p. 88 and 90
[532] ElBaradei 2004

[533] New York Times 2005, Washington Post 2005b, IAEA 2006, and IAEA 2006b
[534] Chomsky 1993 p. 144-154 and Bamford 2001 p. 62-63 and 70-91
[535] UNICEF 2003
[536] Chomsky 1992 p. 199, Global Policy Forum, and Boucher 2004
[537] Baruch 1946
[538] Gaddis 1972 p. 332-335, Hewlett and Anderson 1990 p. 580-619, and Holloway 1994 p. 161-166
[539] For a review of the U.S. and NATO nuclear weapons policies, including their potential use against non-nuclear weapons states, see [Smith 2003] and [Makhijani and Deller 2003].
[540] The countries participating in the PSI as of 2004 were: Australia, Canada, Denmark, France, Germany, Italy, Japan, the Netherlands, Norway, Poland, Portugal, Singapore, Spain, Turkey, the United Kingdom, and the United States. [Chaffee 2004 p. 4]
[541] White House 2003b
[542] For a further discussion of the PSI and its legality under the International Law of the Sea see [Chaffee 2004].
[543] Chaffee 2004 p. 5
[544] Clary 2004 p. 35
[545] State Dept 2004 and State Dept 2005
[546] Sanders 2005 p. 10-11
[547] Chaffee 2004 p. 4, 6, and 8
[548] CIA World Factbook
[549] London Times 1981, Washington Post 1981b, and New York Times 1981
[550] Washington Post 2002
[551] White House 2002b p. 15
[552] White House 2002c p. 3
[553] White House 2006 p. 23
[554] See for example Los Angeles Times 2003, Jerusalem Post 2003, Washington Post 2003, MSNBC 2005, and Washington Post 2006d
[555] Toronto Star 2003b
[556] ElBaradei 2004c
[557] For an analysis of the legal obligation of the five nuclear weapons states to bring to a conclusion negotiations on disarmament under the Nuclear Nonproliferation Treaty see [Deller, Makhijani, and Burroughs 2003 p. 19-40]
[558] ElBaradei 2005b
[559] Cohen and Graham 2004 p.43
[560] ElBadardei 2004b
[561] OTA 1993b p. 26
[562] ElBaradei 2004b
[563] Washington Post 2004e
[564] Washington Post 2004e and Guardian 2004b
[565] Deller, Makhijani, and Burroughs 2003 p. 24
[566] New York Times 2004

[567] Deller, Makhijani, and Burroughs 2003 p. 48 and Washington Post 1999
[568] For a further discussion see Smith 2003, Deller, Makhijani, and Burroughs 2003, and Makhijani and Deller 2003.
[569] DOD 2001
[570] DOD 2001
[571] White House 2006 p. 22
[572] UCS 2004, ANA Factsheet, and ANA 2005
[573] as quoted in NTI 2004
[574] NATO 2001 p. 160
[575] Chirac 2006
[576] New York Times 2004
[577] The Committee on Atomic Energy was commissioned in 1946 by then Under-Secretary of State Dean Acheson. The committee was chaired by David Lilienthal, then the Chairman of the Tennessee Valley Authority and later the first Chairman of the Atomic Energy Commission. The committee also included the presidents of the New Jersey Bell Telephone Company, Monsanto, and General Electric, as well as Robert Oppenheimer who had headed the bomb design work at Los Alamos Laboratory during the Manhattan Project. [Acheson and Lilienthal 1946 p. 4]
[578] MIT 2003 p. 48
[579] Ruckelshaus 1984 p. 157-158
[580] Meserve 2002 (emphasis in the original)
[581] In addition to the Exxon Valdez, which spilled an estimated 38,000 tons of oil into Prince William Sound in 1989, a number of other large oil spills have occurred around the world. Examples include:

1967	the Torrey Canyon spilled 119,000 tons off the coast of the Scilly Isles
1978	the Amoco Cadiz spilled 220,000 tons off the coast of Brittany, France
1991	the Haven spilled 50,000 tons off the coast of Genoa, Italy
1991	the ABT Summer spilled 260,000 tons off the coast of Angola
1992	the Aegean Sea spilled 80,000 tons near La Coruna, Spain
1993	the Braer spilled 85,000 tons off the Shetland Islands
1996	the Sea Empress spilled 72,000 tons near Milford Haven, U.K.

[BBC 2002]
[582] Weinberg 1994 p. 133-134
[583] MIT 2003 p. 9 and 49
[584] Weinberg 1994 p. 141
[585] Makhijani and Saleska 1999 p. 140-141
[586] UCS 1990 p. 3-5
[587] Makhijani, Hu, and Yih 2000 p. 417-418 and Kemeny Commission 1979 p. 31
[588] Makhijani and Saleska 1999 p. 145 and 147
[589] The releases from the July 1959 SRE accident were estimated by IEER. [IEER 2005]

[590] Makhijani and Saleska 1999 p. 93
[591] Bergeron 2002 p. 59-60
[592] For an excellent review of the early history surrounding the debate over the effectiveness of the ECCS and the NRC's response to these concerns see Daniel Ford's *Cult of the Atom*. [Ford 1982 p. 90-130]
[593] Bergeron 2002 p. 54-55, 61, and 72
[594] Bergeron 2002 p. 60-61, 64-70, and 72-74
[595] See [Lochbaum 2000] for a further discussion.
[596] Table adapted from [Makhijani and Saleska 1999 p. 152-153] with additional information from [Fuller 1975 p. 103-115, 128, and 194-206, Kemeny Commission 1979 p. 110-161, Jedicke 1989, Medvedev 1990 p. 12-36, Wise 1999 p. 4-5, Makhijani, Hu, and Yih 2000 p. 417-419, Stacy 2000 p. 138-149, EIA 2002, and IEER 2005]
[597] Ford 1982 p. 217-224
[598] Monju Reactor Factsheet and Monju Accident Factsheet
[599] NRC 2006
[600] GAO 1999 p.4
[601] EIA 2005 p. 89
[602] NRC 2005 p. 55
[603] Lochbaum 2000 p. 9-10
[604] CNN 2006
[605] Davis-Besse Factsheet
[606] NEI 2004 p. 3 and GAO 2004 p. 20
[607] A typical refueling shutdown would last 42 days and cost $60 million ($37 million for operations, repairs, and maintenance and $23 million for replacement power). [GAO 2004 p. 20] The total overnight capital cost of the Davis-Besse Plant in 2003 dollars was $1.94 billion. [EIA 1986 p. 109]
[608] NRC 2002, UCS 2003 p. 1, 11 and GAO 2004 p. 29-30
[609] OIG 2002 p. 6 and UCS 2003 p. 3
[610] The 12 plants in Categories One and Two were Arkansas Nuclear One Unit 1, D.C. Cook Unit 2, Davis-Besse, North Anna Units 1 and 2, Oconee Units 1, 2, and 3, Robinson Unit 2, Surry Units 1 and 2, and Three Mile Island Unit 1. [UCS 2003 p. 3]
[611] OIG 2002 p. 6-9
[612] OIG 2002 p. 9-10 and GAO 2004 p. 10
[613] OIG 2002 p. 10
[614] OIG 2002 p. 10-11 (brackets supplied by OIG)
[615] OIG 2002 p. 10-11
[616] OIG 2002 p. 12
[617] GAO 2004 p. 21 and 80
[618] UCS 2003 p. 9
[619] OIG 2002 p. 19
[620] The five NRC safety principles are that: "(1) current regulations are met, (2) NRC's defense-in-depth philosophy is maintained, (3) sufficient safety margins are maintained, (4) minimal increase in risk of core damage, and (5) risk meas-

urement is monitored using performance measurements strategies." The only safety principle the staff believed was met by their decision to allow Davis-Besse to continue to operate until mid-February was number four. [OIG 2002 p. 13-14]

[621] OIG 2002 p. 14, GAO 2004 p. 16, and NRC 2002
[622] GAO 2004 p. 19
[623] GAO 2004 p. 18 and 20 and NRC 2002
[624] Meserve 2002
[625] GAO 2004 p. 28
[626] FENOC 2002 p. 10
[627] OIG 2002 p. 23
[628] GAO 2004 p. 54-55
[629] NRC 2005b
[630] GAO 2005b p. 12
[631] OIG 2002 p. 14
[632] NRC 2002 and GAO 2004 p. 31-32
[633] The term "normalization of deviance" was coined by sociologist Diane Vaughan to indicate a situation in which there is an ongoing acceptance of unexpected operating conditions. [Gehman et al. 2003 p. 130]
[634] Kemeny Commission 1979 p. 28-29 and 43-44 and Perrow 1999 p. 20-23
[635] Gehman et al. 2003 p. 196
[636] GAO 1999, p. 12
[637] GAO 2004 p. 57
[638] GAO 1997 p. 2-3, 5, 8, and 10
[639] GAO 2004 p. 5
[640] Holt 2005b p. CRS-3 to CRS-5
[641] White House 2005
[642] Unless otherwise noted, the following discussion of the impacts of the Chernobyl disaster is drawn from the work of Arjun Makhijani and Scott Saleska [Makhijani and Saleska 1999 p. 153-164] which was based in large part upon the work of Zhores Medvedev [Medvedev 1990].
[643] Anspaugh, Catlin, and Goldman 1988 p. 1516-1517
[644] NRC 1987 p. 8-11 to 8-12
[645] Anspaugh, Catlin, and Goldman 1988 p. 1517-1518
[646] NRC 1987 p. 8-10 to 8-11 and 8-13 to 8-14
[647] United Nations 2005
[648] NAS/NRC 2006c p. 9-10 and 14-18
[649] In 2000, there were 10 reactors at 6 sites located within 30 km (18.6 miles) of urban centers with populations greater than 100,000 people and 23 reactors at 15 sites located within 40 km (24.8 miles) of such cities. The inclusion of the population in the greater metropolitan area of these cities would add significantly to the numbers of people at potential risk from an accident at one of these plants. [NRC 2004b and Census 2004 p. 26-32] We have chosen to use the range of 30 to 40 kilometers in our consideration of reactors with large populations living nearby, however, it is important to note that areas of intense fallout

at Chernobyl extended out to roughly 60 km or more from the reactor. Due to the multi-day fire at the reactor, the fallout traveled in several directions stretching towards the towns of Narovlia and Bragin to the north and northwest and the towns of Narodichi, Polesskoye, and Bober to the west and southwest. Hotspots were found to extend out to distances of as much as 100 to 300 kilometers from the site of the plant. [Makhijani and Saleska 1999 p. 156-157]

[650] For instance, the population of New York City increased by 13 percent between 1980 and 2000, while the population of Philadelphia decreased by 10 percent over that same time. In addition, some areas of the U.S. have undergone very rapid growth over these two decades. Phoenix, Arizona for instance, which is near the largest nuclear plant in the country, experienced a 67 percent increase in its population between 1980 and 2000. [Census 2004 p. 31 and Palo Verde Factsheet]

[651] The analysis supporting the Sandia report entitled Calculation of Reactor Accident Consequences for U.S. Nuclear Power Plants (CRAC-2) was completed in 1981. This report, prepared for the Nuclear Regulatory Commission, was not initially released to the public. The Union of Concerned Scientists, working with the House Subcommittee on Oversight and Investigation, eventually obtained the CRAC-2 report, and the information it contained on the peak consequences of reactor accidents was finally published in 1982. [Riccio 2001]

[652] Riccio 2001

[653] Riccio 2001

[654] The Committee on Health Effects of Exposure to Low Levels of Ionizing Radiations of the U.S. National Academy of Science's Board on Radiation Effects Research conducts periodic reviews of the latest available scientific evidence regarding the risks from exposure to low dose radiation. The BEIR VII (Biological Effects of Ionizing Radiation) report published in 2005 is the latest of these assessments. [NAS/NRC 2006c] The BEIR VII report updates the previous assessment (BEIR V) which was published in 1990.

[655] EPA 1999 p. 179 and 182 and NAS/NRC 2006c p. 15

[656] NAS/NRC 2006c p. 154

[657] NAS/NRC 2006c p. 152

[658] NAS/NRC 2006c p. 10

[659] Pediatrics 2003 p. 1461 and Bromet et al. 2002 p. 627-629

[660] Perrow 1999 p. 355-360

[661] FAA 2001

[662] Kemeny Commission 1979 p. 13

[663] Pediatrics 2003 p. 1461

[664] New York Times 2005d

[665] GAO 1997 p. 4

[666] New York Times 2000

[667] GAO 2004 p. 20

[668] Riccio 2001

[669] Riccio 2001 and GAO 1986 p. 15

[670] U.S. Congress 2005 p. 186-189

[671] The estimates of the accidents' costs have been converted to year 2000 dollars for ease of comparison. [GAO 1986 p. 6 and 24, Riccio 2001, and Sahr 2004]
[672] See for example New York Times 1986, New York Times 1986b, Davis 1994 p. 50, NRC 1996 p. 7, IAEA 1999 p. 1, 3, 11-12, 27, and 29, and CNN 2000
[673] GAO 2005 p. 8
[674] Andrews 2004 p. CRS-5
[675] Alvarez et al. 2003 p. 10
[676] Alvarez et al. 2003 p. 7, 10, and 39-40
[677] Alvarez 2002
[678] NRC 1987 p. 6-6, Anspaugh, Catlin, and Goldman 1988 p. 1517, and Alvarez et al. 2003 p. 7
[679] Alvarez et al. 2003 p. 11-16
[680] NAS/NRC 2006 p. 57
[681] NAS/NRC 2006 p. 53-55 and 57
[682] NAS/NRC 2006 p. 35
[683] GAO 2005 p. 9
[684] NAS/NRC 2006 p. 9
[685] Alvarez et al. 2003 p. 2-3 and MIT 2003 p. 42
[686] Makhijani, Hu, and Yih 2000 p. 413-417 and 470-475
[687] Makhijani, Hu, and Yih 2000 p. 220-224 and 249-253
[688] Hu, Makhijani, & Yih 1992 p. 80-94
[689] Hu, Makhijani, & Yih 1992 p. 95-107
[690] OECD 1996 p. 68-70
[691] For a further discussion of the safety implications regarding the use of MOX fuel see for example [Takagi et al. 1997 p. 121-156], [Makhijani 2001 p. 41-43], and [Lyman 2001 p. 33-79].
[692] MIT 2003 p. 51
[693] SDA 1999
[694] MIT 2003 p. 71
[695] IEA 2001b p. 155, 228, and 246 and MIT 2003 p. 21
[696] The countries included in the survey were: Argentina, Australia, Cameroon, Canada, France, Germany, Hungary, India, Indonesia, Japan, Jordan, Mexico, Morocco, Russia, Saudi Arabia, South Korea, United Kingdom, and the United States. [IAEA 2005f p. 18-20]
[697] Shell 2001 p. 34
[698] Washington Post 1981c
[699] Washington Post 1979a and Washington Post 1979b
[700] Washington Post 1981a and Washington Post 1981c
[701] Washington Post 1988 and Komanoff and Roelofs 1992 p. 23
[702] Kemeny Commission 1979 p. 87
[703] NVSR 2002 p. 8, NVSR 2003 p. 8 and 10, and NVSR 2004 p. 5
[704] NRC 2005 p. 38
[705] Assuming that accidents at nuclear plants will occur randomly, we can calculate the probability of having seen less than two accidents over approximately

3,000 hours of operation for different assumptions about the underlying accident rate. The given range represents the 5-95 percent confidence interval for the average accident rate (i.e. there is a 5 percent chance that the actual accident rate is greater than 1 in 633 per year and a 5 percent chance that it is less than 1 in 8,440 per year.) The median accident rate (i.e., the value for which half of the estimates are above and half below) is approximately 1 in 1,800 per year.

[706] MIT 2003 p. 48
[707] Knief 1992 p. 387-397
[708] Kendall et al. 1977 p. 144-145 and Makhijani and Saleska 1999 p. 89-90 and 96
[709] Kendall et al. 1977 p. 145-148
[710] In January 1975, the Atomic Energy Commission was broken up into the Nuclear Regulatory Commission and the Energy Research and Development Administration (ERDA), which would later become the Department of Energy.
[711] Lewis 1978 p. 10
[712] Wilson et al. 1985 p. S11
[713] For a detailed discussion of the problems inherent in the Rasmussen Report and a revision of its risk estimates to correct for some of its more serious omissions see *The Risks of Nuclear Power Reactors: A Review of the NRC Reactor Safety Study WASH-1400 (NUREG-75/014)* published by the Union of Concern Scientists. [Kendall et al. 1977]
[714] Lewis 1978 p. vi
[715] Lewis 1978 p. ix
[716] Lewis 1978 p. vii
[717] NRC 1979, p. 3
[718] NRC 1979, p. 3 (emphasis added)
[719] Lewis 1978 p. ix
[720] Lewis 1978 p. 6
[721] Murray and Teare 1993, p. 2551-2552
[722] NRC 1998 p. 72
[723] GAO 1999 p. 2
[724] Lochbaum 2000 p. 7 and 17
[725] NRC 2002
[726] GAO 2004 p. 68 and 73
[727] GAO 2004 p. 74-75
[728] Kendall et al. 1977 p. 44- 51 and Lochbaum 2000 p. 10
[729] London Times 1981, New York Times 1982, and Washington Post 1982
[730] This concern was discussed by the authors of the MIT report as well. [MIT 2003 p. 22]
[731] EIA 2001 p. 25
[732] The number of different types of attacks that might seriously damage a reactor and its containment dome or a spent fuel pool are likely to be very small which aids in trying to make an accurate risk assessment. However, even within this more limited range, there are still an unknown number of actions possible.
[733] NRC 2005 p. 42-43, NRC 2005c, and NRC 2005d

[734] In addition to raising questions over completeness, design defects pose other challenges to the PRA methodology as well. In particular, the fault trees used in these risk assessments are most appropriate for accident scenarios in which cause and effect are linear and unidirectional and where each individual failure is random and independent of all others. Accidents where different failures are interconnected or dependent upon the existence of other, seemingly unrelated failures within distant parts of the same fault tree or on failures in entirely different fault trees are less suited to the PRA methodology. The most important class of such failures are known as common mode/common cause failures. In these kinds of accidents, a single event or condition may simultaneously affect a variety of different components or subsystems which may be functionally related or un-related adding to the complexity of properly diagnosing the causes of an accident. In addition to design or construction problems, other types of common mode/common cause failures can be caused by such things as human error (see Section 4.3.3), failures in systems controlled by software (see Section 4.3.4), adverse environmental conditions such as unusually high temperatures and humidity or excessive levels of electronic noise, and natural disasters such as earthquakes, floods, or fires. For a further discussion of these issues as they relate to the safety of nuclear power plant see for example [Hagen 1980] and [Perrow 1999 p. 15-100].
[735] Lewis 1978 p. 48
[736] Lochbaum 1998 p. 1 and 3
[737] Lochbaum 2000 p. 8
[738] Hagen 1980 p. 185-186
[739] Hagen 1980 p. 189-191
[740] NRC 2001 p. 23
[741] Hagen 1980 p. 191
[742] Lewis 1978, p. 31
[743] Lochbaum 1998 p. 1 and 3
[744] NRC 2000b
[745] NRC 2001 p. 21
[746] NRC 2000b
[747] Boston Globe 2003 and UCS 2003 p. 1 and 5
[748] Dumas 1999 p. 145
[749] as quoted in [Lochbaum 2000 p. 11]
[750] Kemeny Commission 1979 p. 8
[751] NAS/NRC 1997 p. 40-41
[752] Le Monde 2006
[753] NRC 2000b
[754] NAS/NRC 1997 p. 7
[755] Fan and Chen 2000 p. 29
[756] Lions 1996 p. 1, 3, and 11
[757] NAS/NRC 1997 p. 49
[758] NAS/NRC 1997 p. 48
[759] Leveson 2004 p. 566

[760] NAS/NRC 1997 p. 50
[761] Leveson 2004 p. 574
[762] NAS/NRC 1997 p. 7
[763] NAS/NRC 1997 p. 44
[764] Leveson 1994 p. 69
[765] NAS/NRC 1997 p. 55
[766] Leveson 2004 p. 572-573
[767] NAS/NRC 1997 p. 55
[768] NAS/NRC 1997 p. 7
[769] Perrow 1999 p. 50-52
[770] Feynman 1988 p. 220-222
[771] Harwood 1996
[772] Feynman 1988 p. 229
[773] NRC 1983 p. II-2
[774] U.S. Congress 1985 p. 413
[775] NRC 1996b p. 3-2 to 3-3
[776] Lochbaum 2000 p. 14-16
[777] Ruckelshaus 1984 p. 157-158
[778] The given range represents the 5-95 percent confidence interval for the average accident rate (see the introduction to Section 4.3).
[779] MIT 2003 p. 48
[780] MIT 2003 p. 48 (emphasis added)
[781] MIT 2003 p. 48
[782] MIT 2003 p. 85 (emphasis added)
[783] GAO 2004 p. 41
[784] U.S. Congress 1985 p. 413
[785] U.S. Congress 1985 p. 378
[786] MIT 2003 p. 48
[787] Figures 4.1 and 4.2 retain the MIT report's assumption of a 10 fold increase in safety at new plants compared to those in operation today.
[788] New York Times 2005d
[789] Feynman 1988 p. 223
[790] MIT 2003 p. 85 (emphasis added)
[791] Hagen 1980 p. 189-191
[792] Ruckelshaus 1984 p. 157-158
[793] MIT 2003 p. 61
[794] Weinberg 1994 p. 183
[795] At a burnup of 50 GWD per ton, a typical 1,000 MW reactor would discharge approximately 20 metric tons of spent fuel per year containing nearly 265 kilograms of Pu-239. If chemically separated through reprocessing, this would be enough plutonium to make 33 nuclear weapons (assuming eight kilograms per bomb). [MIT 2003 p. 120]
[796] For further discussion of the impacts of uranium mining on Indigenous Peoples see for example [Brugge and Goble 2002] and Chapter 5 of *Nuclear Wastelands* [Makhijani, Hu, and Yih 2000 p. 105-168]. In the U.S., the site of the

proposed Yucca Mountain repository is on land claimed by the Western Shoshone, while the Private Fuel Storage facility, that has been proposed by a consortium of nuclear utilities, is planned for construction on the land of the Skull Valley Band of Goshute Indians (see Sections 5.2.3 and 5.4.1).

[797] SDA 1999

[798] The DOE and EPA definitions of TRU waste differ slightly, but both set a threshold of 100 nanocuries per gram for long-lived, alpha-emitting transuranic radionuclides.

[799] One current exception to this rule is the high-level reprocessing wastes that the DOE now plans to leave behind in the tank farms at the Idaho and Savannah River sites. This waste would be covered with cement and left in these near-surface tanks. The DOE's plan, which was authorized in the FY05 Defense Authorization Act, must be analyzed by a National Research Council committee, and be approved by the states. For more details about the proposal to use grout to immobilize these high-level wastes see [Smith 2004] and [NAS/NRC 2005c p. 1-10, 33-34, and 50-51].

[800] See [Makhijani and Smith 2004] and [Makhijani and Smith 2005] for a further discussion of the issues surrounding the management and disposal of large quantities of depleted uranium.

[801] Makhijani and Saleska 1999 p. 111

[802] There are currently ten low-level waste compacts (Northwest, Southwestern, Rocky Mountain, Midwest, Central, Texas, Appalachian, Central Midwest, Atlantic, and Southeast) which have been joined by 44 states. The remaining six states plus the District of Columbia and Puerto Rico are not affiliated with any compact and must make individual arrangements for disposal of their low-level waste. [LLW Compact Factsheet]

[803] Holt 2005 p. CRS-12 to CRS-13

[804] [Washington State and Holt 2005 p. CRS-12 to CRS-13] For a discussion of the weaknesses of the WCS license application see [Makhijani and Smith 2005 p. 8-22].

[805] MIT 2003 p. 117-119

[806] DOE 1999 p. 1-18

[807] IAEA/NEA 2001 p. 7

[808] Makhijani and Smith 2004 p. 51

[809] [MIT 2003 p. 63] This calculation assumes a 90 percent capacity factor and a burnup of 50 MWd per metric ton.

[810] Andrews 2004 p. CRS-3 and NRC 2005 p. 26

[811] Andrews 2004 p. CRS-3 and CRS-5 and Holt 2005 p. CRS-6

[812] The calculation assumes that each ton of spent fuel discharged contains 13.3 kilograms of plutonium and that 63,000 metric tons of spent fuel would be disposed of in the repository. [MIT 2003 p. 120] In addition, the percentage of Pu-239 in the spent fuel's plutonium is assumed to be approximately 55 percent.

[813] Uncertainties remain about the precise timeline, but it is believed that Homo sapiens evolved roughly 200,000 to 400,000 year ago, while modern humans

(homo sapiens sapiens) are believed to have emerged only about 130,000 years ago. [Hominid Timeline]

[814] Agriculture Factsheet and Lawler 2001 p. 2418-2420
[815] NAS/NRC 1995 p. 11
[816] See SDA 1999 and Makhijani and Makhijani 2006
[817] NAS/NRC 2001 p. 100-101
[818] MIT 2003 p. 157
[819] It is important to note that the involvement of the public should not be viewed as a delay or hindrance to repository development. Just as public intervention in the licensing process of nuclear power plants has led to safer reactors, so to will public involvement in repository development.
[820] MIT 2003 p. 158 and Holt 2005 p. CRS-1
[821] New York Times 2006
[822] MIT 2003 p. 158, Long and Ewing 2004 p. 367, and Andra
[823] NAS/NRC 2001 p. 91
[824] NAS/NRC 2001 p. 91
[825] NAS/NRC 2001 p. 92
[826] NAS/NRC 2001 p. 93-94 and Long and Ewing 2004 p. 391
[827] NAS/NRC 2001 p. 93 and Long and Ewing 2004 p. 391
[828] NAS/NRC 2001 p. 95-96 and Long and Ewing 2004 p. 381
[829] Kersting et al. 1999 p. 56
[830] Unless otherwise noted, the following discussion of the nuclear waste disposal program in the U.S. is drawn from [Macfarlane 2003 p. 785-788], [Long and Ewing 2004 p. 365-367], and [Makhijani and Saleska 1999 p. 115-121 and 126-127]
[831] Currently 80 percent of the installed nuclear capacity in the U.S. is located east of the Mississippi river. [NRC 2005 p. 99-113]
[832] Holt 2005 p. CRS-3 and GAO 2001 p. 17
[833] GAO 2001 p. 4
[834] GAO 2001 p. 7-8 and 20
[835] NWTRB 2002 p. 2 (emphasis added)
[836] NWTRB 2002 p. 2 (emphasis added)
[837] Long and Ewing 2004 p. 367 and Holt 2005 p. CRS-4
[838] GAO 2001 p. 7
[839] Holt 2005 p. CRS-2 to CRS-4
[840] NAS/NRC 2001 p. 100
[841] Long and Ewing 2004 p. 369-372 and Peterson 2003 p. 28
[842] NWTRB 2003 p. 1-2 and MIT 2003 p. 160
[843] DOE 2005 p. A-5, Long and Ewing 2004 p. 387, and Holt 2005 p. CRS-12
[844] DOE 2005 p. 1-4
[845] NRC 2005e and Holt 2005b p. CRS-2
[846] Peterson 2003 p. 30 and Holt 2005 p. CRS-6,
[847] UNHCR 2001
[848] IAEA 2003b p. 6
[849] Macfarlane 2003 p. 787-788 and 795 and Long and Ewing 2004 p. 382-384

[850] IAEA 2003b p. 6
[851] MIT 2003 p. 56
[852] At higher burnups than typically achieved today (50 GWd per ton versus 30 to 40 GWd per ton) the fission products would amount to just over 5 percent of the spent fuel mass while the transuranics would amount to nearly 1.5 percent of the mass. [MIT 2003 p. 118 and 120]
[853] Long and Ewing 2004 p. 387
[854] IAEA 2003b p. 50 and MIT 2003 p. 160
[855] Long and Ewing 2004 p. 373-374 and 376
[856] NAS/NRC 2001 p. 95-96, Macfarlane 2003 p. 789-790, Campbell et al. 2003 p. 43, 59-60, and Long and Ewing 2004 p. 374-375
[857] Long and Ewing 2004 p. 374 and 379-380
[858] NWTRB 2003 p. 14
[859] NWTRB 2003 p. ii, 2, and 12
[860] NWTRB 2003 p. 3-4 and 14 and NWTRB 2004 p. 4
[861] NWTRB 2003 p. 13-14 and 17 and NWTRB 2004 p. 4
[862] Sharpe 2003 p. 62 and Sharpe 2003b p. 6-71 to 6-72
[863] Long and Ewing 2004 p. 377
[864] Macfarlane 2003 p. 790
[865] Sharpe 2003b p. 6-69 to 6-70
[866] See for example Runde et al. 2002 p. 837 and 849-850, Toulhoat 2002 p. 978 and 980-981, Schäfer et al. 2003 p. 1528 and 1533, and Lu et al. 2003 p. 713 and 719.
[867] Long and Ewing 2004 p. 381
[868] NAS/NRC 1995 p. 6-7 (emphasis in the original)
[869] Court of Appeals 2004 p. 22, 29, 31, and 74
[870] For a further discussion of the proposed EPA Yucca Mountain standard and the IEER recommendations for a more protective regulation to govern repository development see [Makhijani and Smith 2005b].
[871] EPA 2005 p. 49063
[872] 10 CFR 20 2005 p. 331
[873] Règle N° III.2.f
[874] NAS/NRC 2006c p. 15
[875] EPA 2005 p. 49041-49047
[876] DOE 2002e p. I-48 to I-49 and I-77 to I-78
[877] DOE 2002e p. I-77 to I-78
[878] DOE 2002e p. I-48 to I-49
[879] NAS/NRC 2006c p. 15
[880] EPA 2005 p. 49063
[881] EPA 2005 p. 49064 (emphasis added)
[882] Macfarlane 2003 p. 791-793
[883] Macfarlane 2003 p. 793
[884] Macfarlane 2003 p. 802
[885] NWTRB 2003 p. 2, Long and Ewing 2004 p. 385, Craig 2004 p. 12, and MIT 2003 p. 160

[886] Craig 2004 p. 12, Long and Ewing 2004 p. 384-385, and NWTRB 2003 p. 2
[887] NWTRB 2003 p. 6, 8, and 16-17
[888] NAS/NRC 2001 p. 88
[889] NWTRB 2003 p. 6
[890] NWTRB 2003 p. 16
[891] NWTRB 2003 p. i
[892] NWTRB 2003 p. 11
[893] NWTRB 2003 p. 18
[894] NWTRB 2003 p. 11
[895] NWTRB 2004 p. 3-5 and NWTRB 2005 p. 3
[896] NWTRB 2003 p. 15
[897] NWTRB 2003 p. 17
[898] Craig 2004 p. 16
[899] IAEA 2003b p. 70 and MIT 2003 p. 15
[900] MIT 2003 p. 158-159
[901] Peterson 2003 p. 30
[902] Long and Ewing 2004 p. 390 and GAO 2004b p. 6
[903] Peterson 2003 p. 28
[904] NWTRB 2005b p. 149-150
[905] In the case of denser waste packaging, the peak temperature could be maintained at its current level or even reduced if a ventilation system was included in the repository. This system, however, would need to be designed, manufactured, and tested and even then important questions would likely remain about its ability to function properly over the long times necessary.
[906] ACNW 2001
[907] ACNW 2001
[908] ACNW 2001
[909] GAO 2004b p. 2-3, 7-8
[910] GAO 2004b p. 8-9
[911] NRC 2004 p. 5
[912] GAO 2004b p. 22
[913] New York Times 2005c
[914] Los Angeles Times 2005
[915] The current safe drinking water standards limit the maximum dose to a member of the public to 4 millirem per year from beta/gamma emitters and limits the maximum concentration of all alpha emitters combined (except radium and radon) to 15 pCi per liter. Uranium is limited to 30 micrograms per liter based on its chemical toxicity. IEER and other groups have requested that the EPA lower the allowable limit for long-lived alpha emitting transuranic elements to a combined total of 0.15 pCi per liter to take into account the most recent science on health risks posed by these materials. For more information see [Makhijani 2005].
[916] EPA 2001b p. 32117
[917] NAS/NRC 2006b p. 2.12 - 2.22

[918] Much of the analysis in this section is drawn from the review by Matthew Lamb and Marvin Resnikoff of Radioactive Waste Management Associates prepared on behalf of Clark County, Nevada. [Lamb and Resnikoff 2000]
[919] Lamb and Resnikoff 2000 p. 2 and 23
[920] Lamb and Resnikoff 2000 p. 23
[921] The Modal Study was a transportation risk assessment for spent fuel conducted by Lawrence Livermore National Laboratory for the NRC. It was published in February 1987 as NUREG/CR-4829. [Lamb and Resnikoff 2000 p. 2]
[922] Lamb and Resnikoff 2000 p. 5-6 and 10-11
[923] This Lawrence Livermore analysis was published in 1987 as UCID-21246. [Lamb and Resnikoff 2000 p. 11]
[924] Lamb and Resnikoff 2000 p. 10-11 and 22
[925] NRC 2000 p. 5-31
[926] CNN 2002, New York Times 2002, and Washington Post 2002b
[927] Daily Yomiuri 2005
[928] NAS/NRC 2006b p. 2.28
[929] Lamb and Resnikoff 2000 p. 12-13 and 22
[930] Lamb and Resnikoff 2000 p. 8 and NAS/NRC 2006b p. 2.28
[931] FEMA 2001 p. 2-4, 6, 12
[932] NIST 2002 p. 28
[933] FEMA 2001 p. 2, 12 and NIST 2002 p. 28
[934] FEMA 2001 p. 20
[935] FEMA 2001 p. 17
[936] NRC 2003c
[937] Washington Post 2004d and Baltimore Sun 2004
[938] New York Times 2004d
[939] NAS/NRC 2006b p. 2.25
[940] NAS/NRC 2006b p. 2.26
[941] U.S. Congress 2005 p. 206-207
[942] Independent 1992
[943] According to information compiled from Department of Defense announcements, more than 30 percent of U.S. deaths in Iraq between March 2003 and March 2006 were caused by improvised explosive devices. [Iraq Casualties]
[944] NAS/NRC 2006b p. 5.3
[945] NAS/NRC 2006b p. 5.3 (emphasis added)
[946] Washington Post 2002b
[947] FEMA 2001 p. 5-6
[948] Baltimore Sun 2004 and New York Times 2004d
[949] MIT 2003 p. 61 (emphasis added)
[950] Juhlin and Sandstedt 1989 Part II p. 73-74 and 80
[951] MIT 2003 p. 62
[952] GAO 2005 p. 4-5
[953] GAO 2005 p. 3, 11-15
[954] Holt 2005 p. CRS-9 and NRC 2005f
[955] NRC 2005f

[956] NRC 2006c
[957] Deseret Morning News 2006
[958] Hatch 2005 and Hatch 2005b
[959] MIT 2003 p. 55
[960] Committee on Appropriations 2005 p. 87-88, 105-108, and 125-126
[961] The onsite storage options considered would need to include both current storage strategies as well as the type of hardened onsite storage discussed in Section 5.5.
[962] MIT 2003 p. 58 and 148
[963] Bunn et al. 2003 p. 65
[964] Bunn et al. 2003 p. 65 and 34-45 and MIT 2003 p. 151
[965] DOE 1997b p. 2-23
[966] This assumes that a regulation football field is approximately 110 meters long by 50 meters wide.
[967] Makhijani, Hu, and Yih 2000 p. 413-417 and 470-475
[968] Hu, Makhijani, & Yih 1992 p. 80-94
[969] Hu, Makhijani, & Yih 1992 p. 95-107
[970] Bunn et al. 2003 p. 39-40
[971] MIT 2003 p. 59 and Bunn et al. 2003 p. 3
[972] see Zerriffi and Makhijani 2000 p. 103-105, 144-145, and 170-174, Makhijani and Smith 2004, and Makhijani and Smith 2005
[973] WIPP Factsheet
[974] Makhijani and Smith 2004 p. 38
[975] Bunn et al. 2003 p. 3, 40, and 62
[976] MIT 2003 p. 60 (emphasis in the original)
[977] For a further discussion of the costs, risks, and vulnerabilities associated with separation and transmutation schemes see *The Nuclear Alchemy Gamble: An Assessment of Transmutation as a Nuclear Waste Management Strategy* [Zerriffi and Makhijani 2000].
[978] For a review of the deep borehole concept for radioactive waste disposal see [Nirex 2004].
[979] MIT 2003 p. 56
[980] MIT 2003 p. 56-57 and 163-164
[981] Juhlin and Sandstedt 1989 Part II p. 18, 21, and 80-81 and MIT 2003 p. 163-164
[982] MIT 2003 p. 57
[983] Juhlin and Sandstedt 1989 Part I p. 2-3, 71-78, and 88-90
[984] NAS/NRC 2001 p. 112
[985] NAS/NRC 2001 p. 123
[986] Juhlin and Sandstedt 1989 Part II p. 83 and Harrison 2000 p. 43-44
[987] Gibb 2000 p. 35 and Harrison 2000 p. 20 and 45
[988] NAS/NRC 2001 p. 123
[989] Nirex 2004 p. vi
[990] Washington Post 2006c

[991] Dr. Arjun Makhijani, the President of the Institute for Energy and Environmental Research, has long advocated for an alternative strategy for the management of long-lived radioactive waste and for the development of hardened onsite storage. For a further discussion of the IEER proposals put forth by Dr. Makhijani see [Makhijani and Saleska 1992], [SDA 1999], [IEER 2002], [Makhijani 2003], and [Makhijani and Smith 2005b].
[992] Millennium Assessment 2005 p. 17
[993] MIT 2003 p. 22 (emphasis added)
[994] Meier 2002 p. i-iii and 25-27
[995] A number of alternate visions have been proposed for how to reduce greenhouse gas emissions and simultaneously phase out nuclear power in the United States. See for example: *Energy Innovations* from the Alliance to Save Energy, American Council for an Energy-Efficient Economy, Natural Resources Defense Council, the Tellus Institute, and the Union of Concerned Scientists [ASE et al. 1997], *Securing the Energy Future of the United States* from the Institute for Energy and Environmental Research [Makhijani 2001b], and the *Clean Energy Blueprint* from the Union of Concerned Scientists [Clemmer et al. 2001]. While the recent increases in the price of oil and natural gas, the advances in lower cost thin-film solar cell technology, and the experience that has been gained with carbon sequestration projects make some of the conclusions in these analyses out of date, they continue to provide many valuable insights into how the energy system could evolve and improve over the coming decades to address climate change without the construction of new nuclear power plants. For a more recent analysis showing how even a country as reliant on nuclear power as France could reduce emissions and phase out nuclear power see *Low-Carbon Diet without Nukes in France* from the Institute for Energy and Environmental Research [Makhijani and Makhijani 2006b].
[996] A typical 1000 MW light-water nuclear power plant requires approximately 100 to 120 MTSWU per year in enrichment services to provide its fuel. For simplicity in this calculation we have assumed 110 MTSWU per year would be required for future reactors. However, the increased demand for uranium under the global or steady-state growth scenario would make the higher enrichment levels more likely since the amount of uranium feed material and the amount of enrichment services needed are inversely related for a fixed percentage of U-235 in the tails. The assumptions made in the MIT report are consistent with this conclusion. In that report the authors assume the equivalent of nearly 125 MTSWU per year of enrichment services will be required for each reactor. [MIT 2003 p. 30 and 145-146]
[997] For simplicity this calculation assumes 110 MTSWU per year is required for each reactor and that 25 kilograms of HEU is required per warhead.
[998] MIT 2003 p. 12
[999] New York Times 2004
[1000] Riccio 2001
[1001] New York Times 2005d

[1002] At a burnup of 50 GWD per ton, a typical 1,000 MW reactor would discharge approximately 20 metric tons of spent fuel per year containing nearly 265 kilograms of Pu-239. If chemically separated through reprocessing, this would be enough plutonium to make 33 nuclear weapons (assuming eight kilograms per bomb). [MIT 2003 p. 120]

[1003] If the number or reactors was held constant at 1000 GW beyond 2050, the rate of waste generation would be roughly three times the present level and a new repository the size of Yucca Mountain would be needed every three to four years as noted by the authors of the MIT study. [MIT 2003 p. 61] The waste problem under the steady-state growth scenario would be proportionally larger.

[1004] For a further discussion of the proposed EPA Yucca Mountain standard and the IEER recommendations for a more protective regulation to govern repository development see [Makhijani and Smith 2005b].

[1005] Given the poor economics of nuclear power, a number of large new subsidies have been proposed to help prop up the industry. The authors of the MIT study favor a government subsidy of $4.85 billion over the next 10 years for a variety of "analysis, research, development, and demonstration" programs. In addition the authors proposed a $200 million per plant production credit to be given to the first 10 new plants to be built. [MIT 2003 p. 8, 13-16, and 94] The Energy Policy Act of 2005 contains even larger subsidies. The law allows the DOE to offer loan guarantees that the Congressional Budget Office estimates could carry a subsidy value of up to $600 million per plant. In addition, the law authorizes up to $1 billion in production tax credits that can be offered to each of the first six new plants. Finally, the DOE can offer insurance against regulatory delays totaling up to $500 million per plant for the first two new plants and up to $250 million for next four. [U.S. Congress 2005 Sections 638, 1306, 1702, and 1703] An additional $650 million in federal subsidies not included in the Energy Policy Act has been requested by two utility consortiums to help pay for preparing the paperwork associated with submitting combined construction and operating licenses. [Holt 2005b p. CRS-3 to CRS-5]

[1006] The importance of the fact that the cost of all of the alternatives tend to cluster around six to seven cents per kWh was originally noted by Dr. Arjun Makhijani.

[1007] DTI 2003 p. 11

[1008] Dr. Arjun Makhijani has long advocated for changes to the U.S. energy system. For a discussion of the IEER recommendations put forth by Dr. Makhijani for how to best facilitate the expansion of energy efficiency programs and the development of renewable energy resources, including actions at the state and local level, see [Makhijani and Saleska 1999 p. 181-195], [Makhijani 2001b p. 48-57], and [Makhijani et al. 2004 p. 7-10]

[1009] MIT 2003 p. 132

[1010] IEA 2003b

[1011] IPCC 2001 p. 159

[1012] Makhijani 2001b p. 21

[1013] The analogy of balancing the tradeoffs between the impacts of transition technologies and those of climate change to the use of chemotherapy in fighting cancer was originally put forth by Dr. Arjun Makhijani.
[1014] MIT 2003 p. 3 and 42 and EIA 2004e p. 224 and 339
[1015] MIT 2003 p. 43, EIA 2005 p. 3 and 97, EIA 2003d p. 35, Zaidi 2005, and Jensen 2003 p. 20 and 27
[1016] Freund 2003 p. 2
[1017] IPCC 2005 p. 5-9
[1018] MIT 2003 p. 22 (emphasis added)
[1019] Bunn et al. 2005 p. 223
[1020] Bunn et al. 2005 p. 223-224
[1021] IEA 2001 p. 368
[1022] IAEA 2001 p. 2-3
[1023] IAEA 2001 p. 1, 3, and 63, OECD/IAEA 2002 p. 21-23 and 27, OECD/IAEA 2004 p. 15-16 and 21
[1024] IAEA 2001 p. 63 and 65, OECD/IAEA 2004 p. 21-22, MIT 2003 p. 153, and Bunn et al. 2003 p. 114
[1025] IAEA 2001 p. 65 and OECD/IAEA 2004 p. 50
[1026] Bunn et al. 2003 p. 110-111
[1027] OECD/IAEA 2002 p. 21-23 and 27, OECD/IAEA 2004 p. 15-16 and 21 and Bunn et al. 2003 p. 110
[1028] Bunn et al. 2003 p. 109-110, Cameco Reserves, and Cameco McArthur History
[1029] MIT 2003 p. 154 and Bunn et al. 2003 p. 107-109
[1030] Bunn et al. 2003 p. 111-113
[1031] Bunn et al. 2003 p. 116 and MIT 2003 p. 91
[1032] MIT 2003 p. 91 and 94
[1033] OECD/IAEA 2004 p. 23 and 25
[1034] OECD/IAEA 2002 p. 42-43 and 52
[1035] OECD/IAEA 2002 p. 69-70 and OECD/IAEA 2004 p. 63-64
[1036] IEA 2000 p. 289
[1037] IAEA 2001 p. 61, OECD/IAEA 2002 p. 76, and OECD/IAEA 2004 p. 66
[1038] IEA 2001 p. 376, IAEA 2001 p. 5, 33, and 72, OECD/IAEA 2002 p. 69, and OECD/IAEA 2004 p. 64
[1039] IAEA 2001 p. 72
[1040] Makhijani, Hu, and Yih 2000 p. 110-168
[1041] IEA 2001 p. 375-376
[1042] OECD/IAEA 2002 p. 41 and OECD/IAEA 2004 p. 35
[1043] IAEA 2001 p. 47, 67-68, and 70
[1044] IEA 2001 p. 368
[1045] OECD/IAEA 2002 p. 22, 23, and 27
[1046] IAEA 2001 p. 70
[1047] OECD/IAEA 2002 p. 36 and IAEA 2001 p. 70
[1048] IAEA 2001 p. 40
[1049] IAEA 2001 p. 1 and 3

[1050] IAEA 2001 p. 64
[1051] This calculation assumes that one-third of the core is replaced per refueling. [MIT 2003 p. 119]
[1052] IAEA 2001 p. 24
[1053] See Makhijani and Smith 2004 and Makhijani and Smith 2005
[1054] Ux Weekly 2004 p. 3 and 6 and Ux Weekly 2005 p. 6
[1055] MIT 2003 p. 151 and Bunn et al. 2003 p. 54-55
[1056] IAEA 2001 p. 33 and 57
[1057] The proposed NEF would have a capacity of 3000 metric tons SWU per year. This is 12 times the size of the enrichment plant proposed to be built by Iran at Natanz which has been the subject of intense international debate. [Makhijani, Chalmers, and Smith 2004 p. 15-25]
[1058] IAEA 2001 p. 22 and 60
[1059] NRC 2005 p. 137
[1060] Guardian 2003
[1061] IAEA 2001 p. 1 and 61
[1062] MIT 2003 p. 42-43 and 118
[1063] IAEA 2001 p. 1
[1064] MIT 2003 p. 119 and 152
[1065] U Chicago 2004 p. 7-14 and MIT 2003 p. 154
[1066] ElBaradei 2004 (emphasis added)
[1067] IAEA 1997 p. 2-3 (emphasis in original)
[1068] The calculations assumes the creation of highly enriched uranium with 90% U-235, depleted uranium tails with 0.25% U-235, a 2 percent loss in the processing and enrichment stages, and total of 25 kilograms of HEU per weapon.
[1069] DOE 2001b p. 6-25 to 6-31, OECD/IAEA 2002 p. 28 and 75, and DOE 2003b p. 1 and 7
[1070] DOE 2002c p. 17-18 and DOE 2003b p. 3-5
[1071] MIT 2003 p. 148
[1072] Bunn et al. 2003 p. 18-19
[1073] Bunn et al. 2003 p. 112-116, OECD/IAEA 2002 p. 28, and OECD/IAEA 2004 p. 22
[1074] IAEA 2001 p. 14-21 and 66-67, Albright, Berkhout, and Walker 1997 p. 80, and USEC 2005
[1075] NRC 2005 p. 135, OECD/IAEA 2002 p. 52, OECD/IAEA 2004 p. 50, and USEC 2005
[1076] Makhijani, Chalmers, and Smith 2004 p. 17-18
[1077] ElBaradei 2005

Index

Abraham, Spencer, 249

Acheson - Lilienthal Report, 9, 100, 131-132, 138-140, 399

Acheson, Dean, 9, 131, 399

Acid gas, 91, 93, 304, see also Carbon sequestration

Additional Protocol, 133-138, 141, 163, 321, 396

Afghanistan, 74, 212

Alaska, 16-17

Alloy-22 (C-22), 242, 252, 263-265, 268, 294, see also Yucca Mountain

American Centrifuge Commercial Plant (ACP), 323-324

Angola, 386, 399

Antarctica, 21-22, 380

Arctic, 16-18

Argentina, 127, 388, 403

Arkansas, 66, 201, 275, 400

Arrhenius, Svante, 13

Asselstine, James, 222-223, 226

Atomic Energy Act, 2

Atomic Energy Commission (AEC), iii-v, 1-3, 5-7, 9, 114-115, 131-132, 142, 205-206, 246, 251, 399, 404, , see also U.S. Department of Energy and U.S. Nuclear Regulatory Commission

Atomics, iii

Atoms for Peace, 5, 9, 103, 114-115, 132, 139-140, 142, 305

Australia, 4, 23, 75, 91, 125, 310-314, 377, 388, 391, 398, 403

Babcock and Wilcox, 177, 204, 207

Bacher, Robert, iv

Bahamas, 77

Bangladesh, 15

Baruch, Bernard, 139-140, 148, 150

Bathtub curve, 171, 173, 182, see also Reactor safety

BBC China, 109, 151

Bechtel Corporation, v, 88-89, 94, 248

BEIR VII, see Committee on Health Effects of Exposure to Low Levels of Ionizing Radiations

Belarus, 184, 186

Belgium, 118, 140, 145, 201, 384

Belize, 152

Bethe, Hans, 221

Bhopal, India, 184, 191

Biomass, 68, 85, 98, 303

Blair, Tony, 11-12, 341

Bodman, Samuel, 240, 292

Boiling water reactors (BWR), see Light-water reactors

Bradford, Peter, 192, 230, 298

Brazil, 111-112, 127, 137, 140, 142-143, 384, 386, 397

Britain, 4, 11-12, 14, 20-22, 25, 40, 58, 60, , 65, 68-69, 72, 76, 83, 88, 95, 102, 109, 112-114, 117, 119-120, 123, 131, 136, 139-140, 149, 151, 154, 156-157, 160-161, 167-168, 172, 185, 198, 200, 209, 221, 240-241, 279, 286, 290, 301-302, 318, 378, 384, 388, 392, 394, 398-399, 403

British Nuclear Fuels Ltd. (BNFL), 118, 394

Brookhaven National Laboratory, 197, 205-206

Bryant, Edward, 23

Burk, Susan, 137

Bush, George W., 1, 12, 47, 49, 50,

417

88, 110, 116-118, 140-144, 146-147, 149, 154, 158-159, 163, 183, 249-250, 267, 270, 377, 379, 397

California, 13, 24, 59-60, 77, 168, 186, 236, 247, 282, 389

Cameroon, 93, 386, 388, 403

Canada, 9, 17, 37, 71, 75, 90-91, 93, 101, 103, 105, 114, 125, 131, 172, 217, 221, 240, 244, 310, 314, 384, 388, 398, 403

Capps, Thomas E., 45

Carbon capture and storage, see Carbon sequestration

Carbon dioxide emissions, see Greenhouse gas emissions

Carbon sequestration, 27-28, 34, 60, 73, 85-87, 89-96, 98-99, 304, 391, 413
   Great Plains Synfuels coal gasification plant, 90, 93
   Sleipner and In Salah natural gas fields, 91, 94, 304
see also Acid gas and Enhanced oil recovery

Carter, Jimmy, 9, 115, 196, 202, 246, 307

Castro, Fidel, 149

Cesium-134, 185

Cesium-135, 238, 291, 299

Cesium-137, 169, 185-186, 190, 197, 284-285

Cheney, Dick, 12, 116, 153

Chernobyl, vii, 10-11, 168, 172, 184-187, 190-191, 197, 200-202, 230, 307, 401-402

China, 1, 25, 86, 89, 97, 99, 102, 109, 111-112, 123, 126, 128, 135-136, 150-152, 157, 169, 191, 224, 296, 384, 386, 391-392, 397

Chirac, Jacques, 160

Clary, Christopher, 152

Climate change, iii, 1-2, 11-12, 27-28, 30, 34, 52, 54-56, 60, 62, 73-74, 93, 97-99, 126-127, 164, 201, 232, 294-297, 305-306, 379, 413, 415,
   effect on repository performance, 242, 244-245, 257-259, 261-262
   impacts, 12-26

Clinton, William J., 62,116

Coal gasification, 27-28, 34, 60, 73, 85-90, 93-96, 303-304, 386, 390-391

Coal, iv, 1, 11, 26-29, 33-38, 42-43, 49-50, 52-60, 68, 73-74, 79, 82, 85-90, 92-95, 97-99, 296, 303-304, 379, 382, 384, 386, 388-391, see also Carbon sequestration, Coal gasification, and Integrated Gasification Combined Cycle

Cole, Sterling, iv

Colorado, 31, 167, 201, 387

Combined construction and operating licenses (COL), see Energy Policy Act of 1992

Combined cycle gas turbines (CCGT), 2, 35-36, 42, 55, 73-74, 79-80, 82, 303-304, 386-387, see also Natural gas

Combined heat and power (CHP), 64

Combustion Engineering, 30, 204

Committee on Health Effects of Exposure to Low Levels of Ionizing Radiation (BEIR), 189-190, 259-260, 402

Congressional Budget Office (CBO), 42, 48, 414

Congressional Office of Technology Assessment (OTA), 62, 108-109, 111, 115-116, 143, 145, 156

Connecticut, 13, 197, 199, 201, 278, 281, 286

Conservation, see Efficiency

Cook, James, 29

Cooling pools, 28, 187, 196-198, 202-203, 212, 229-230, 234, 238, 283-284, 294, 404, see also Spent fuel

CRAC-2 (Calculation of Reactor Accident Consequences for U.S. Nuclear Power Plants), 188-189, 193-195, 203, 229, 298, 402

Croatia, 152, 377

CSX Railroad Tunnel Fire (Baltimore), 276-277, 280

Cuba, 149

Cult of the Atom, 400

Cyprus, 152

Czech Republic, 88, see also Czechoslovakia

Czechoslovakia, 185, see also Czech Republic

Davidson, Ward, iii

Davis-Besse, 174-182, 193, 210-211, 400-401

Deep boreholes, 283, 287-290, 293, 300, 412

Democratic Republic of the Congo, 386

Denmark, 398

Department of Trade and Industry, U.K. (DTI), 21-22, 65, 72, 95, 301-302

Depleted uranium, 235, 237, 287, 291, 293, 316-318, 393, 407, 416

Deutch, John, 157, 292, 382

Direct containment heating (DCH), 170

Disarmament, 9, 103, 155-160, 163, 323-324, 398, see also Nuclear proliferation and Nuclear weapons

District of Columbia, 13, 248, 407

Dominion, 45, 183

Duke Power, 113, 183

Early Site Permit (ESP), 45, 183

Earth source heat pump, 77, 302

Economics of nuclear power, 29-55, 96-97, 300-301, 414

Egypt, 25, 127-128

Eisenhower, Dwight D., 2, 5-6, 9, 132, 139-140, 142, 305

ElBaradei, Mohamed, 104, 110, 129, 134, 143-144, 147-150, 155-156, 160, 297-298, 321, 324

Electric Power Research Institute (EPRI), 84-85, 118, 186, 221

Emergency Core Cooling System (ECCS), 169, 212, 400

Energy efficiency and conservation, 27-28, 34-35, 60-65, 71-73, 75, 77, 97-99, 301-304, 380, 413-414

Energy Information Administration (EIA), 36, 39, 41, 49, 53, 63, 70, 77-78, 85, 89, 174, 212, 238, 382, 388

Energy Policy Act of 1992, 45, 182

Energy Policy Act of 2003, 48

Energy Policy Act of 2005, 48-50, 117, 183, 194, 278, 414

Energy Research and Development Agency (ERDA), 246, 251, 404, see also Atomic Energy Commission and U.S. Department of Energy

Energy security, vii, 62-63, 72-73, 98, 142-143, 301-303, 413, see also Terrorism

Enhanced oil recovery (EOR), 90-91, 93, 95, 304, see also Carbon sequestration

Enrichment, see Uranium enrichment

Entergy, 113, 183

Ethiopia, 386

Eurodiff, 140, 142, see also Uranium enrichment

European Wind Energy Association, 67

Ewing, Rodney, 257

Exelon, 113, 250

Extinctions and biodiversity loss, 16-18, 20, 56, 295, 381

Exxon-Mobil, 78, 166, 399

Falcone, Giovanni, 279

Fallout, 4-5, 101, 184-187, 197, 199-200, 205, 255, 401-402

Farhar, Barbara, 71

Fast breeder reactors, vi, 115, 118, 123, 172-173, 221-222, 307-308, 322, 394, see also Plutonium

Federal subsidies, see Subsidies

Fermi I (fast breeder reactor), 115, 172, 221-222

Fermi, Enrico, v

Feynman, Richard, 230

Finland, 42-43, 46, 240-241, 268, 286, 384

FirstEnergy Nuclear Operating Company (FENOC), 174-175, 177-179, see also Davis-Besse

Fissile materials, 101-102, 104-106, 114-116, 118, 121-122, 126, 129-130, 146, 392, 394, see also Plutonium and Uranium

Florida Power and Light, 283

Florida, 15, 77, 88, 187

Forbes, 10, 29, 96

Ford, Gerald R., 9, 115, 129, 196 246, 307

Fourier, 12-13

France, 10, 25, 33, 40, 43, 102, 111-115, 117-120, 123, 136, 140, 145, 151, 153, 156-157, 160-161, 175, 180, 198, 217, 240-241, 259, 284, 286, 318, 384, 388, 392, 394, 398-399, 403, 413

Fukuda, Yasuo, 123

Gas centrifuges, 105-106, 108-113, 122, 134-135, 138-140, 147, 161, 318, 321, 323-324, see also Uranium enrichment and Urenco

Gas-cooled reactors, 30-31, 105-106, 166-168

Gaseous diffusion, 108, 112-113, 140, 161,
 Oak Ridge, Tennessee, 106, 108, 392,
 Paducah, Kentucky, 108, 112-113, 392,
 Portsmouth, Ohio, 112, 392,
see also Eurodiff and Uranium enrichment

Gasoline, 87, 276, 278

General Accounting Office (GAO), see Government Accountability Office

General Electric, v, 7, 30, 48, 88, 132, 204, 221, 399

Generation IV, 29-30, 146, 311, 322

Geologic repositories, 32, 116, 196-197, 200, 233-235, 237-274,
 alternatives to, 280-294
 changes in human behavior, 238-239, 242, 245
 climate change, 242, 244-245, 257-259, 261-262
 conceptual models, 242-245, 251, 255-256, 266-267

Georgia, 15, 76, 201, 389

Germany, 25, 31, 60, 66, 86, 88, 109, 112-113, 118, 127, 139, 145, 160-161, 185, 201, 240-241, 384, 388, 398, 403

Ghana, 386

Glaciers, 14, 22-23

Global North, 97-99, 281, 301, 303, 381

Global Nuclear Energy Partnership (GNEP), 117, 141

Global South, 21, 82-83, 97, 201, 296, 301, 303, 305, 381

Global Warming, see Climate change

Government Accountability Office (GAO), 179-180, 182, 210, 226, 248-250, 271-272, 282

Graphite moderated reactors, 166-168, 172, 184, 230, 318

Greenhouse gas emissions, iii, vi-vii, 1-2, 11-14, 18-20, 26-28, 30, 34, 37, 54-61, 64-66, 70, 72-74, 82, 85-87, 89-95, 97-99, 166, 195-196, 202, 212, 232, 296-299, 301-306, 377, 379, 382, 386, 388, 391, 395-396, 413

Greenland, 14, 22-23

Greenpeace, 67

Groves, Leslie, 9, 131

Gummer, John, 12

Hagen, Edward, 213-214, 231

Hanford, v, 4, 114, 117, 199, 200, 236, 243-244, 247, 285-286, 395

Hansen, James, 11

Harvard University, 62, 285, 311, 316-317, 322

High-level waste, 116, 124, 198-200, 202-203, 233-238, 240, 243-244, 246, 250-252, 259, 268, 279, 285-286, 288, 290-294, 394, 407
  vitrification, 116, 124, 198-199, 285-286, 293, 394
see also Reprocessing and Spent fuel

Holifield, Chet, 5

Hu, Howard, 313

Hungary, 185, 388, 403

Hurricane Katrina, 15, 25, 191

Hurricanes, 15, 25, 74, 191

Hussein, Saddam, 10, 110, 149

Hutchins, Robert, iv, 3

Hydroelectric power, 34, 54, 59, 68, 80-83, 98, 108, 166, 303, 389

Hydrogen, 19, 31, 86-87, 89-91, 170

Idaho National Engineering and Environmental Laboratory (INEEL), see Idaho National Laboratory

Idaho National Laboratory (INL), 216, 243, 285, 407

Illinois, 59, 86, 183, 201

India, 1, 9-11, 15, 25, 37, 43, 79, 86, 97, 99, 102-104, 111, 114-115, 119-120, 126-128, 133, 140, 143, 146-147, 155, 157, 184, 191, 224, 296, 304, 307, 378, 384, 386, 388, 392, 294, 403

Indiana, 59

Indonesia, 388, 403

Institute for Energy and Environmental Research (IEER), 36, 59, 69-70, 97, 200, 234, 237, 239, 262, 292, 294, 317, 319, 324, 399, 409-410, 413-414

Integrated Gasification Combined Cycle (IGCC), 85-90, 93-96, 304, 386, 390-391, see also Coal gasification

Interdiction, 109, 151-152, 164, 392, see also Proliferation Security Initiative

Intergovernmental Panel on Climate Change (IPCC), 2, 13-16, 19-22, 58, 66, 70, 74, 84, 89, 92, 94, 302, 377

Intermittency, 68-69, 79-83, 85, 303, 389, see also Solar power and Wind power

International Atomic Energy Agency (IAEA), 6, 9, 40, 72, 104, 112-113, 119, 122, 130, 132-138,

421

140-141, 143-146, 148, 153, 155, 163, 201, 253-254, 268, 308-315, 317-323, 384, 394, 396-397

International Energy Agency (IEA), 42, 70-71, 83, 92, 95, 302, 312, 390

Iodine-129, 238, 286, 291, 299

Iodine-131, 101, 168-169, 185

Iowa, 13, 59, 66, 387

Iran, 25, 74, 104, 106-107, 109, 111, 114, 127-129, 134-135, 140, 143, 147, 149-150, 154, 164, 416

Iraq, 10-11, 74, 103, 109-110, 129, 133, 145, 149-150, 153-154, 212, 279, 411

Ireland, 198, 286

Israel, 10-11, 25, 102-103, 11, 127-129, 133, 149-150, 153-156, 164, 211, 378, 392

Ivory Coast, 386

Japan, 4-5, 25, 30, 32-33, 40, 42-43, 76, 79, 89, 93, 111-112, 115-120, 123, 125-128, 140, 143, 150, 173-174, 199-201, 240-241, 275, 284, 304, 318, 384, 388, 394, 398, 403

Jordan, 388, 403

Kansas, 59, 66, 223, 246, 387

Kazakhstan, 4, 102, 314, 377

Kennedy, John F., 102

Kenya, 386

Khan, Abdul Qadeer, 106, 109-110, 135, 140, 147, 150

Kissinger, Henry, 157

Kuboyama, Aikichi, 5

Kurchatov, Igor, 7

Kyoto Protocol, 1-2, 12-13, 377, 379

Lake Nyos, 93-94

Lamb, Matthew, 411

Lambert, Sheldon, 61

Larsen B ice shelf, 22

Lawrence Livermore National Laboratory (LLNL), 104, 147, 186, 275-276, 411

Leveson, Nancy, 219-220

Lewis Commission, 207-209, 213-214, see also Probabilistic Risk Assessment

Liberia, 152

Libya, 102, 109, 127, 136, 147, 151

Light-water reactors, 29-33, 37, 42, 101-102, 105, 107-108, 112, 120-122, 126, 139, 160, 165-168, 172, 174, 187, 204, 224, 226-227, 230, 232-233, 280, 312, 316, 318, 323, 383, 392, 413
  boiling water reactors, 30, 42, 105, 118, 172-173, 183, 196, 204, 207, 221, 223, 276
  pressurized water reactors, 30, 105, 118, 172, 174-176, 196, 204, 207, 221, 223, 276, 316

Lilienthal, David, 3, 9, 131, 399

Liquefied natural gas (LNG), 28, 34-35, 73-79, 82, 86, 97-99, 196, 303-305, 388-389

Lochbaum, David, 171

Long, Jane, 257

Los Alamos National Laboratory (LANL), iv-v, 124, 132, 399

Los Angeles Times, 125

Louisiana Energy Services (LES), see also National Enrichment Facility

Louisiana, 15, 66, 76, 113, 187, 246

Lovelock, James, 13

Low-level waste, 234-237, 282, 290-291, 407

Macfarlane, Allison, 263

Maine, 13

Makhijani, Arjun, i, vii, 6-7, 97, 302, 313, 377, 413-415

Manhattan Project, iv-v, 3, 9, 131-132, 167, 234, 399

Marshall Islands, 5, 152

Marshall, Andrew, 23-24

Maryland, 76, 206, 276, 278

Massachusetts Institute of Technology (MIT), 26-27, 31-40, 42-44, 46-49, 52-60, 64, 71-72, 74, 79, 88-89, 95-97, 109-110, 112, 120, 123, 125-128, 130, 135-136, 148, 155-157, 162, 166, 195, 200-201, 205, 215, 219, 224-227, 230-232, 240, 254, 263, 268, 280-281, 283, 285, 287-290, 292-293, 297, 300, 302, 304-305, 311, 314-315, 317, 319-320, 322, 382, 384-385, 388-389, 395-396, 404, 406, 413-414

Massachusetts, 13, 76, 201, 211, 215, 389

Meacher, Michael, 12

Medvedev, Zhores, 401

Meserve, Richard, 165, 179

Methane, 17, 86, 95, 377

Mexico, 77, 101, 388, 403

Michigan, 59, 201, 389

Millennium Ecosystem Assessment, 16, 295, 380

Minnesota, 59, 66, 88, 187, 387

Mississippi, 15, 183, 246, 408

Missouri, 59, 66, 223, 389

Mixed-oxide plutonium fuel (MOX), 32-33, 116-120, 162-163, 199-200, 283-286, 292-293, 308, 318, 322-323, 383, 394, 403

Monaco, 377

Monitored Retrievable Storage (MRS), see Private Fuel Storage and Spent fuel

Moniz, Ernest, 125, 157, 292, 382

Montana, 387

Morocco, 12, 388, 403

Murray, Thomas, iv

Musharraf, Pervez, 110

Myanmar, 386

National Aeronautics and Space Administration (NASA), 11, 222

National Cancer Institute (NCI), 101

National Enrichment Facility, 318, 323-324, 416, see also Louisiana Energy Services

National Renewable Energy Laboratory (NREL), 69-72, 83

National Research Council of the U.S. National Academies of Science (NAS/NRC), 21, 39-40, 63, 87, 189-190, 197-198, 218-221, 239, 242-243, 247, 251, 259, 261, 265, 278-279, 289-290, 378, 402, 407

Natural gas, 1, 11, 26-29, 33-38, 42, 49-50, 52-60, 67, 70, 72-80, 82, 86-99, 143, 196, 296, 301, 303-305, 379, 382, 387-389, 391, 413, see also Combined cycle gas turbines and Liquefied natural gas

Nebraska, 59, 66, 187, 236, 387

Nepal, 15, 386

Netherlands, 88, 109, 112-113, 139-140, 160-161, 201, 384, 398

Nevada Test Site (NTS), 101, 244, 247, 253, 273, 292

Nevada, 236, 243, 246-247, 249, 257-258, 272, 275, 411

New Hampshire, 11, 201

New Jersey, 13, 131-132, 278, 399

New Mexico, 4, 66, 69, 114, 275, 286, 291, 293, 323-324, 387

New York City, 13, 101, 130, 164,

187-188, 197, 402

New York Times, 45, 157

New York, iii, 13, 115, 187-188, 201, 278, 389

New Zealand, 201

Nigeria, 78

Nishimura, Shingo, 123

Nixon, Richard M., 142, 165, 223, 232

Non-Proliferation Treaty (NPT), see Treaty on the Non-Proliferation of Nuclear Weapons

North - South: A Program for Survival, 381

North Anna, 45, 183, 189, 400

North Atlantic Treaty Organization (NATO), 110, 145-146, 150, 157-160, 398

North Dakota, 59, 66, 90, 93, 387

North Korea, 25, 102-103, 109, 126-128, 133, 147, 150, 152, 154-155, 378, 392, 395-396

Norway, 198, 286, 384, 398

Nuclear Energy Agency (NEA), 42-43, 309, 312

Nuclear Energy Institute (NEI), 11

Nuclear Power and Non-Proliferation, 397

Nuclear Power Deception, i, 6-7, 377

Nuclear Waste Policy Act (NWPA), 246-247, 249, 268, see also Yucca Mountain

Nuclear Waste Technical Review Board (NWTRB), 248-249, 256-257, 264-267, 270

Nuclear waste, see High-level waste, Low-level waste, Spent fuel, and Transuranic waste

Nuclear Wastelands, 313, 406

Nuclear weapons proliferation, i, iii, vi-vii, 1-2, 9, 25, 28-29, 32, 34, 71, 99-165, 196, 200, 229, 233, 246, 281, 284, 293, 296-300, 305, 307-308, 318, 321-324, 393, 416

Nuclear weapons states, 9, 102-103, 106, 128, 134-139, 145, 147, 149, 155, 157, 163-164, 284, 297, 393, 398

Nuclear weapons, iv, 1-7, 9-11, 100-164, 167, 189, 191, 199, 201, 234, 237-238, 243-244, 246, 251-252, 272-274, 284, 291, 293-294, 297-300, 305, 307, 313, 318, 321-322, 324, 392-394, 396, 399, 406, 414, 416

  Hiroshima, 4-5, 101, 106, 189-190, 392

  hydrogen bomb, iv, 4-6, 12, 124, 295

  Nagasaki 4, 101, 114, 189-190, 392, 394

Nucleonics, v

NuStart, 183

Oak Ridge National Laboratory (ORNL), vi, 31, 166, 213-214, 231, 233

Ocean acidification, 18-20, 92

Ohio, 59, 174, 187, 236, 324, 391-392

Oil, 1, 11, 61-63, 68, 71, 75, 90-95, 166, 196, 289, 296, 302, 304, 307, 379. 399, 413

Oklahoma, 66, 201, 212, 275, 387

Oppenheimer, J. Robert, v, 104, 132, 399

Oregon, 59

Organization for Economic Cooperation and Development (OECD), 42-43, 81, 89, 309-312, 322-323, 384

Ozawa, Ichiro, 122-123

Pacific Northwest Laboratory (PNL), 66

Pakistan, 15, 25, 102-103, 106, 109-111, 126-129, 135, 140, 142, 150, 155, 378, 392

Panama, 152

Papua New Guinea, 386

Pennsylvania, 2, 7, 31, 167, 187, 215, 389

Performance and Innovation Unit of the Cabinet Office, U.K. (PIU), 58, 65, 69

Perle, Richard, 154

Peterson, Russell, 202

Petroleum, see Oil

Photovoltaics, see Solar power

Plutonium, iv-vii, 3-4, 9-10, 32-33, 101-103, 114-124, 133, 140, 144-145, 162-163, 167, 172, 199-200, 222, 233, 235-238, 243-244, 246, 252, 258, 273, 281, 284, 286-287, 290-294, 299-300, 307, 317-318, 322, 392-395, 406-407, 414,
   material unaccounted for (MUF), 118-119, 123-124
   suitability for nuclear weapons, 115-116, 121-123, 393-394
see also Fissile materials, Mixed-oxide plutonium fuel, Reprocessing, and Transuranic waste

Poland, 185, 398

Portugal, 82, 398

Pressurized water reactors (PWR), see Light-water reactors

Price-Anderson Act, 7, 194-195, 206

Private Fuel Storage (PFS), 282-283, 406-407, see also Spent fuel

Probabilistic Risk Assessment (PRA), 165, 168, 202-224, 226-227, 231-232, 242, 404-405
   common mode/common cause failures, 219, 226, 405
   computers and digital systems, 216-221, 226, 231, 405
   design defects, 212-214, 218, 231, 405
   expert judgment, 221-223
   human factors, 214-216, 218, 223, 226, 229, 231, 405
   issues with completeness, 208-209, 212-214, 216, 220, 231, 242, 405,
see also Reactor safety

Proliferation Security Initiative (PSI), 151-152, 398

Proliferation, see Nuclear weapons proliferation

Public opposition, 44, 49-51, 165, 187-188, 192, 200-202, 224-225, 229, 232, 239-241, 298-299

Puerto Rico, 407

Pumped hydro power, see Hydroelectric power

PUREX, 121-122, 163, 285, see also Reprocessing

Pyroprocessing, 117, 121-122, 163, 293, 300, see also Reprocessing

Qatar, 78

Radiation effects, 186, 189-190

Radioactive waste, see High-level waste, Low-level waste, Spent fuel, and Transuranic waste

Randall, Doug, 24

Rapid climate change, 21-25

Rapley, Chris, 22

Rasmussen Report, see Probabilistic Risk Assessment and WASH-1400

Rasmussen, Norman, 206-207

Reactor safety, vii, 1-2, 7, 11, 28-30, 34, 51, 71, 98-100, 115, 165-196, 200-232, 296-299, 305, 307, 393, 399, 400-406,
   loss of coolant accidents (LOCA),

167-169, 172, 175-176, 204, 211-212
see also Chernobyl, Davis-Besse, Direct containment heating, Emergency Core Cooling System, Probabilistic Risk Assessment, Sodium Reactor Experiment, Three Mile Island, and Windscale

Reactor security, see Terrorism

Reagan, Ronald, 116, 165, 223, 232

Renewable energy, 27-28, 34, 54-55, 59, 65-73, 77, 79-85, 98-99, 301-303, 382, 414, see also Biomass, Hydroelectric power, Solar power, and Wind power

Reprocessing, iii, vi, 25, 32-33, 101, 103-104, 109, 114-123, 125, 128-129, 133, 136, 138-141, 143-146, 162-163, 196, 198-200, 233-235, 237-238, 246, 250, 268, 281, 283-287, 290-291, 293, 300, 307-308, 311, 318, 322, 394, 406-407, 414
   Chelyabinsk, 120, 199-200, 286
   La Hague, 119-200, 198
   Thorp (Sellafield), 119-120, 394
   Rokkasho, 33, 119-120
see also Plutonium, PUREX, Pyroprocessing, and UREX+

Resnikoff, Marvin, 411

Rhode Island, 13

Richardson, Bill, 124

Romania, 185

Royal Society, U.K., 20

Ruckelshaus, William, 165, 223, 232

Rule of Power or Rule of Law, 377

Russia, iv, 25, 102, 111-113, 117-120, 136, 142-143, 151-152, 157, 159, 163, 184, 287, 293, 307, 314, 323, 377, 384, 386, 388, 392-394, 403, see also Soviet Union

Russian roulette, 230, 261

Saleska, Scott, i, 7, 377, 401

Sandia National Laboratories, 170, 188-189, 193-194, 203, 274, 298

Saudi Arabia, 388, 403

Savannah River Site (SRS), 117, 124-125, 199, 285-286, 407

Schwartz, Peter, 24

Schwarzenegger, Arnold, 13

Science, iv, 9

Scowcroft, Brent, 157

Seaborg, Glenn, iv-vi, 3-4

Senegal, 386

Sequoyah, 124-126, 170, 223

Shaker, Mohamed, 137

Shalom, Silvan, 156

Sharon, Ariel, 156

Shell Oil, 24, 61, 68

Siberia, 16-17, 199

Siemaszko, Andrew, 180

Singapore, 398

Skull Valley Band of Goshute Indians, 282, 407

Sodium Reactor Experiment (SRE), 168, 172, 399

Solar power,
   photovoltaics, 34, 54-55, 69, 72, 83-85, 98, 303, 382, 390, 413
   solar water heating, 64, 302

South Africa, 29-30, 102-103, 106, 110-111, 127-128, 140, 211-212, 393

South Carolina, 124, 175, 236, 282, 389

South Dakota, 59, 66, 387

South Korea, 25, 76, 79, 126-128, 140, 142, 150, 201, 304, 384, 388, 403

Southern Company, 283, 379, 391

Soviet Union (U.S.S.R.), iv, 3-7, 10, 102, 108, 114, 117, 139, 150, 185-187, 200, 230, 286, 314, see also Russia

Space shuttle, 181, 222, 230

Spain, 43, 66, 88, 140, 384, 398, 399

Spent fuel, vi, 32-53, 101, 104, 114-116, 119-121, 123, 135-136, 144, 162, 196-198, 200, 202-203, 212, 233-235, 237-240, 246-248, 250, 252, 254, 262, 264, 268-269, 273-294, 299-300, 393-394, 404, 406-407, 409, 411, 414,
  dry cask storage, 197-198, 238, 281-283, 294
  hardened on-site storage (HOSS), 198, 294, 412-413
  transportation, 2, 234-235, 240, 274-280
see also Cooling pools, Geologic repositories, High-level waste, Reprocessing, and Yucca Mountain

Standard & Poor's, 44-45, 50-51, 97, 384

Stanford University, 66, 80

Statoil, 91

Strauss, Lewis, 1, 3, 5-6, 9

Strontium-90, 169, 190, 197, 284-285

Subsidies, vi, 7, 34, 37, 48-52, 54-55, 96-97, 117, 183-184, 300, 311, 385, 414, see also Energy Policy Act of 2005 and Price-Anderson Act

Sudan, 386

Suits, C.G., v

Sunday Times, 10

Sutcliffe, William, 104, 147

Sweden, 22, 180, 185, 201, 240-241, 268, 281, 288-289

Switzerland, 118, 185, 240-241

Taiwan, 76, 126-128, 136, 201

Tanzania, 386

Technetium-99, 238, 254, 286, 291, 299

Tennessee Valley Authority (TVA), 3, 31, 46, 108, 125, 131, 183, 223, 379, 399

Tennessee, 10, 106, 108, 124, 174, 223, 389, 392

Terrorism, vii, 34, 116, 121-122, 155, 165-166, 191, 195-198, 202-203, 211-212, 229-230, 232, 274, 278-280, 294, 298-299, 302-303, 404

Texas, 66, 236, 246-247, 387, 407

The Nuclear Alchemy Gamble, 412

The Risks of Nuclear Power Reactors, 404

Thermohaline circulation (THC), 18-19, 21-24, 380

Thorium, 138-139, 321, 392

Three Mile Island (TMI), 11, 31, 170, 172-173, 180-181, 188-189, 191-192, 200-202, 204, 207, 216, 230, 232, 298, 307, 400

Total System Performance Assessment (TSPA), 249, 260, 269-271, see also Yucca Mountain

Transmutation, 117, 284-287, 412, see also Reprocessing

Transportation sector, 60, 215, 302-303, 386

Transuranic waste (TRU), 235, 286-287, 291, 293, 407

Treaty on the Non-Proliferation of Nuclear Weapons, 9, 102-104, 106, 127-129, 132-137, 139, 141, 145-149, 152, 155-157, 163-164, 378, 395-396, 398

Tritium, 104, 124-126, 158, 243, 395

Truman, Harry S., 131, 139

Turkey, 384, 398

U.S. Congress, iv, 2, 5, 7, 9, 49, 116, 131, 137, 146, 159, 197, 222-223, 226, 240-241, 247, 249-250, 270, 377
  House of Representatives, 49, 116, 249, 283-284, 403
  Senate, 48, 131, 137-138, 157-158, 247, 249,
see also Congressional Budget Office and Congressional Office of Technology Assessment

U.S. Department of Defense (DOD), 23-24, 159, 411

U.S. Department of Energy (DOE), 11, 40-42, 48, 50, 62-63, 84-85, 87-88, 90, 116, 121, 124-125, 157, 183, 222, 234, 236-238, 240, 243-244, 246-252, 255-273, 283, 287, 292, 294, 299, 322, 382, 391, 393,-394, 404, 407, 414

U.S. Enrichment Corporation (USEC), 113

U.S. Environmental Protection Agency (EPA), 13, 165, 186, 189, 223, 232, 246-247, 259-262, 270, 272-273, 299-300, 407, 409-410, 414

U.S. Federal Aviation Administration (FAA), 191

U.S. Federal Emergency Management Agency (FEMA), 277

U.S. Food and Drug Administration (FDA), 185

U.S. Geological Survey (USGS), 272, 310

U.S. National Academies of Science, 32, 245-246, 248-249, 290, see also National Research Council

U.S. Nuclear Regulatory Commission (NRC), 30, 39, 42, 45, 51, 67, 113, 124-126, 165, 170, 173-184, 186, 192, 197, 200, 206-211, 213-215, 217, 222-223, 226, 229-230, 246-252, 259-260, 262-263, 271-272, 274-278, 281-283, 292, 298, 402, 404, 411

Uganda, 386

Ukraine, 102, 143, 168, 172, 184, 186

Union of Concerned Scientists (UCS), 167, 188, 206, 402, 404, 413

United Kingdom, see Britain

United Nations (U.N.), 5-6, 10, 13, 16, 102-103, 135-136, 139, 148-151, 153, 186, 253, 295, 377-378, 380

United States, iii-iv, vi, 1-5, 7-13, 15, 24-27, 29-34, 37-47, 49, 53-54, 58-60, 63-69, 71, 74-83, 86-91, 93, 96, 99, 101-106, 108-113, 115-117, 123-126, 128-131, 134, 136-142, 145-147, 149-159, 161, 163-164, 167, 169-170, 172-175, 182-183, 187, 189, 191, 195-197, 200-201, 203-205, 212, 215, 242-243, 224-229, 232-238, 240-241, 245-247, 249, 253-254, 263, 267-268, 274-277, 279-281, 285, 290-292, 294, 298, 301-302, 304-307, 311, 313-314, 317, 323-324, 377, 379, 382, 384, 386-389, 392-395, 397-398, 402-403, 408, 411, 413-414

University of California, Davis, 186

University of Chicago, iv, 3, 33, 36-38, 40, 43, 45-47, 50, 52-54, 57-58, 79, 86, 88-89, 94, 96-97, 129-130, 385

University of Michigan, 257

University of Nevada, Reno, 257

Uranium enrichment, iii, 1, 25, 101-114, 122, 128-130, 133-147, 150-151, 154, 160-164, 196, 235, 237, 263, 291, 296-297, 316-319, 321, 323-324, 392-393, 397, 413, 416, see also Eurodiff, Gas centrifuges, Gaseous diffusion, and Urenco

Uranium, iii, vi, 9, 32-33, 85, 101-115, 117-118, 122, 124-125, 128-129, 133-140, 142-147, 151, 154, 161-164, 168, 196, 199-200, 234-

235, 237, 253-254, 263-264, 268, 285-287, 290-291, 293, 297, 307-324, 392-394, 397, 406-407, 410, 413, 416,
  highly enriched uranium (HEU), 105-111, 114, 124, 135, 161-162, 297, 322-324, 392, 413, 416
  low enriched uranium (LEU), 32-33, 105-108, 110-111, 114, 118, 162, 285, 316, 318, 323-324, 383, 392, 397,
  mining and milling, 1, 85, 138, 233-235, 253, 296, 308-309, 312-316, 406
  McArthur River mine, 310
  supply and demand, 85, 113, 117, 307-324
see also Depleted uranium, Fissile materials, Reprocessing, and Uranium enrichment

Urenco, 106, 109, 112-113, 139-140, 142, 160-161, see also Uranium enrichment

UREX+, 117, 121, 163, 293, 300, see also Reprocessing

Utah, 236, 246, 274-275

Vanunu, Mordechai, 10, 133, 156

Vaughan, Diane, 401

Venezuela, 78, 305

Vermont, 13, 282

Virginia, 45, 174, 183, 201, 389

Wall Street, vi, 43, 46-47, 51, 192, 230, 298, see also Standard & Poor's

WASH-1400, 205-209, 211, 216, 404, see also Probabilistic Risk Assessment

WASH-740, 205-206

Washington State, v, 4, 59, 114, 236, 243-244, 246, 247, 395

Waste Isolation Pilot Plant (WIPP), 286-287, 291, 293

Waste, see High-level waste, Low-level waste, Spent fuel, and Transuranic waste

Watts Bar, 10, 124-126, 174, 223, 395

Weinberg, Alvin, vi, 31, 166, 233

Western Shoshone, 253, 272, 292, 406-407

Westinghouse, 7, 30, 48, 204, 207, 221

Wilson, Charles, 5

Wind power, 34, 54-55, 60-61, 65-73, 77, 79-83, 88, 97-98, 301-303, 382, 387, 389, see also Intermittency

Windscale, 167-168, 172

Wisconsin, 13, 59

World Health Organization (WHO), 21

Wyoming, 387

Xcel Energy, 283, 379

Yemen, 152, 386

Yih, Katherine, 313

Yucca Mountain, 233-234, 238, 240-242, 244-245, 247-274, 280-281, 283, 286-287, 291-294, 299, 406-407, 409, 414,
  capacity of, 238, 246, 251-252, 268-272, 281, 291, 407
  engineered barriers, 238-239, 242, 245, 249, 252, 254-256, 262-268, 270
  EPA standards, 246-247, 259-262, 270, 272-274, 299-300, 409-410, 414
  history, 245-250
  quality assurance problems, 270-272, 292
  water infiltration, 251, 254-258, 262, 263
see also Geologic repositories, Nevada Test Site, Spent fuel, and Total System Performance Assessment

Zambia, 377